011800747

PLANT MEMBRANES — STRUCTURE, ASSEMBLY AND FUNCTION

A meeting organized by the Phytochemical Society of Europe, the
Membrane Group of the Biochemical Society and the Plant Metabolism
Group of the Society for Experimental Biology and held at
University College Cardiff in April 1988

Edited by
J.L. HARWOOD and T.J. WALTON

1988
LONDON : THE BIOCHEMICAL SOCIETY

The Biochemical Society
7 Warwick Court
London WC1R 5DP, U.K.

Order from: The Biochemical Society Book Depot
P.O. Box 32, Commerce Way
Colchester CO2 8HP, Essex, U.K.

ISBN: 0 904498 23 9

Printed in Great Britain by Whitstable Litho Ltd, Whitstable, Kent

CONTENTS

Preface vii

Acknowledgements viii

PART 1 - MEMBRANE STRUCTURE

Recent structural studies on biomembranes   1-16
    P.I. Harris and D. Chapman

Identification and isolation of photosystem I and   17-31
photosystem II pigment-proteins from higher plants
    G.F. Peter, O. Machold and J.P. Thornber

Structure, function and biogenesis of nuclear-encoded   33-45
proteins of photosystem II
    B. Andersson and R.G. Herrmann

Lipid topology and the role of lipids in plant   47-64
membrane structure
    W.P. Williams

PART 2 - MODIFICATION OF MEMBRANE STRUCTURE BY DIFFERENT FACTORS

Use of yeast lipid-synthesis mutants in establishing   65-72
membrane function
    K.D. Atkinson

Surface electrical charges and their role in membrane   73-83
function
    J. Barber

Catabolic regulation of thylakoid membrane structure   85-95
and function during senescence
    H. Thomas

Heat stress and membranes   97-112
    K.A. Santarius and E. Weis

PART 3 - SYNTHESIS OF COMPONENTS AND MEMBRANE ASSEMBLY

Glycerolipid synthesis   113-127
    J.L. Harwood

Processing peptidases of higher plant chloroplasts   129-132
    P.D. Elderfield, J.E. Musgrove, P.M. Kirwin and
    C. Robinson

The synergistic effect of light and heat stress on the   133-138
inactivation of photosystem II
    G. Schuster, S. Shochat, N. Adir, D. Even,
    D. Ish-Shalom, B. Grimm, K. Kloppstech and
    I. Ohad

The molecular genetics of thylakoid proteins   139–147
    T.A. Dyer

PART 4 – MOLECULAR FUNCTION OF MEMBRANE COMPONENTS

Properties of the photosystem II quinone binding region   149–158
    M.C.W. Evans, J.H.A. Nugent, J.A.M. Hubbard,
    C. Demetriou, C.J. Lockett and A.R. Corrie

A veteran's look at the chloroplast H^+-ATPase and   159–167
photosystem I reaction center
    N. Nelson

Lipid–protein interactions and membrane function   169–177
    K. Gounaris, D.J. Chapman and J. Barber

Plasma membrane H^+-ATPase   179–188
    R.T. Leonard

Intracellular cannibalism in higher plant cells   189–199
    R. Douce, R. Bligny, A.-J. Dorne and C. Roby

PART 5 – PHYSIOLOGICAL ASPECTS OF MEMBRANES

Herbicide action on photosynthetic membranes   201–208
    A. Trebst

Effects of water stress on photosynthesis and related   209–221
processes
    M. Speer, J.E. Schmidt and W.M. Kaiser

Chilling sensitivity and phosphatidylglycerol   223–230
biosynthesis
    N. Murata, O. Ishizaki and I. Nishida

Calcium, protein kinase and the plasma membrane   231–237
    S. Gilroy, D. Blowers, M. Collinge, H. Harvey
    and A.J. Trewavas

Transport across membranes   239–251
    J.A. Raven

PREFACE

The meeting on **Plant Membranes — Structure, Assembly and Function** was held at University College Cardiff in April 1988 under the auspices of the Phytochemical Society of Europe. We were able to organize the meeting in conjunction with the Membrane Group of the Biochemical Society and the Plant Metabolism Group of the Society for Experimental Biology. By so doing we were able to invite an impressive caste of internationally known biochemists. This, we felt, provided those attending with a valuable insight into the latest views on different aspects of plant membranes.

The programme was divided into five sessions. These developed a theme by beginning with structural aspects, moving on to modifications of structure and synthesis of components and closing with functional and physiological aspects. We have retained the same order in this book. It would be invidious to single out any particular contribution — all set out useful background information together with recent (in some cases unpublished) data. We would like to thank all the authors for their efforts.

Finally we would like to write of you, the reader. Whether you be an expert, a tyro or a worker from another area of plant biochemistry, we're sure that you will all find the descriptions of these aspects of plant membranes fascinating. It is an area which is very wide-ranging and one which is an extremely active subject for research. One predicts that it will not be long before another meeting on plant membranes will be needed.

John L. HARWOOD and Terry J. WALTON
Cardiff and Swansea
June 1988

ACKNOWLEDGEMENTS

The meeting was organized with financial backing from the Phytochemical Society of Europe, the Biochemical Society and the Society for Experimental Biology, for which we are very grateful.

We would also like to acknowledge additional financial support from Amersham International plc, Cadbury-Schweppes plc, May and Baker Ltd and the Plant Protection Division of ICI plc.

The cover picture of chloroplast thylakoid membranes from Agrostis stolonifera was kindly provided by Dr Helgi Opik, Biochemistry Research Group, School of Biological Sciences, University College, Swansea.

Recent structural studies on biomembranes

Parvez I. HARIS and Dennis CHAPMAN

Department of Protein & Molecular Biology, Royal free Hospital School of Medicine, University of London, Rowland Hill Street, London NW3 2PF, U.K.

INTRODUCTION

The fluid dynamics of membrane lipids are emphasized in the popular 'fluid mosaic model' of membrane structure (Singer and Nicolson, 1972). This summarizes the results of many previous investigations and envisages membrane proteins as floating in a two-dimensional lipid bilayer matrix. In this model it is pictured that the lipids and proteins diffuse freely within the plane of the bilayer. This emphasis on intrinsic mutability accommodates the great variety of molecular species present in biological membranes, and permits the selective modification (via protein insertion) of the bilayer permeability barrier without profound structural alterations. This model has proved useful for emphasizing the dynamic character of biological membranes.

There are, however, many exceptions to the simple picture provided by this model. Thus (i) proteins in many biomembranes are indeed arranged in a random fashion within the membrane plane; however, differentiated regions of some membranes contain specific proteins arranged in an organized two-dimensional matrix. The spatial organization of biomembrane components can have different degrees of order, ranging from essentially complete disorder to the quais-crystalline order exhibited by bacteriorhodopsin in the purple membrane (Henderson & Unwin, 1975). The electron micrographs of whole cells and thylakoids of the phototrophic bacterium, Rhodopseudomonas viridis demonstrate the remarkably regular array which the proteins in some biomembranes can adopt (Welte et al., 1981; Welte & Kreutz, 1982). (ii) In many biomembranes the intrinsic proteins are not free to diffuse readily in the plane of the membrane, but are fixed in position as a result of either (a) high protein concentration, (b) protein aggregation (Naqvi et al., 1973), (c) lipid domain formation, or (d) the interaction of intrinsic membrane proteins with the underlying cytoskeleton (Nicolson, 1976). (iii) Whilst the lipid hydrocarbon chains in biological membranes are often in a fluid, disordered state in some

---

ABBREVIATIONS: FTIR, Fourier transform infrared; nmr, nuclear magnetic resonance; esr, electron spin resonance; CD, circular dichroism.

We would like to thank the SERC (P.I.H., CASE award sponsored by Smith Kline and French Research Ltd.) and the Wellcome Trust for financial support.

biomembranes (e.g. <u>Acholeplasma</u> <u>laidlawii</u> and <u>Halobacterium</u>
<u>halobium</u>) large amounts of ordered lipid are present (Steim et al.,
1969; Chapman & Urbina, 1971; Jackson & Sturtevant, 1978).

Biological membranes exhibit obvious functional asymmetries in
the homeo-static maintenance of their surrounded volume; the inside
of any cell or organelle is different from the outside. Transport
of molecules across the membrane must be directional and any given
receptor, enzymatically active site or immunological determinant is
found on only one side of a membrane. In contrast to the absolute
asymmetry of membrane proteins, membrane lipids exhibit partial
asymmetry in that most lipid species are found in both halves of the
bilayer, but often at different concentrations (Bretscher, 1973;
Rothman & Lenard, 1977; Op den Kamp, 1979). The origins, mechanisms
for maintenance and functions of lipid asymmetry are poorly
understood. However, it is known that the lipids normally situated
on the inner membrane surface, for example phosphatidylserine can
cause the blood coagulation processes to occur.

LIPID ORGANIZATION

Most phospholipids such as the lecithins spontaneously form
bilayers upon dispersion in aqueous media; and this is the basic
organization of the lipid matrix of biomembranes. A number of model
lipid membrane systems such as liposomes have been developed and
used to study a variety of transport processes. Studies of
anaesthetics, drugs and ion transport processes have been
investigated using these model lipid membrane systems. There are,
however, a variety of non-lamellar structures which can be formed
with specific lipids under specific conditions. The individual
structures of non-bilayer conformations may be identified
unequivocally by X-ray diffraction (Luzzati & Husson, 1962).
Recently, the inverted hexagonal (H_{II}) phase has been the subject of
much attention because it is suggested that it has a role in
secretion, membrane fusion and in transbilayer lipid transport
(Cullis & de Kruijff, 1979). ^{31}P nuclear magnetic resonance (nmr)
spectra yield an essentially diagnostic pattern for membrane systems
containing hexagonal phase lipid. The added dimension of rotational
motion that occurs along the cylindrical axis of the H_{II} phase gives
rise to a slightly sharpened line shape but with an axial powder
pattern that is in the sense opposite to that of the lamellar powder
pattern (Cullis & de Kruijff, 1978). The unequivocal biological
relevance for hexagonal phase lipids is not yet clear but the
speculations about these transitions have emphasized the importance
which non-lamellar structures may play in various transport
functions.

Recently there has been considerable interest in a particular
lipid class i.e. the phosphatidylinositols. These lipids are
considered to play a role as a second messenger. The
inositoltriphosphate polar group is considered to be released from
the phospholipid upon activation by a receptor mechanism and causes
release of Ca^{2+} ions from calcium vesicles.

CHOLESTEROL

Perhaps the best studied component of mammalian membranes,
cholesterol, is the most abundant sterol in animal tissues and an

important determinant of their fluid properties and permeability. Its amphiphilic structure consists of a hydroxyl group oriented at the aqueous interface and capable of hydrogen bonding to the glycerol ester oxygen of phospholipids, a rigid planar ring that renders a monomolecular film virtually incompressible (Chapman et al., 1969), and a short alkyl side-chain that may contribute to the increased fluidity of the membrane interior (Taylor & Smith, 1980).

Many studies have been made of the modulation of lipid chain ordering by the presence of cholesterol in lipid bilayers. The proton nmr studies (Chapman & Penket, 1966) demonstrated that, above the lipid phase transition T_c, addition of choleserol inhibits the motion of the hydrocarbon chains of fluid lipid bilayers. In the ^{2}H nmr experiment, this results in a 2-fold increase in the quadrupolar splitting at equimolar ratios of cholesterol to phosphatidylcholine (Oldfield & Chapman, 1972; Rice et al., 1979). Below the T_c, however, cholesterol prevents lipid hydrocarbon from crystallizing into the more ordered crystalline or gel phases. Deuterium nmr studies show this effect as a maintenance of the 50kHz quadrupolar splitting in the presence of cholesterol upon cooling below the T_c as compared with the broad, featureless spectrum that is obtained for gel-state phospholipids in the absence of cholesterol.

Perturbations by cholesterol may also be observed by using infrared spectroscopy. The more commonly used infrared spectra parameters are the frequency maxima and the band widths at half-height of the individual vibrational modes. Frequency maxima are determined by the nature of the vibrational mode and the vibrating group. Infrared analyses of phospholipid/water systems have shown that the gel-to-liquid crystal transition is accompanied by an abrupt change in the methylene band parameters (Cameron et al., 1980; Cortijo & Chapman, 1981). The C-H and C-^{2}H stretching frequencies provide information on the proportions of trans and gauche rotamers and have been used to monitor changes in lipid conformation. One of the advantages of applying infrared spectroscopy for biomembrane studies is the non-perturbing nature of the technique. The addition of an external probe molecule is not required and the absorptions of lipid and protein groupings reflect their genuine environments. This is in contrast to the other techniques commonly used, such as esr and fluoresence spectroscopy, where perturbations induced by the added probe molecules may confuse the conclusions. An example of this can be seen in studies of protein-lipid interactions (Restall & Chapman, 1986). A further advantage of infrared spectroscopy is that it's time scale, which is about $10^{13}s^{-1}$, ideally complements the esr and nmr time scales of 10^8s^{-1} and 10^5s^{-1}.

MEMBRANE PROTEIN STRUCTURE

Crystallographic techniques, which reveal the greatest structural detail for soluble proteins, are now being applied to the problem of membrane protein structure. A major problem in the application of these techniques to membranes is that integral proteins are in contact with both polar and non-polar environments. This bi-polar environment must be reproduced in the crystal in order to maintain conformation and thereby complicates crystallization (see Michel (1983) for a discussion of crystallization of membrane proteins).

Matrix Porin (Gravito & Rosenbusch, 1980) and photosynthetic reaction centres from purple bacteria (Deisenhofer et al., 1985; Michel et al., 1986; Allen et al., 1987a,b; Yeates et al., 1987) have been crystallized in forms suitable for x-ray diffraction analyses.

Matrix porin, a major component of the outer membrane of Escherichia coli (Di Rienzo et al., 1978) in which it forms ordered, two-dimensional arrays, was crystallized from detergent solutions (Garavito et al., 1983). Large quantities of detergent remained associated with the crystalline protein; however, the structural resolution was to within 0.29nm and reproduced the hexagonal arrangement found for the protein in phospholipid bilayers (Dorset et al., 1983).

The reaction centre is an integral membrane-pigment complex that mediates the primary process of photosynthesis i.e. the light-induced electron transfers from a donor to a series of acceptor species. The photosynthetic reaction centres from purple bacteria are the first membrane proteins for which the three-dimensional structures are available. The structure of the reaction centre from the photosynthetic bacterium Rhodopseudomonas viridis was determined by x-ray diffraction at a resolution of 2.9Å (Deisenhober et al., 1985; Michel et al., 1986). More recently the structure of the reaction centre from Rhodobacter sphaeroides has been obtained to a resolution of 2.8Å (Allen et al., 1987a,b; Yeates et al., 1987). The structure of the reaction centre from Rhodobacter sphaeroides is very similar to the one described for Rhodopseudomonas viridis. The similarity between their three-dimensional structures is consistent with the similarity in their primary structures.

Three-dimensional crystals of bacteriorhodopsin have also been obtained (Henderson & Shotton, 1980; Michel & Oesterhelt, 1980), although their small size and the presence of structural defects preclude high-resolution X-ray crystallographic studies. The results of crystallographic attempts suggests that only those proteins which are known to form two-dimensional crystal-like arrays within membranes have a sufficient propensity towards self-ordering to form three-dimensional crystals.

For those proteins that do not form ordered arrays, X-ray diffraction has been used to obtain a profile of electron density in the direction perpendicular to the membrane plane. Such profiles are generally prepared from wet pellets.

In the absence of three-dimensioal crystals, a great deal of effort has been focused on the study of pre-formed two-dimensional arrays of membrane proteins. Some of these arrays occur naturally in differentiated membrane regions (e.g. bacteriorhodopsin in purple membrane) while others have been induced in model systems (e.g. vesicles isolated from sarcoplasmic reticulum (Dux & Martonosi, 1983a,b) and acetylcholine receptor (Klymkowsky & Stroud, 1979). The diffraction of X-rays, which are scattered by electrons, is not possible with such a thin crystal. Under these circumstances, diffraction patterns are best obtained from neutrons and electrons, for which the scattering centres are the atomic nuclei. The three-dimensional image reconstruction of the purple membrane, obtained by Fourier analysis of electron micrographs prepared from unstained samples, has contributed greatly to our knowledge of the static structure of a membrane protein (Henderson & Unwin, 1975; Unwin &

Henderson, 1975; Tsygannik & Baldwin, 1987).

The amount of structural detail revealed by this method is dependent on structural preservation of the sample while in the electron beam. The use of electron-opaque stains must be avoided in order to access the non-surface structure of the protein. Extremely low electron doses limit the damage induced by the electron beam. The damage associated with dehydration of membranes in vacuo is prevented by immersion of the sample in a glucose solution before dehydration. This is essential, as X-ray diffraction patterns showed that drying causes shrinkage as well as disordering (Blaurock, 1975) of the purple membrane lattice.

The electron density contour maps for bacteriorhodopsin show seven rods of density, which may be recognized as α-helices that span the membrane (Henderson & Unwin, 1975; Tsygannik & Baldwin, 1987). Three of the rods are perpendicular to the plane of the lipid bilayer, while the other four are tilted slightly. Their relative tilts are consistent with a structural stabilization by helix-helix interactions; this notion finds support in the capacity to regenerate the native structure from proteolytic fragments of bacteriorhodopsin (Liao et al., 1983). Attempts have been made to fit the amino acid sequence (Ovchinnikov et al., 1979; Khorana et al., 1979) to the three-dimensional structure (Engelman et al., 1980; Jap et al., 1983).

In their present work, Henderson and co-workers are further improving the resolution of the structure of bacteriorhodopsin. Recently they determined the 2.8Å projection structure by recording micrographs at 4°K and carrying out image processing of the entire area of the micrograph (Baldwin, J.M., Henderson, R., Beckmann, E. & Zemlin, F., (1988) J. Mol. Biol. in press). The exact mechanism by which bacteriorhodopsin translocates proton is likely to be solved once its structure is determined at an atomic resolution.

Various spectroscopic techniques are being applied to the study of membrane proteins. One of these new techniques is the application of Fourier transform infrared (FTIR) spectroscopy. This technique now makes it possible to obtain good spectra of membranes in H_2O dispersions.

Early infrared spectroscopic studies (Susi et al., 1967; Timasheff et al., 1967) of the amide I bands have demonstrated the sensitivity of the band maximum frequency to the different secondary conformation of proteins. Measurements performed in H_2O and ^2H_2O enable discrimination among α-helical, parallel and anti-parallel β-sheet and random coil conformations.

Studies conducted in our laboratory have shown that FTIR spectroscopy is sensitive enough to detect very small changes in protein structure (Haris et al., 1986a; Alvarez et al., 1987a). We have also applied the technique in conjunction with secondary structure prediction methods to determine the structures of various proteins for eg. Factor H, which is a complement protein (Perkins, et al., 1988). Studies on a large number of membrane proteins are also being carried out. FTIR spectroscopy appears to be particularly useful for probing the structure of membrane proteins. Other spectroscopic techniques such as circular dichroism and nmr are not readily applicable to large membrane-bound proteins; CD suffers from light scattering problems whereas nmr is restricted by line broadening effects.

Infrared spectroscopy has been used by various workers to obtain structural information on a number of membrane proteins which includes, matrix porin (Kleffel et al., 1985), human erythrocyte glucose transporter (Alvarez et al., 1987b), bacteriorhodopsin (Cortijo et al., 1982; Jap et al., 1983; Lee et al., 1985), Rhodopsin (Downer et al. 1986; Haris, P.I., Coke, M. & Chapman, D., unpublished data), H^+/K^+-ATPase , Na^+/K^+-ATPase (Haris et al 1986b; Mitchell, R.C., Haris, P.I., Fallowfield, C., Keeling, D.J. & Chapman, D. (1988) Biochem. Biophys. acta., in press), Cytochrome C oxidase (Bazzi & Woody, 1985; Grahn et al., 1987).

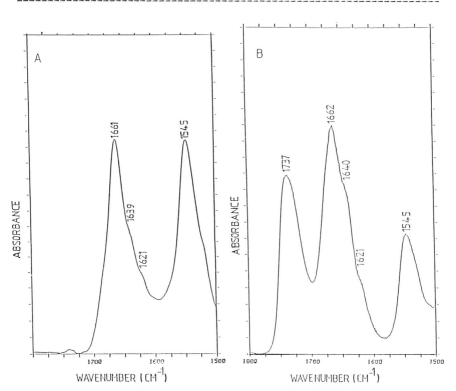

FIGURE 1: FTIR spectra of (a) bacteriorhodopsin and (b) alamethicin in dimyristoylphosphatidylcholine, at $20^{\circ}C$. The spectra were obtained after subtraction of the aqueous background (Haris, P.I. & Chapman D. (1988) Biochem. Biophys. Acta. In press).

Investigations of the purple membrane using conventional and polarized infrared spectroscopy by various workers have focused on the helical nature of the protein and its orientation relative to the membrane plane (Rothschild & Clark, 1979; Cortijo et al., 1982; Krimm & Dwivedi, 1982). The unusually high frequency of the amide I band has been interpreted to indicate the presence of α_{II}-helices (Krimm & Dwivedi, 1982). In an α-helix the plane of the peptide group is essentially parallel to the helix axis, whereas in α_{II} it

is tilted with the N-H bond pointing inward to the axis. However
our recent studies suggest that bacteriorhodopsin may not have α_{II}-
helices as another membrane associated polypeptide, alamethicin,
also shows an unusually high amide I maxima (Haris, P.I. & Chapman,
D.(1988) Biochem. Biophys. Acta., in press). The x-ray structure of
alamethicin (Fox & Richards, 1982) shows that it has regular α-
helices. The infrared spectrum of alamethicin and bacteriorhodopsin
exhibited marked similarity suggesting common folding arrangement of
their polypeptide chains. A comparison between the spectra of
bacteriorhodopsin and alamethicin is shown in Fig. 1.

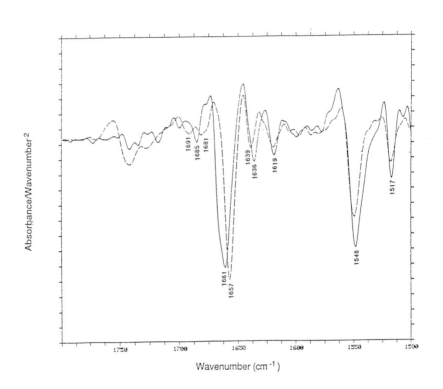

FIGURE 2: Second-derivative FTIR spectra of bacteriorhodopsin (———) and bovine rhodopsin (----)
in H_2O buffer. The spectra were obtained after
the subtraction of the aqueous background
(Haris, P.I., Coke, M. & Chapman, D.,
unpublished data).

Rhodopsin is an integral membrane protein that shares a common
property with bacteriorhodopsin in its ability to bind a cis-retinal
chromophore at an active site in its membrane domain (for a recent
review see Findlay & Pappin, 1986). The biological functions of the
two proteins, however, are quite different. Bacteriorhodopsin
serves as a light-driven photon pump, transducing energy from

photons into electrochemical energy in the form of a pH gradient
across the cell membrane of Halobacterium halobium. Rhodopsin on
the other hand serves as a photon detector in vertebrate rod
photoreceptors, utilizing the energy of a photon to trigger an
increase in the Ca^{2+} concentration of rod cytoplasm and to initiate
a biochemical signal cascade involving cGMP. The structure of
rhodopsin is much less well characterized than that of
bacteriorhodopsin. However, the amino acid sequence of rhodopsin
has been determined, and it contains seven relatively apolar
stretches that would constitute a membrane-bound domain of rhodopsin
analogous to the structure proposed for bacteriorhodopsin. Although
there is no discernible sequence homology, the structural parallels
are striking. For example, the two proteins occupy roughly the same
cross sectional areas in the membrane (Corless et al., 1982) and the
retinal attachment lysine is located on the C-terminal transmembrane helix (Eliopoulos et al., 1982; Pappin et al., 1984).
Furthermore, infrared spectroscopic studies conducted in our
laboratory (Haris P.I, Coke, M. & Chapman, D., unpublished results)
and others (Downer et al., 1986) show that they have similar
infrared spectra. The latter workers obtained their infrared
spectra of rhodopsin and bacteriorhodopsin in ^2H_2O. We have
obtained spectra of these membrane proteins in both ^2H_2O and H_2O and
have applied second-derivative and deconvolution procedures for
detailed spectral analysis. Fig.2 shows the second-derivative
spectra of bacteriorhodopsin and rhodopsin recorded in H_2O. It is
clear that there are some structural similarities evident from the
number and positions of the amide I components. However an
important difference is the position of the amide I maxima, which is
unusually high in the case of bacteriorhodopsin.

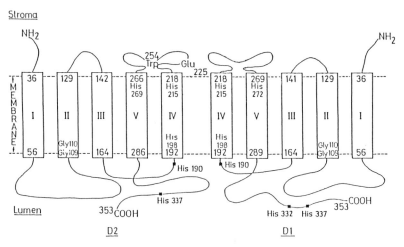

FIGURE 3: A diagrammatical representation of the organization of the D1, D2 polypeptides in
photosystem II (Adapted from Barber, 1987).

With the three-dimensional structures of two bacterial reaction
centres known in detail attempts are now being made to elucidate the
structure of the photosystem II reaction centre from higher plants.
How water is split in "Oxygen-evolving complex" of photosystem II is

one of the most important questions to be answered. Photosystem II reaction centre is a complex consisting of peripheral and integral membrane proteins, several chlorophyll A molecules, two pheophytin A molecules, two or three plastoquinone molecules, and one non-heme iron atom. The amino acid sequence of this reaction centre has been determined (for a recent review on photosynthetic reaction centres see Barber, 1987). A model for the organization of the D1 and D2 polypeptides of photosystem II is presented in Fig.3. On the basis of amino acid sequence comparisons with the more structurally well characterised Rhodopseudomonas viridis, some workers have already made some structural predictions for photosystem II (Michel & Deisenhober, 1988).

We ourselves, are collaborating with Prof. Barber's group at Imperial College to gain structural information on the photosystem II reaction centre. We have made some preliminary comparisons between the secondary structure of the photosystem II reaction centre with the reaction centre from Rhodobacter sphaeroides by obtaining their infrared spectrum under identical conditions (Haris, P.I., Newell, W., Barber, J., & Chapman, D. - unpublished data). Some similarities in their structure are clearly evident from the position of the major amide I and amide II maximas (see Fig.4). Further studies are in progress in order to investigate, in detail, the structure of the photosystem II.

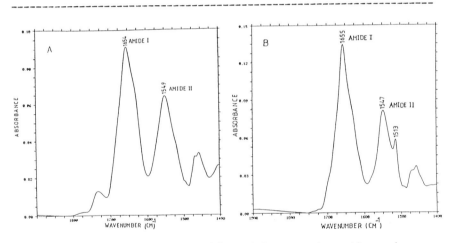

FIGURE 4: FTIR spectra of (a) Rhodobacter sphaeroides and (b) photosystem II reaction centre from higher plants, after subtraction of the aqueous background. The spectra were recorded in H_2O buffer at $2^{\circ}C$ (Haris, P.I., Newell, W., Barber, J. & Chapman, D., unpublished data).

Glycophorin (MN glycoprotein) is the major glycoprotein of the human erythrocyte membrane and is composed of 131 amino acid residues with 16 attached oligosaccharide chains (Segrest et al., 1973; Tomita et al., 1978). The N-terminus has been shown to contin a stretch of about 30 residues with a high proportion of hydrophobic amino acids and not a single charged residue (Segrest et al., 1973). A recent spectroscopic investigation (Mendelsohn et al., 1984a) of glycophorin reconstituted in phosphatidylserine has revealed the presence of significant β-sheet conformation (10% of total). The major vibration of the amide I band occurred at 1653cm^{-1}, which indicates a primarily α-helical and random coil backbone. This study supports an earlier suggestion (Schulte & Marchesi, 1979), based upon circular dichroism studies, that the protein contains 27% α-helix, 10% β-sheet and 63% random coil conformations.

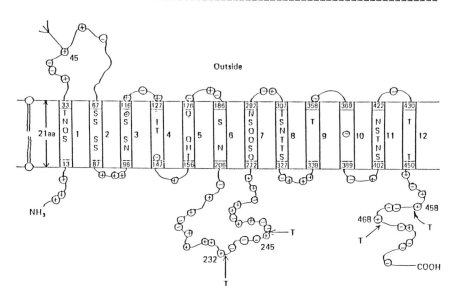

FIGURE 5: Proposed model for the human glucose transporter. The 12 putative membrane spanning domains are numbered and shown as rectangles. The relative positions of the acidic and basic amino acid residues are indicated by circled (+) and (−) signs respectively (after Meuckler et al., 1985).

A model for the structure of the glucose transporter which operates using a facilitated diffusion transport process has been proposed based upon a DNA clone method for determining the amino acid sequence. The model is shown in Fig.5. It can be seen that it involves 12 putative helices embedded in the lipid matrix (Mueckler et al., 1985).

FTIR spectroscopy has been used to study the secondary structure of the human erythrocyte glucose transporter after purification and reconstitution in erythrocyte lipids (Alvarez et al., 1987b). The spectra indicate that the glucose transporter contains, in addition to the predominant α-helical structure, an appreciable amount of β-structure and random coil conformation. A study of the time dependency of H-^{2}H exchange revealed that more than 80% of the polypeptide backbone is readily accessible to the solvent (Alvarez et al., 1987b). This result is interpreted to indicate that a portion of the intramembrane-spanning region of the membnrane protein is exposed to the solvent, suggesting the existence of an intraprotein aqueous channel. The residual (10-20%) portion of the protein which exchanges slowly includes some α-

helical structure, probably situated in a hydrophobic environment
inside the membrane. The infrared spectra of transporter
preparations were also examined after incubation with substrate and
substrate analogues.

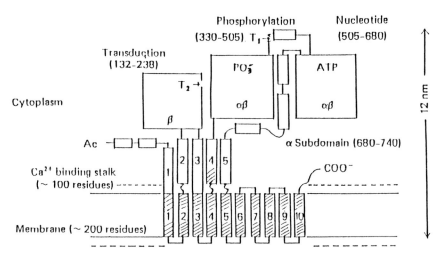

FIGURE 6: Assembly of Ca^{2+}-ATPase domains. The predicted
arrays are laid out in a planar diagram (after
MacLennan et al., 1985).

A recent structural model for the Ca^{2+}-ATPase of sarcoplasmic
reticulum based on the amino acid sequence determined from a DNA
clone (MacLennan et al., 1985; Brandl et al., 1986) is considered to
contain transmembrane α-helices together with domains of parallel
and anti-parallel β-sheet on the outside of the membrane. The
proposed structure of the Ca^{2+}-ATPase is indicated in Fig.6. It can
be seen that it is proposed that there are 10 helical segments
embedded in the lipid matrix. There is also a considerable portion
of the protein outside the lipid and connection by a polypeptide
stalk. Phosphorylation nucleotide and transaction segments are
indicated. Recently the three-dimensional structure of Ca^{++}-ATPase
was reconstructed from electron micrographs of negatively stained
and frozen hydrated P2 type crystals at $=25$ A resolution (Taylor et
al., 1986; Ho et al., 1986). Analysis of Ca^{++}-ATPase at higher
resolution requires large and well ordered three-dimensional
crystals that are suitable for X-ray diffraction studies.

Information on the structure of the Ca^{2+}-ATPase of sarcoplasmic
reticulum from rabbit striated muscle has also been obtained by
infrared spectroscopy (Cortijo et al., 1982; Mendelsohn et al.,
1984b; Lee et al., 1985). The position of the amide I and amide II
maxima in H_2O and ^2H_2O buffers indicated that the Ca^{2+}-ATPase
contains unordered as well as α-helical conformation (Cortijo et
al., 1982). Fourier deconvolution (Mendelsohn et al., 1984),
second-derivative (Lee et al., 1985) and fourth-derivative (Lee &
Chapman, 1986) of the amide I region revealed a minor amount of β-
structure.

Research in our laboratory has also been extended to investigate the structure of two other closely related ion transport enzymes namely Na^+/K^+-ATPase and H^+/K^+-ATPase (Haris et al., 1986b). The infrared spectra of the three different ATPases displayed marked similarities suggestive of secondary structural homology.

Our infrared studies show that the α-helical conformation is the predominant secondary structural motif in membrane proteins. We have studied a range of membrane proteins isolated from both plants and animals. The matrix porin, however, has been reported to be predominantly β-sheet in structure (Kleffel et al., 1985 and our own unpublished data).

SIGNAL SEQUENCE PEPTIDES

Secretory signal peptides

Secreted proteins are generally synthesized as precursors which contain an extra peptide extension that is termed the signal sequence. Signal sequences appear to be required for the export of proteins but the molecular mechanism by which proteins are able to cross biomembranes is still under discussion. Several models have been proposed to explain the translocation of proteins but none of them is in complete agreement with the experimental data. Although signal peptides exhibit little sequence homology, some common structural features have been identified, which include a hydrophobic core that is bordered by basic residues at the N-terminal and by acidic residues at the C-terminal end. At the cleavage site there is invariably an amino acid with a small uncharged side-chain. Fidelio et al.,. (1986) reported that signal sequences share another common property, namely an ability to support a high-lateral pressure at the air-water interface. In order to obtain information about the conformation of these signal sequence in membrane systems, three signal sequence polypeptides (the pre-trypsin gen signal sequence, putative signal sequence of ovalbumin and a synthetic 'consensus' signal sequence) and dioleoylphosphatidylcholine have been studied in lipid-water dispersions using FTIR spectroscopy (Haris et al., 1987). The presence of strong absorptions at $1695cm^{-1}$ and $1627cm^{-1}$ indicate that the conformation adopted by these peptides in aqueous phospholipid dispersions is anti-parallel β-sheet. It can be suggested that this conformation may be responsible for the high surface stability observed with these signal peptides. Further studies are in progress to confirm this.

Plant chloroplast transit peptides

Many chloroplast proteins are known to be nuclear encoded. These proteins are synthesised on cytoplasmic ribosomes and imported post-translationally into the chloroplast by an energy-dependent process, where they are eventually segregated into their functional sub-compartments, either in the envelope membranes, the stroma, the thylakoid membranes or the thylakoid space (Chua & Schmidt, 1978). Various investigations involving the construction of chimeric proteins have shown that the information required for transport of the proteins to the chloroplast, translocation across the chloroplast envelope and subsequent localisation within a sub-compartment is contained in the transient amino-terminal transit peptide of the precursor (Robinson & Ellis, 1984; Smeekens et al., 1986).

The cell has unique mechanisms for targetting proteins synthesised in the cytoplasm to their precise subcellular destination. Mechanisms exist which differentiate the signal sequences that target secretory and membrane proteins to the endoplasmic reticulum membrane from the leader and transit sequences which direct proteins to their respective organelles.

The precursors of the small subunit of ribulose-1,5-bisphosphate carboxylase (RUBISCO) contain transit peptides of between 44 and 57 residues in length, depending on the plant species of origin (Van den Broeck et al., 1985). Despite variation in length, there are common recognisable features in the primary sequence (Karlin-Neumann & Tobin, 1986). In order to see if transit peptides exhibit secondary structural features which could be important in the way they interact with chloroplast membranes, we conducted FTIR spectroscopic studies on the synthetic transit peptide of the small subunit of ribulose-1,5-bisphosphate carboxylase from wheat (Austen, B.M., Chapman, D., Fidelio, G.D., Haris, P.I., Kaderbhai, M.A. & Lucy, J.A. - unpublisehed data). Preliminary results indicate that unlike the secretory peptides we studied, the transit peptide adopts a predominatly α-helical structure. This is further supported by secondary structure prediction methods of other workers (Schmidt et al., 1979).

CONCLUSION

The bilayer arrangement of lipids is a common organizational feature to most biological membranes. Despite the regularity of its structure, it appears remarkably mutable. For technological reasons, lipid structure and dynamics have been the focus of most biophysical studies over the last two decades. However, new developments, particularly in the fields of molecular biology, electron microscopy, electron diffraction and spectroscopic methods now permit examination of the membrane proteins. Molecular details of ion transport are still largely unknown although model membrane studies using ionophores such as valinomycin and gramicidin have provided insight into these transport processes. Combined studies, utilizing some of these new methods for determining the static and dynamic structure of membrane proteins, hold great promise for the delineation of enzyme molecule and ion transport mechanisms.

REFERENCES

Allen, J.P., Feher, G., Yeates, T., Komiya, H. & Rees, D.C. (1987a) Proc. Natl. Acad. Sci. USA **84**, 5730-5734.
Allen, J.P., Feher, G., Yeates, T., Komiya, H. & Rees, D.C. (1987b) Proc. Natl. Acad. Sci. USA **84**, 6162-6166.
Alvarez, J., Haris, P.I., Lee, D.C. & Chapman, D. (1987a) Biochem. Biophys. Acta. **916**, 5-12.
Alvarez, J., Lee, D.C., Baldwin, S.A. & Chapman, D. (1987b) J. Biol. Chem. **262**, 3502-3509.
Barber, J. (1987) Trends. Biochem. Sci. **12**, 321-326.
Bazzi, M.D. & Woody, R.W. (1985) Biophys. J. **48**, 957-966.
Blaurock, A. (1975) J. Mol. Biol. **93**, 139-157.

Brandl, C.J., Green, N.M., Korczak, B. & MacLennan, D.M. (1986) Cell 44, 597-607.

Bretscher, M. (1973) Science 181, 622-629.

Cameron, D.G., Casal, H.L. & Mantsch, H.H. (1980) Biochemistry 19, 3665-3672.

Chapman, D. & Penket, S.A. (1966) Nature (London) 211, 1304-1305.

Chapman, D. & Urbina, J. (1971) FEBS lett. 12, 169-172.

Chapman, D., Owen, N., Phillips, M. & Walker, D. (1969) Biochem. Biophys. Acta. 183, 458-465.

Chua, N.H., & Schimdt, G.W. (1978) Proc. Natl. Acad. Sci. USA 75, 6110-6114.

Corless, J.M., McCaslin, D.R., & Scott, B.L. (1982) Proc. Natl. Acad. Sci. (USA) 79, 1116-1120.

Cortijo, M. & Chapman, D. (1981) FEBS Lett 131, 245-247.

Cortijo, M., Alonso, A., Gomez-Fernandez, J. & Chapman, D. (1982) J. Mol. Biol. 157, 597-618.

Cullis, P.R. & De Kruijff, B. (1978) Biochem. Biophys. Acta. 513, 31-42.

Cullis, P.R. & De Kruijff (1979) Biochem. Biophys. Acta 559,399-420.

Deisenhofer, J., Epp, O., Miki, K., Huber, R. & Michel, H. (1985) Nature (London) 318, 618-624.

Di Rienzo, J., Nakamur, K. & Inouye, M. (1978) Annu. Rev. Biochem. 47, 481-532.

Dorset, D.L., Engel, A., Massalski, A. & Rosenbusch, J.P. (1983) J. Mol. Biol. 165, 701-710.

Downer, N.W., Bruchman, T.J. & Hazzard, J.H. (1986) J. Biol. Chem. 261, 3640-3647.

Dux, L. & Martonosi, A. (1983a) J. Biol. Chem. 258, 1011-10115.

Dux, L. & Martonosi, A. (1983b) J. Biol. Chem. 258, 11896-11902.

Eliopoulos, E., Geddes, A.J., Brett, M., Pappin, D.J.C. & Findlay, J.B.C. (1982) Int. J. Biol. Macomol. 4, 263-268.

Engelman, D., Henderson, R., MacLachlan, A & Wallace, B.A. (1980) Proc. Natl. Acad. Sci. USA 77, 2023-2027.

Fidelio, G.D., Austen, B.M., Chapman, D. & Lucy, J.A. (1986) Biochem. J. 238, 301-304.

Findlay, J.B.C. & Pappin, D.J.C. (1986) Biochem. J.238, 625-642.

Fox, R.O., Jr. & Richards, F.M. (1982) Nature (London) 300, 325-330.

Grahn, M.F, Haris, P.I., Wrigglesworth, J.M., & Chapman, D. (1987) International Symposium on "Cytochrome Systems: Molecular Biology and Bioenergetics" IUB symposium No. 159. Plenum.

Gravito, R.M., Jenkins, J., Jansonius, J.N., Karlsson, R. & Rosenbusch, J.P. (1983) J. Mol. Biol. 164, 313-327.

Gravito, R.M. & Rosenbusch, J.P. (1980) J. Cell Biol. 86, 327-329.

Haris, P.I., Fidelio, G.D., Austen, B.M., Lucy, J.A. & Chapman, D. (1987) Biochem. Soc. Trans. 15, 1129-1131.

Haris, P.I., Lee, D.C. & Chapman, D (1986a) Biochem. Biophys. Acta. 874, 255-265.

Haris, P.I., Mitchell, R.C., Falllowfield, C., Keeling, D.J. & Chapman, D. (1986b) Biochem. Soc. Trans. 14, 1126-1127.

Henderson, R. & Shotton, D., (1980) J. Mol. Biol. 139, 99-109.

Henderson, R. & Unwin, P. (1975) Nature (London) 257, 28-32.

Ho, M.H., Taylor, K.A. & Martonosi, A.N. (1986) Biophys. J. 49, 570a.

Jackson, M.B. & Sturtevant, J.M. (1978) Biochemistry 17, 911-15.

Jap, B.K., Maestre, M.F., Hayward, S.B. & Glaeser, R.M. (1983) Biophysical Journal 43, 81-9.

Karlin-Neumann, G.A.S. & Tobin, E.M. (1986) EMBO J. 5, 9-13.

Khorana, H.G., Gerber, G.E., Herlihy, W.C., Gray, C.P., Andregg, R.J., Bienmann, K. & Nihei, K. (1979) Proc. Natl. Acad. Sci. (USA) 76, 5046-50.

Kleffel, B., Gravito, R.M., Baumeister, W. & Rosenbusch, J.P. (1985) EMBO J. 4, 1589-1592.

Klymkowsky, M.W. & Stroud, R.M. (1979) J. Mol. Biol. 128, 319-34

Krimm, S. & Dwivedi, A.M. (1982) Science 216, 407-8.

Lee, D.C. & Chapman, D. (1986) Bioscience Rep. 6, 235-56.

Lee, D.C. Hayward, J.A. Restall, C.J. & Chapman, D. (1985) Biochemistry 24, 4364-73.

Liao, M.J., London, E. & Khorana, H.G. (1983) J. Biol. Chem. 258 9949-55.

Luzzati, V. & Husson, F. (1962) J. cell Biol. 12, 207-19.

MacLennan, D.H., Brandl., C.J., Korczak, B. & Green, N.M. (1985) Nature (London) 316, 696-700.

Mendelsohn, R., Dluhy, R.A., Crawford, T. & Mantsch, H.H. (1984a) Biochemistry 23, 1498-504.

Mendelsohn, R., Anderle, G., Jaworsky, M., Mantsch, H.H. & Dluhy, R.A. (1984b) Biochimica et Biophysica Acta 775, 215-24.

Michel, H. (1983) Trends Biochem. Sci. 8, 56-9.

Michel, H. and Deisenhofer, J. (1988) Biochemistry 27, 1-7.

Michel, H. & Oesterhelt, D. (1980) Proc. Natl. Acad. Sci. (USA) 77, 1283-5.

Michel, H., Epp, O. & Deisenhofer, J. (1986) EMBO J. 5, 2445-2451.

Mueckler, M., Caruso, C., Baldwin, S.A., Panico, M., Blench, I., Morris, H.R., Allard, W.J., Lienhard, G.E. & Lodish, H.F. (1985) Science 229, 941-5.

Naqvi, K., Gonzalez-Rodriguez, J., Cherry, R. & Chapman, D. (1973) Nature, London 245, 249-51.

Nicholson, G.L. (1976) Biochimica et Biophysica Acta 457, 57-108.

Oldfield, E. & Chapman, D. (1972) FEBS Letters 23, 285-97.

Op den Kamp, J. (1979) Annu. Rev. Biochem. 48, 47-81.

Ovchinnikov, Y., Abdulaev, N., Feigira, M., Kieselev, A. & Labanov, N. (1979) FEBS Letters 100, 219-24.

Pappin, D.J.C., Eliopoulos, E., Brett, M. & Findlay, J.B.C. (1984) Int. J. Biol. Macromol. 6, 73-76.

Perkins, S.J., Haris, P.I., Susi, R.B. & Chapman, D. (1988) Biochemistry 27, 4004-4012.

Restall, C.J. & Chapman, D. (1986) in Lipids and Membranes. Past Present and Future (Op den Kamp, J.A.F., Roelofsen, B. and Wirtz, K.W.A., eds.), pp. 61-92, Elsevier, Amsterdam.

Rice, D., Meadows, M., Scheinmann, A., Goni, F., Gomez-Fernandez, J., Moscarello, M., Chapman, D. & Oldfield, E. (1979) Biochemistry 18, 5893-903.

Robinson, C. & Ellis, R.J. (1984) Eur. J. Biochem. 142, 343-346.

Rothman, J. & Lenard, J. (1977) Science 195, 743-53.

Rothschild, K.J. & Clark, N.A. (1979) Biophys. J. 25, 473-88.

Schmidt, G.W., Devillers-Thiery, A., Desruisseaux, H., Blobel, G. & Chua, N.H. (1979) J. Cell. Biol. 83, 615-622.

Schulte, T.H. & Marchesi, V.T. (1979) Biochemistry 18, 275-80.

Segrest, J.P., Kahane, I., Jackson, R.L. & Marchesi, V.M. (1973) Arch. Biochem.and Biophys. 155, 167-83.

Singer, S.J. & Nicholson, G.L. (1972) Science 175, 720-31.

Smeekens, S., Baurerie, C., Hageman, J., Keegstra, K. & Weisbeck, P. (1986) Cell. 46, 365-375.

Steim, J., Tourtelotte, M., Reinert, J., McElhaney, R. & Rader, R. (1969) Proc. Natl. Acad. Sci. (USA) 63, 104-9.

Susi, H., Timasheff, S.N. & Stevens, L. (1967) J. Biol. Chem. 242, 5460-6.

Taylor, K., Dux, L. & Martonosi, A. (1986) J. Mol. Biol. 187, 417-427.

Taylor, M.G. & Smith, I.C.P. (1980) Biochimica et Biophysica Acta 599, 140-9.

Timasheff, S.N., Susi, H. & Stevens,L. (1967) J. Biol. Chem. **242**, 5467-73.

Tomita, M., Furthmayr, H. & Marchesi V. (1978) Biochemistry **17**, 4756-70.

Tsygannik, I.N. and Baldwin, J.M. (1987) Eur. Biophys. J. (1987) **14**, 263-272.

Unwin, P.N.T. & Henderson, R. (1975) J. Mol. Biol. **94**, 425-40.

Van den Broeck, G., Timko, M.P., Kausch, A.P., Cashmore, A.R., Van Montagu, M. & Herrera-Estrella, L. (1985) Nature (London) **313**, 358-363.

Welte, W., Hodapp, N., Aehnelt, C. & Kreutz, W. (1981) Biophys. Struct. Mech. **7**, 209-12.

Welte, W. & Kreutz, W. (1982) Biochimica et Biophysica Acta **692**, 479-88.

Yeates, T.O., Komiya, H., Rees, D.C, Allen, J.P. & Feher, G. (1987) Proc. Natl. Acad. Sci. (USA) 84, 6438-6442.

Identification and isolation of photosystem I and photosystem II pigment-proteins from higher plants

Gary F. PETER*, Otto MACHOLD[+] and J. Philip THORNBER*

*Dept of Biology, University of California, Los Angeles, California 90024, USA., and [+] Zentralinstitut fur Genetik und Kulturpflanzenforschung der Akademie der Weissenschaften, DDR-4235 Gatersleben, DDR.

SYNOPSIS

Various glycosidic surfactants have been used to extract photosynthetic pigment-proteins from higher plant thylakoid membranes. The pigment-proteins showed greatly improved resolution and stability when separated with a modified Deriphat-PAGE system; in particular, they retained their associated chlorophyll molecules during fractionation. The polypeptide composition of various PS I fractions and the biogenesis of PS I in the absence of cytoplasmic protein synthesis have been studied. The association of LHC I with the P700-containing core complex is described.

INTRODUCTION

Genetic, biophysical, and biochemical studies of photosynthetic bacteria, algae, and higher plants have made it clear that each photosystem (PS) is composed of multiple chlorophyll (chl) and carotenoid-binding proteins (Thornber, 1986). A PS is organized into two multiprotein components: a core complex (CC) which is composed of those polypeptides and cofactors essential for a stable complex *in vivo* that performs a photochemical oxidation/reduction reaction, and a light-harvesting complex (LHC) where much of the antenna pigments are organized, but which is not essential for a stable and functional CC *in vivo* (Thornber, 1986; Thornber et al., 1986). The first descriptions of multiple pigment-binding proteins in higher plant thylakoids were made by extracting chloroplast lamellae with anionic surfactants (see Thornber (1986) for review). The first two pigment-proteins described were purified by polyacrylamide gel electrophoresis (PAGE) of sodium dodecyl sulfate (SDS) or sodium dodecyl benzene sulfonate extracts. These chlorophyll-proteins were designated CP I and CP II (Ogawa et al., 1966; Thornber et al., 1966). CP II is the major LHC component associated with PS II and is the most abundant pigment-protein in green plants (Thornber, 1986). CP I is the second most abundant plant pigment-protein, and contains the polypeptides that bind, among other things, the reaction center chl molecule(s), P700, of PS I (Dietrich & Thornber,1971; Shiozawa et al., 1974).

ABBREVIATIONS: chl, chlorophyll; cmc, critical micelle concentration; PS, photosystem; CC core complex; LHC, light-harvesting complex; PAGE, polyacrylamide gel electrophoresis.

Research was supported by NSF grants (DMB 84-17760 and CHE 85-09657) to JPT. GFP was supported by a USPHS predoctoral training grant and the McKnight Foundation.

Improved extraction and SDS-PAGE conditions led later to the discovery and delineation of several more pigment-proteins, as well as to an elucidation of the associations between some pigment-proteins in each photosystem (cf. Thornber, 1986). Anderson et al. (1978) detected an additional chl a-containing pigment-protein termed, CPa, and showed that CP I is associated with colorless polypeptides in PS I, and that CP II could exist in slower migrating, presumably oligomeric forms, termed by them LHCP 1 - 3 (see also Markwell et al., 1978). Machold et al. (1979) resolved yet another pigment-protein, chl a/b-P 1, which they suggested to be part of the PS II LHC, but distinct from CP II. Replacement of SDS with lithium dodecylsulfate (LiDS) allowed Delepelaire and Chua (1979) to separate CPa into two different chl a-proteins, termed CP III and CP IV.

The major advantage of separation by SDS- or LiDS-PAGE is the fine resolution allowing rapid purification of several pigment-proteins; however, a serious limitation is that this procedure disrupts the noncovalent interactions between proteins as well as between pigments and proteins. As a result any photochemical activity is often lost, and 10 to 33% of the total chl, depending on the exact system, is released from the proteins. Markwell et al. (1979) overcame part of this problem by replacing SDS during PAGE with the zwitterionic surfactant, Deriphat 160 (lauryl-iminodiproprionate). Using a Deriphat-PAGE system the authors separated SDS-extracted pigment-proteins while generating only 3% free chl, strongly suggesting that all chl molecules in higher plant thylakoid membranes are associated with protein (Markwell et al., 1979). The results also showed that when SDS is added to thylakoids it does not dissociate chl from proteins but rather disruption of the pigment-protein complexes occurs during the electrophoresis.

Extraction of thylakoid pigment-proteins by nonionic surfactants (e.g. Triton X-100) was pioneered by Vernon et al. (1966) who found it superior to SDS for obtaining _active_ pigment-proteins, but while fractionation can be achieved by ultracentrifugation or column chromatography of such extracts, Triton does not permit the extracts to be separated on the more highly resolving PAGE. Nevertheless, extracting higher plant thylakoids with Triton X-100, led later to Bengis and Nelson (1975) purifying a multiprotein core complex of PS I in a form that photochemically oxidizes plastocyanin and reduces $NADP^+$ when supplemented with ferredoxin-$NADP^+$ reductase. Useful PS I and PS II holoenzymes have continued to be prepared from Triton X-100 extracts (see, for example, Mullet et al. (1980a) and Berthold et al. (1981), respectively).

The potential of surfactants other than SDS, LiDS, and Triton X-100 to solubilize and fractionate thylakoid pigment-proteins has been investigated. Camm and Green (1980) first showed that octyl-glucoside mildly solubilizes thylakoids, and that when such extracts are separated by SDS-PAGE, the pigment-proteins show increased stability and improved resolution. This view was supported by the separation of chl a/b-P1, termed by them CP 29, from monomeric CP II, and fractionation of the two PS II core chl a-proteins, termed by them CP 47 and CP 43, in higher yields than previously obtainable. _The number of pigment-proteins and the names by which they are known have grown substantially during the 1980s. In Table I we have attempted to help the reader understand this rather complex situation by correlating each pigment-protein with the alternative names by which it is sometimes known._ Dunahay and Staehelin (1985) and Bassi and Simpson (1987), using octyl-glucoside together with low amounts of SDS, have separated an oligomeric CP I* that contains LHC I pigment-protein(s) and provided evidence for another LHC II pigment-protein, CP 24 (Bassi et al., 1987; Dunahay & Staehelin, 1986). Thus, a combination of octyl-glucoside and SDS represents the only method for solubilizing and electrophoretically resolving all of the known pigment-proteins (Bassi & Simpson, 1987; Bassi et al., 1987; Dunahay & Staehelin, 1985; 1986; Machold, 1984). Yet, substantial amounts of free pigment are still generated by such a system. More recent work with octyl-glucoside and dodecyl-maltoside indicates that these surfactants have favorable properties for solubilizing membranes and separating more native complexes by velocity sedimentation, e.g. an oxygen-evolving PS II core complex (Ikeguchi et al., 1985) and a PS I holoenzyme (Nechushtai et al., 1987).

A detailed biochemical and biophysical description of *intact* photosynthetic pigment-proteins is essential for us to understand how light energy is trapped and converted to chemical energy. In order to achieve this goal we have studied several glycosidic surfactants which have either glucose, maltose, or glucamide as polar groups, for their use in extracting pigment-proteins from higher plant thylakoids in their native states. We demonstrate that Deriphat-PAGE (Markwell et al., 1979) is an excellent method with which to separate pigment-proteins from glycosidic surfactant extracts of thylakoid membranes. Table I gives some details of the pigment-proteins that have been resolved so far, using such a procedure. All of the pigment-proteins can be isolated in association with other pigment-proteins or individually without the loss of chl or probably carotenoid. In the second half of this paper we compare fractionation of glycosidic surfactant extracts on Deriphat-PAGE with that on the highly resolving SDS-PAGE system pioneered by Machold et al. (1979). The data enable us to make some conclusions about how LHC I is associated with the PS I core complex. Finally, we examine the effect of inhibition of cytoplasmic protein synthesis on the assembly of PS I.

Table I. Characteristics of barley thylakoid pigment-protein complexes[1]

| | APPARENT SIZE(kDa) OF: | | | % TOTAL | ALTERNATIVE |
	HOLOCOMPLEX	APOPROTEIN(S)	CHL a/b	CHL	NAMES
PS I	230	multiple	5.5	38	
CC I	120	58 + others	a only	20	CPI, CHLa-P1
LHC Ia	25	24	1.4		LHCPIa
				18	
LHC Ib	65	21	2.3		LHCPIb
CC II-RC	80	32(D2), 30(D1), 9, 4	a + pheo	1	
CC IIa	55	47	a only		CPIII, CHLa-P2, CP47
				10	
CC IIb	50	43	a only		CPIV, CHLa-P3,
LHC IIa	35	31	2.25	4	CHLa/b-P1, CP29,
LHC IIb	72	28, 27, 25	1.33	40	CHLa/b-P2, CP2
LHC IIc	30	29, 26.5	1.80	4	CP27
LHC IId	24	21	0.93	4	CP24

----

[1]Unpublished data; see also Thornber et al. (1986); Peter and Thornber (1988)

METHODS

Thylakoid membrane isolation

Barley leaves were cut near the soil and chopped into 1-2cm lengths and 50-75 grams of leaves were mixed with 150 ml chilled grinding buffer (0.4M sorbitol, 10mM tricine-NaOH pH 7.6, 10mM MgCl) and homogenized 3-4X for 5 sec at full power in a Wareing blendor . The homogenate was filtered through four layers of Miracloth. This filtrate was spun at 3020 xg for 2 min. The pellet was washed with 2M NaBr (Nechushtai & Nelson, 1984) and EDTA buffer (Markwell et al., 1978), and then resuspended to 1.1 mg chl/ml final concentration in extraction buffer (6.2mM tris-48mM glycine pH 8.3, and 10% vol/vol glycerol) (Markwell et al., 1978; 1979), and frozen at -70°C.

Membrane solubilization

Surfactant stock solutions were prepared; typically 1% (w/v) SDS and 9% (w/v) of a glycosidic surfactant were combined to make a 10% (w/v) stock. Membranes were thawed quickly at 20°C but not allowed to reach room temperature and were chilled on ice. The membrane and surfactant stocks were then mixed, 1 part surfactant and 9 parts membranes to yield a final 10:1 weight ratio of surfactant to chl. The surfactant extracts were spun in a microfuge for 2 minutes to remove starch, and the green supernatant was immediately applied to PAGE. Solubilizations and incubations were carried out on ice.

Deriphat polyacryamide gel electrophoresis

Gels were composed of 12.4 mM tris-48mM glycine pH ~8.3 and 8% acrylamide (33.5% acrylamide 0.3% bis-acrylamide). These gels were polymerized with 0.1% ammonium persulfate and 0.005% TEMED. Typically 25 ug chl was loaded per lane for two millimeter thick gels. The electrophoresis resevoir buffer was 12.4 mM tris-96 mM glycine pH ~8.3 and 0.2% Deriphat 160 (Markwell et al., 1979), which was precooled on ice and, when used, SDS was added to 0.01% final concentration immediately before use to prevent precipitation of Deriphat before the sample enters the gel. The SDS is not needed in the resevoir buffer when purified complexes are rerun, or when the maltoside surfactants are used. Gels were electrophoresed at 100V constant voltage (Markwell et al., 1979) for 35 min; after longer times the complexes were not as sharply resolved due to decreasing pH. The current started at 20 mA and decreased to about 15 mA after 35 min.

SDS-PAGE

"Non-denaturing" SDS-PAGE was performed on 7.5-20% linear gradient acrylamide gels (2.5% bis-acrylamide) with System IV buffers (Machold et al., 1979). The samples were solubilized with 20:1:1 octyl-glucoside/SDS/ Chl, applied and electrophoresed overnight in gels rigidly maintained at 4°C.

Denaturing SDS-PAGE (Figs 3 & 4) was on 10-20% linear gradient acrylamide gels (3% bis-acrylamide) plus 6M urea with System IV buffers; electrophoresis was overnight at room temperatures. Alternatively (Fig. 2) linear gradient gels of 10-15% acrylamide were cast according to the Laemmli (1970), except that the separating gel contained 4 M urea and twice the published ionic strength, 0.755 M Tris-Cl pH 8.8. Gel slices were incubated for 30-60 minutes in 75 mM Tris-HCl pH 6.8, 5.0 mM EDTA, 100 mM DTT, and 4% SDS at 25°C. Gels were run overnight at a constant current of 20-25 mA. They were fixed and stained for 2 hr in 50% methanol, 10% acetic acid, 40% water, 0.1% coomassie brilliant blue, and destained in the same solution without coomassie blue.

RESULTS

Analysis of glycosidic detergents

We have tested the effectiveness of some nonionic glycosidic surfactants for extracting pigment-protein complexes from thylakoid membranes. We judged their efficacy by resolving the pigment-proteins in such extracts by Deriphat-PAGE. First, we confirmed that octyl-glucoside/SDS extracts of thylakoids could be separated by Deriphat-PAGE. We then varied the nature of the gel buffer and the acrylamide percentage to optimize the resolution without the loss of pigment from proteins. We concluded that higher ionic strengths, increased acrylamide to bisacrylamide ratios and the use of impure Deriphat 160 increased the amount of free pigment generated; however, resolution of the pigment-proteins was still excellent. Since previous work showed that complete solubilization of higher plant thylakoids is achieved with 1% SDS at 1 mg/ml chl (10:1, w/w) final concentration (Markwell & Thornber, 1982), this surfactant/chl ratio was used as a guide in our studies. Using optimal PAGE conditions, we tested the efficacy of octyl-, nonyl-, and decyl-glucosides; octanoyl-, nonanoyl-, and decanoyl-glucamides; octylthio- and heptylthio-glucoside; as well as decyl- and dodecyl-maltosides for the complete solubilization of thylakoids and separation of pigment-proteins. A representative PAGE separation of barley thylakoids extracted with various glycosidic surfactants used singly or in mixtures is shown (Fig. 1A). Submicellar concentrations of SDS were included in the solubilizations and in the resevoir buffer when monosaccharide-containing surfactants were used, since this addition decreased the slight background smear of chl.

The pattern of pigment-protein complexes observed after Deriphat-PAGE depended upon the surfactant(s) used to solubilize whole membranes (Fig. 1A). For example, nonyl-glucoside/SDS mixtures solubilized all the pigment-proteins, converting some LHC IIb into its monomeric form, disrupting CC II subunit associations, and dissociating substantial amounts of LHC I from PS I (Fig. 1A), whereas octyl-glucoside/SDS mixtures neither completely solubilized thylakoids nor disrupted LHC IIb oliogomers. Importantly, upon longer incubations with octyl- or nonyl-glucoside only small amounts of pigment were released, even though more pigment-proteins migrated in their monomeric forms. A combination of octyl- and nonyl-glucosides, each at their critical micelle concentrations (cmc), gave less disruption of the PS I, CC II, and LHC IIb multimeric complexes than when nonyl-glucoside was used alone. Both decyl- and dodecyl-maltoside have different properties and produced different results than the monosaccharide surfactants. The maltosides did not disrupt subunit interactions but rather they stabilized these associations, particularly those between CC IIa and CC II-RC. Importantly, we detected little dissociation of CC II subunit interactions or increases in free pigment during the longer incubations required to completely solubilize thylakoid membranes. However, these surfactants consistently released slightly more carotenoid and chl from their associations with protein than the monosaccharide surfactants (Fig. 1A). Interestingly, the size of surfactant micelles determined the rate of migration through the gel (Fig. 1A); solubilizations with decyl- and dodecyl-maltoside as well as SDS showed decreased mobility of all the pigment-protein complexes with respect to the glucosides, even though SDS is negatively charged.

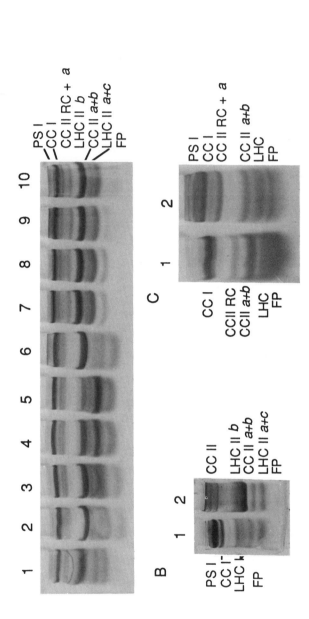

Figure 1. Separation of barley whole membranes (WM), stroma lamellae, Ps II, and chl b-less chlorina f2 (F2) pigment-proteins from glycosidic surfactant extracts by nondenaturing Deriphat-PAGE. All samples except for lane 2 in part A had a 1% final concentration of surfactant and 1 mg chl/ml; i.e. surfactant:chl=10:1 wt/wt. Gels are unstained, the rationale for labeling is described in Table 1; FP is free pigment.

A- Various solubilizations of thylakoid membranes of barley, lane: 1) 9:1 octyl-glucoside/SDS, 2) 19:1 octyl-glucoside/SDS, 3) 6:3:1 octyl-/nonyl-glucoside/SDS, 4) 9:1 nonyl-glucoside/SDS, 5) 9:1 nonyl-glucoside/SDS incubated for 30 minutes, 6) 9:1 nonyl-glucoside/nes incubated for 30 minutes, 7) 10 dodecyl-maltoside, 8) 10 dodecyl-maltoside incubated for 30 minutes, 9) 9:1 decyl-maltoside/SDS 10) 4.5:4.5:1 heptylthio-glycoside/decyl-maltoside/SDS.

B- separation of 4.5:4.5:1 heptylthio-glycoside/decyl-maltoside/SDS extracted: 1) digitonin-prepared stroma lamellar membrane fragments (Ghanotakis & Babcock, 1983); 2) Triton X-100 PS II particles (Leto et al., 1985).

C- separation of barley chl b-less mutant pigment-proteins: lane 1) 6:3:1 octyl-/nonyl-glucoside/SDS; lane 2) 9:1 decyl-maltoside/SDS.

Conclusions about the glycosidic surfactant-Deriphat 160 PAGE system.

The results demonstrate that pigment-proteins separated by Deriphat-PAGE from glycosidic surfactant extracts of thylakoid membranes have retained their integrity as judged by their minimal loss of pigment. Nonionic surfactants with a carbohydrate polar group prove to be excellent for solubilizing thylakoid membranes: one, they are effective at low surfactant to protein ratios; two, they can be easily removed by dialysis; three, low concentrations of them maintain pigment-proteins in solution; and four, they are available in a variety of chain lengths.

Each available glycosidic surfactant exhibits distinct properties from the others when used in conjunction with Deriphat-PAGE. In general, mixtures of a monosaccharide and a disaccharide surfactant yield mainly multimeric pigment-protein complexes, whereas a long chain monosaccharide surfactant used alone yields monomeric pigment-proteins. For surfactants with the same polar group, shorter hydrocarbon chains are less solubilizing and less disruptive of interactions between pigment-proteins than longer chains. It is possible that the decrease in cmc with increasing chain length accounts for this phenomenon. For example, the efficacy of octylthio-glucoside (cmc = 9mM) is intermediate between heptylthio- or octyl-glucoside (cmc $\geq$ 25mM) and nonyl-glucoside (cmc = 6mM). Nonanoyl-glucamide's properties are much like octyl-glucoside's, although less solubilization occurs on a weight per weight basis. The type of polar group also greatly influences the effectiveness of the surfactant. Decyl-glucoside and decanoyl-glucamide have similar properties, including their cmc's, to decyl- and dodecyl-maltoside. Whereas the former disrupt protein-protein contacts the latter do not. Furthermore, the maltoside surfactants stabilize protein-protein contacts in solution. The maltoside surfactants do, however, consistently release slightly more chl from its association with protein than the shorter, $\leq$ 9, hydrocarbon chain monosaccharide surfactants. The most effective surfactant mixtures with which to solubilize higher plant pigment-protein complexes and maintain their stability during Deriphat-PAGE, proved to be a combination of a short monosaccharide and a long chain disaccharide surfactant at concentrations equal to or slighty above their cmc's (see next section).

Using such combinations of glycosidic surfactants with submicellar concentrations of SDS to solubilize barley thylakoids, we have purified each of the higher plant individual pigment-proteins previously described: CC I RC, LHC Ib, CC IIa, CC IIb, and LHC IIb, and five others that are less well characterized: LHC IIa, LHC IIc, LHC IId, LHC Ia, and CC II RC. Table I gives some details of their characteristics in addition to covering their nomenclature. Our fractionation conditions maintain the pigment-proteins close to their native state: First, each pigment-protein can be isolated in complex with other PS subunits; for example, over 95% of PS I is isolated as a multisubunit holoenzyme, containing LHC I and CC I (Fig. 1A), which indicates that P700 is functionally associated with the antenna chls of LHC I and CC I *in vivo*. In contrast the PS II holoenzyme is disrupted under these same conditions, even though abundant multimeric forms of LHC II, CC II and LHC IIb were isolated. Thus the polar forces that stabilize interactions between CC II and LHC II pigment-proteins are weaker than those between CC I and LHC I subunits. Second, surfactants shift the absorbance and fluorescence of protein-bound chls to shorter wavelengths (Markwell & Thornber, 1982; Nechushtai et al., 1986), which may be due to slightly different protein conformations and/or to surfactant-chl interactions (Markwell & Thornber, 1982). The longer wavelength red maxima of our isolated pigment-proteins and the broadness of their red peaks (data not shown) suggest that they are spectrally closer to the pigment-proteins in the membrane than are most other

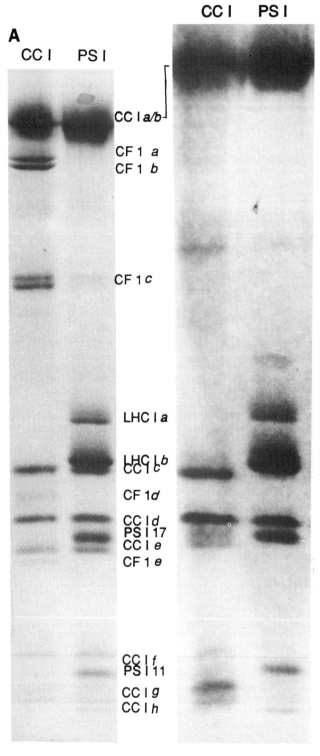

Figure 2. Polypeptide composition of Photosystem I pigment-protein complexes of Hordeum vulgare. A) PS I and CC I complexes were isolated from nonyl-glucoside extracts by Deriphat-PAGE (Fig. 1A, lane 4). The amounts of PS I and CC I obtained from 100 ug and 175 ug whole membrane chl, respectively, were denatured and separated by SDS-PAGE 10-15% gels with 4M urea. B) The PS I complex was purified from a 19:1:1 octyl-glucoside/SDS/chl extract of thylakoids by Deriphat-PAGE using SDS rather than Deriphat in the upper buffer (see text). CC I was purified, denatured and electrophoresed as in A.

preparations. Third, and most significantly, little carotenoid and <u>no chls</u> are dissociated from the pigment-proteins during solubilization and separation under our optimum conditions. Interestingly, increased incubation times result in the dissociation of PS I, CC II, and LHC IIb multimers (Fig. 1), yet the amount of free pigment does not increase. Our data therefore show that protein-protein associations can be disrupted without altering the pigment-protein interactions, and thus, for the most part, chl and carotenoid molecules are bound specifically to individual subunits rather than simultaneously to two subunits, as occurs in the purple bacterial B800-850 antenna (cf. Thornber, 1986).

IDENTIFICATION OF PIGMENT-PROTEINS IN HEPTYTHIO-GLUCOSIDE AND DECYL-MALTOSIDE EXTRACTS OF THYLAKOIDS AND SUBMEMBRANE FRACTIONS

Comparison of PSII and stroma lamellae preparations.

It is now well established that PS I is localized mainly in the nonappressed stroma lamellae and PS II in the appressed grana membranes (see B. Andersson, this volume). We used digitonin to prepare stroma lamellae (Leto et al., 1985), as well as Triton X-100 extraction in the presence of Mg^{++} to obtain PS II particles (Berthold et al., 1981; Ghanotakis & Babcock, 1983). Each preparation was extracted with heptylthio-glucoside plus decyl-maltoside, and the pigment-proteins in them separated by Deriphat-PAGE (Fig. 1B). Stroma lamellae were greatly enriched in the two slowest migrating pigment-protein complexes (Fig. 1B), demonstrating that they represent PS I forms in Deriphat-PAGE gels of whole membranes (Fig. 1A). In comparison with PS II particles, stroma lamellae contain much greater amounts of a yellow-green pigment-protein which migrated slightly faster than LHC IIb, the major pigment-protein of PS II; this pigment-protein is an oligomeric form of LHC I (Nechushtai et al., 1987). Electrophoresis of the PS II particles (Fig. 1B) resolved at least six pigment-protein complexes (cf. Table I).

Photosystem I

The PS I holoenzyme contains 13 major subunits organized into at least two pigment-protein complexes: (a) a core complex (Bengis and Nelson 1975; Nechushtai & Nelson, 1984) which contains the beta-carotene chl a-RC protein complex (CC Ia + CC Ib), and six smaller nonpigmented subunits; (b) an LHC I chl a/b-xanthophyll-protein complex having at least two protein subunits (Mullet et al, 1980b; Lam et al., 1984a); Bassi & Simpson, 1987). When nonyl-glucoside extracts of thylakoids are separated by Deriphat-PAGE (Fig. 1) both a PS I holoenzyme and CC I complex are isolated. Polypeptide analysis (Fig. 2) showed that PS I contained all the CC I subunits as well as other polypeptides, two of which are among those of LHC I (see above). Treatment of the isolated PS I complex with 0.2% nonyl-glucoside and reelectrophoresis allowed us to isolate two LHC I pigment-proteins (LHC Ia and b (Nechushtai et al., 1987; Peter and Thornber, unpublished data) in addition to CC I. LHC Ia had a 77°K fluorescence maximum of 690 nm and a 24 kDa apoprotein, whereas LHC Ib had an emission maximum at 730nm and a single 21 kDa subunit. Since a small percentage of the chl was released during re-electrophoresis of PS I, it is quite possible that there are other LHC I pigment-protein(s). In this connection, the functions of the 17 and 11 kDa subunits of the PS I fraction are unknown, and these polypeptides were not isolated with either the LHC I or CC I pigmented complexes. Our CC I complex had a very similar subunit composition to those reported previously (Nechushtai & Nelson, 1984; Ortiz et al., 1985); the functions of most of these subunits are unknown.

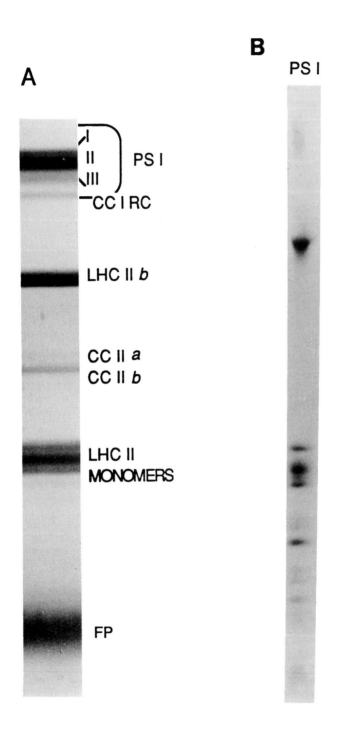

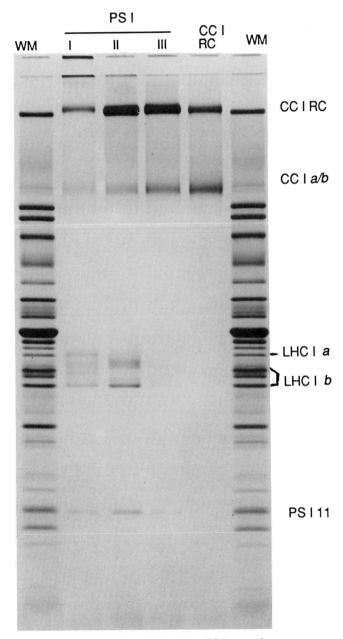

Figure 3. Separation and analysis of *Vicia faba* PS I complexes isolated by a highly resolving "nondenaturing" SDS-PAGE system and Deriphat-PAGE. **A)** "Nondenaturing" SDS-PAGE separation of octyl-glucoside extracts of thylakoid pigment-proteins. The gel is unstained. **B)** Subunit composition of Deriphat-PAGE-purified PS I holoenzyme on 10-15% gradient gels with 4M urea as in Fig 2. **C)** Polypeptide composition of the PS I complexes isolated in A. The complexes were denatured and electrophoresed on 10-20% gradient gels with 6M urea.

In order to understand the organization of PS I and to test if any of the small CC I subunits are involved in binding LHC I to CC I-RC, we have isolated and compared the polypeptide compositions of PS I complexes purified from octyl-glucoside extracted thylakoids by two different systems (Figs 2 and 3): The frequently used Machold et al. (1979) system, and the Deriphat-PAGE system described above.

The association of LHC I with CC I as studied by PAGE

When 0.1% SDS is substituted for Deriphat 160 in the resevoir buffer of the Deriphat-PAGE system, and all other conditions are maintained, the PS I holoenzyme band observed in Deriphat-PAGE was not present. The most abundant PS I component (~80% of the total PS I chl) in such gels was slightly smaller than in the Deriphat system, and had a distinctly different polypeptide composition (Fig. 2): The proteins present were those of CC I-RC, LHC I*a* and *b*, and the 17 and 11 kDa PS I subunits mentioned in the previous section. All of the colorless CC I subunits, except CC I*d*, were conspicuously absent (Fig 2). Interestingly, under the conditions we used, CC I*d* is tightly associated with CC I-RC, whereas other treatments specifically release this subunit (Nechushtai and Nelson 1986; Ortiz et al. 1985). Thus, it cannot be ruled out that CC I*d* is involved in binding LHC I to CC I-RC. However, separation of octyl-glucoside extracted *V. faba* thylakoid membranes on the highly-resolving, "nondenaturing" SDS-PAGE system of Machold (Fig. 3) revealed four, and sometimes five, pigmented PS I complexes. None of them contained any of the small CC I subunits, even though LHC I subunits are present (Fig. 3). We must therefore conclude that LHC I binds directly to CC I-RC, and that the small (<25kDa) CC I subunits do not mediate this binding.

On the association of LHC I with CC I-RC

Three of the four PS I pigment-protein complexes resolved by the SDS-PAGE conditions of Machold et al. (1979) are combinations of LHC I with CC I-RC (i.e., with the 68kDa polypeptides CC I*a* and *b*). Polypeptide analysis of these four pigmented bands (Fig. 3) shows that most of them contain LHC I*b* apoproteins, the 11 kDa PS I subunit, and the CC I-RC apoproteins. LHC I apoproteins are present in all of the fractions except CC I-RC (Chl*a*-P1). Note that in this system and in *V. faba*, LHC I*b* contains four resolvable polypeptides. The largest PS I band contains significant amounts of both LHC I*a* and *b* (Fig 3). The next smaller and most abundant PS I form lacked the 24 kDa LHC I*a* apoprotein and one of the four LHC I*b* polypeptides (Machold et al. 1987), but it still had appreciable amounts of the other LHC I*b* subunits. Thus, it seems likely that LHC I*a* and one of the LHC I*b* polypeptides are more easily released and/or more peripheral to CC I-RC than the other LHC I*b* components. The next smaller PS I band has lost some of each of the three remaining LHC I*b* polypeptides from CC I-RC (Fig. 3). The 17 kDa subunit is not detected in any fraction, and hence cannot act as an internal antenna to LHC I.

CONSEQUENCES OF CYTOPLASMICALLY SYNTHESIZED PS I POLYPEPTIDES ON THE SYNTHESIS AND ASSEMBLY OF CC I-RC and PS I

Strong evidence exists that most of the smaller PS I subunits are synthesized in the cytoplasm, post translationally imported into the plastid, and then assembled with the 68kDa CC I reaction center polypeptides (Mullet et al., 1980b; Nechushtai et al. 1986; Tittgen et al., 1986). CC I*a* and *b*, the 68kDa polypeptides, are synthesized

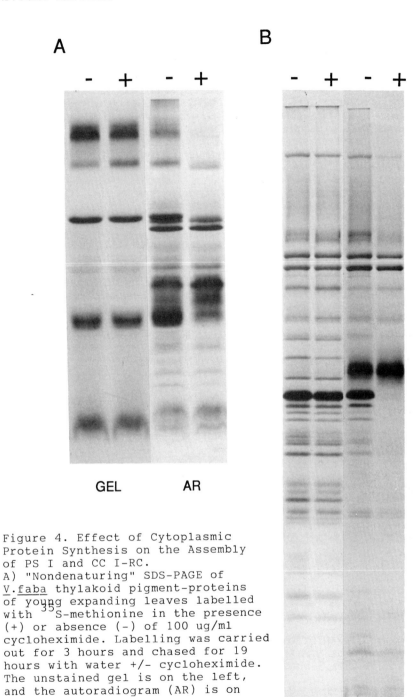

Figure 4. Effect of Cytoplasmic
Protein Synthesis on the Assembly
of PS I and CC I-RC.
A) "Nondenaturing" SDS-PAGE of
V.faba thylakoid pigment-proteins
of young expanding leaves labelled
with ^{35}S-methionine in the presence
(+) or absence (-) of 100 ug/ml
cycloheximide. Labelling was carried
out for 3 hours and chased for 19
hours with water +/- cycloheximide.
The unstained gel is on the left,
and the autoradiogram (AR) is on
the right.
B) Denaturing SDS-PAGE of the
labelled thylakoids showing
the protein synthesis patterns.

on thylakoid-bound polyribosomes, and then complexed with pigments to form the CC I reaction center (Klein et al., 1988). We have looked at the *in vivo* assembly of CC I a/b in young expanding leaves of *Vicia faba* in the absence of cytoplasmic protein synthesis. In control plants, which were not treated with cycloheximide, radioactivity was detected in all of the PS I subunits resolved (Fig. 4). However, when cytoplasmic protein synthesis was blocked with cycloheximide, the pattern of synthesis and assembly of CC I a/b differed. Less CC I a/b accumulated, and very little was detected in higher ordered forms of PS I on PAGE (Fig. 4). Assembly of the radiolabelled CC I a/b was somewhat impaired. We had expected CC I a/b to assemble into a dimeric state, i.e., CC I-RC; however, they did not. A slight decrease in mobility of this complex was reproducibly seen (Fig. 4), the significance of which is unknown, but two possiblities are apparent. One, that this complex is dimeric but lacks some chl's, or two, that other CC I subunits are neccessary for the proper confirmation of CC I-RC.

Why a lack of cytoplasmic synthesis should affect the assembly process in this manner is unclear. A nuclear encoded subunit could be required for stable assembly (Nechushtai & Nelson, 1985), or a decrease in chl synthesis might affect CC I a/b synthesis, thereby providing a control mechanism for CC I a/b assembly and stability. For example, the complex could be unstable in the membrane even though partially formed into the dimer (Fig. 4). Examining the assembly of newly synthesized CC I a/b on milder Deriphat-PAGE showed no detectable assembly of CC I · a/b into CC I or PS I when cytoplasmic synthesis was blocked (Fig. 4). The relationship between assembly and stability remains unclear.

REFERENCES

Anderson, J. M., J. C. Waldron, & S. W. Thorne. (1978) FEBS Lett. 92:227-233.
Bassi, R., & D. J. Simpson. (1987) Eur. J. Biochem. 163:221-230.
Bassi, R., G. Hoyer-Hansen, R. Barbato, G. M. Giacometti, & D. J. Simpson. (1987) J. Biol. Chem. 262:13333-13341.
Bengis, C., & N. Nelson. (1975) J. Biol. Chem. 250:2783-2788.
Berthold, D. A., G. T. Babcock, & C. F. Yocum. (1981) FEBS Lett. 134:231-234.
Camm, E. L., & B. R. Green. (1980) Plant Physiol. 66:428-432.
Delepelaire, P., & N-H. Chua. (1979) Proc. Natl. Acad. Sci. USA 76:111-115.
Dietrich, W. E., & J. P. Thornber. (1971) Biochim. Biophys. Acta 245:482-493.
Dunahay, T., & L. A. Staehelin. (1986) Plant Physiol. 80:429-434.
Dunahay, T. G., & L. A. Staehelin. (1985) Plant Physiol. 78:606-613.
Ghanotakis, D. F., & G. T. Babcock. (1983) FEBS Lett. 153:231-234.
Ikeuchi, M. M., M. Yuasa, & Y. Inoue. (1985) FEBS Lett. 185:316-322.
Klein, R. R., Mason, H. S. & Mullet, J. E. (1988) J. Cell Biol. 106: 289-301
Laemmli, U. K. (1970) Nature London 227:680-685.
Lam, E., W. Ortiz, & R. Malkin. (1984a) FEBS Lett. 168:10-14.
Lam, E., W. Ortiz, S. Mayfield, & R. Malkin. (1984b) Plant Physiol. 74: 650-655.
Leto, K. J., E. Bell, & L. McIntosh. (1985) EMBO J. 4:1645-1653.
Machold, O. (1984) Advances in photosynthesis research Nijhoff/Junk: The Hague-II: 107-114.
Machold, O., D. J. Simpson, & B. Moller-Lindberg. (1979) Carlsberg Res. Commun. 44:235-254.
Machold, O. (1986) Carlsberg Res. Commun. 51:227-238
Markwell, J. P., & J. P. Thornber. (1982) Plant Physiol. 70:633-636.

Markwell, J. P., J. P. Thornber, and R. T. Boggs. (1979) Proc. Natl. Acad. Sci. USA 76:1233-1235.

Mullet, J. E., J. J. Burke, & C. J. Arntzen. (1980a) Plant Physiol. 65:814-822.

Mullet, J. E., Burke, J.J. & Arntzen, C. J. (1980b) Plant Physiol. 65:823-837.

Nechushtai, R., & N. Nelson. (1984) Adv. Photosyn Res. II:85-93.

Nechushtai, R. & Nelson, N. (1985) Plant Mol. Biol. 4:377-384.

Nechushtai, R., S. D. Nourizadeh, & J. P. Thornber. (1986) Biochim. Biophys. Acta 848:193-200.

Nechushtai, R., C. C. Peterson, G. F. Peter, & J. P. Thornber. (1987) Eur. J. Biochem. 164:345-350.

Ogawa, T., F. Obata, & K. Shibata. (1966) Biochim. Biophys. Acta 112:223-234.

Ortiz, W., Lam, E., Chollar, S., Munt, D. & Malkin, R. (1985) Plant Physiol. 77:389-399.

Peter, G.F. & Thornber, J.P. (1988) in Photosynthetic Light-harvesting Systems - Structure and Function (Scheer, H. & Schneider, S., eds.) W. de Gruyter and Co., Berlin, in press.

Shiozawa, J. A., R. S. Alberte, & J. P. Thornber. (1974) Arch. Biochem. Biophys. 165:388-397.

Thornber, J. P. (1986) Encl. Plant Physiol. New Series:19-98.

Thornber, J. P., G. F. Peter, R. Nechushtai, P. R. Chitnis, F. A. Hunter, & E. M. Tobin. (1986) In Regulation of Chloroplast Differentiation (Akoyunoglou G. & Senger H.), Alan Liss, N.Y. :249-258.

Thornber, J. P., C. A. Smith, & J. L. Bailey. (1966) Biochem. J. 100:14-15.

Tittgen, J., J. Hermans, J. Steppuhn, T. Jansen, C. Jansson, B. Anderson, R. Nechushtai, N. Nelson, & R. G. Herrmann. (1986) Mol. Gen. Genet. 204:258-265.

Vernon, L. P., E. R. Shaw, & B. Ke. (1966) J. Biol. Chem. 241:4101-4123.

Structure, function and biogenesis of nuclear-encoded proteins of photosystem II

Bertil ANDERSSON[1] and Reinhold G. HERRMANN[2]

[1] Department of Biochemistry, Arrhenius Laboratories, University of Stockholm, S-106 91 Stockholm (Sweden) and
[2] Botanisches Institut der Ludwig-Maximilian Universität, Menzinger Strasse 67, D-8000 München (FRG)

INTRODUCTION

The components required for the photosynthetic energy conversion of plants have their location in the chloroplast thylakoids. This membrane system has a very high degree of organization which is manifested at several structural levels (Anderson and Andersson, 1982). To begin with, there is a pronounced transverse asymmetry of lipids and protein such that the electron acceptor sides of both photosystems are located at the outside of the thylakoid membrane whereas the donor sides are located near the inner membrane surface. Secondly, the plant thylakoid membrane has a heterogeneous organization in the lateral plane of the membrane. Photosystem I (PS I) and the ATP synthase are excluded from the appressed regions of the thylakoids and confined to the stroma-exposed membrane regions. The appressed membrane region contains most of photosystem II (PS II) and its light-harvesting apparatus, including a number of chlorophyll a/b proteins. Thirdly, most of the proteins of the thylakoid membrane are organized into at least four supramolecular complexes, ATP synthase, cytochrome b/f complex, PS I and PS II.

The thylakoid membrane contains proteins encoded by the plastid genome (plastome) and the nuclear genome, adding to the complexity of the membrane. Thus, when considering thylakoid membrane biogenesis and adaptation one has to consider events such as: the coordinated expression of two intracellularly segregated genomes, the import and processing of precursor proteins from the cytoplasm, the assembly of functional protein complexes containing subunits derived from the two genomes, the incorporation of catalytic ligands and chlorophyll, and the acquisition of the correct location of proteins along the stacked thylakoid membrane.

In this communication we will concentrate on the composition, organization and biosynthesis of PS II, with special emphasis on the nuclear encoded-protein subunits. Just 10 years ago PS II was more or less a black box from a biochemical point of view. However, during this decade there has been rapid progress through the combination of biochemistry and molecular genetics. We know now that PS II, including its total light-harvesting apparatus, is a giant multiprotein

----

We thank professor P. Westhoff and Dr. U. Ljungberg for stimulating collaboration and discussions. The work has been supported by the Swedish Natural Science Research Council and the German Research Foundation.

complex comprised of at least 20 different subunits species.

PLASTID-ENCODED PROTEINS OF PHOTOSYSTEM II

       Before describing the nuclear-encoded proteins of PS II, a short
description of the plastid-encoded proteins (Herrmann et al., 1985)
will be given (Table 1). Currently, a lot of attention is given to
the D_1- and D_2-proteins (Fig. 1). These are two proteins, around 32
kDa in size, which show a high degree of sequence homology with each
other (Alt et al., 1984). Both are integral membrane proteins, each
with five trans-membrane helices as suggested from hydropathy analysis of primary sequence data (Trebst, 1986). For the D_1-protein, the
secondary structure prediction has been supported by comparing the
binding of site-specific antibodies to inside-out and right-side out
thylakoid vesicles (Sayre et al, 1986). The D_1- and D_2-proteins have
a high degree of regional sequence homology with the L and M reaction
center subunits of photosynthetic purple bacteria. By analogy with
the organization of the reaction center of the photosynthetic bacteria, as deduced from crystallographic studies, it is thought that the
D_1- and D_2 proteins are arranged as a heterodimer carrying all the
redox components required for the primary photochemical reactions of
PS II (Michel and Deisenhofer, 1986; Trebst, 1986). This assumption
has been experimentally supported by the isolation of a D_1/D_2/ cytochrome b_{559} protein complex competent in the primary charge separation of PS II (Nanba and Satoh, 1986; Barber et al., 1986). The physiological function of cytochrome b_{559} still remains unknown. It may
donate electrons to the reaction center under conditions of impaired
water oxidation or it may be required for cyclic electron flow around
PS II. Cytochrome b_{559} is probably in itself a heterodimer where the
heme group is bound between two small polypeptides of 9 and 4 kDa
(Herrmann et al., 1984). Each of these two polypeptides are predicted
to have one membrane span, each of which carries a single histidine
residue as a likely heme ligand. PS II contains two plastid-encoded
chlorophyll a binding proteins of 51 and 44 kDa apparent molecular

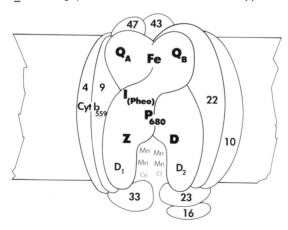

Fig. 1. Schematic illustration of the composition and organization
        of photosystem II. The numbers indicate the apparent molecu-
        lar weight to the proteins subunit. Z and D represent donor
        components. Q_A and Q_B are the primary and secondary quinone
        electron acceptors. P_{680} is the primary electron donor while
        the pheofytin I is the primary acceptor.

_ _

Table 1. Properties of photosystem II polypeptides from spinach deduced from nucleotide sequence analysis

Protein	Coding compartment	Precursor resid./MolWt	Mature Protein resid./MolWt	Transit Sequence resid./MolWt	Predicted membrane spans
51 kDa	p	–	508/56.3	–	5-7
44 kDa	p	–	473/51.9	–	5-7
32 kDa (D1)	p	–	353/39.0	–	5
32 kDa (D2)	p	–	353/39.5	–	5
cyt. b-559/1	p	–	83/9.4	–	1
phosphoprot.	p	–	73/7.8	–	1
cyt. b-559/2	p	–	39/4.4	–	1
33 kDa	n	331/33.0	247/26.5	84/8.5	0
23 kDa	n	267/28.5	186/20.2	81/8.3	0
16 kDa	n	232/24.9	149/16.5	83/8.4	0
10 kDa	n	140/14.4	99/10.2	41/4.2	1
LHCII	n	267/28.4	233/25.1	34/3.3	3
20 kDa	n	261/27.7	210/22.7	51/5.0	2

weight. Both are integral membrane proteins predicted to have 5-7 membrane spans. They are involved in light-harvesting as the core antenna of PS II but they may also fulfill catalytic or structural functions. A 9 kDa phosphoprotein is present in PS II and has been identified as a gene product of the plastid chromosome (Westhoff et al., 1986). PS II contains several small polypeptides (3-7 kDa) of unknown function (Ljungberg et al., 1986). One of these polypeptides has very recently been shown to be plastid-encoded (N. Murata, personal communication).

BIOCHEMICAL ASPECTS ON NUCLEAR-ENCODED PROTEINS OF PHOTOSYSTEM II

The 33, 23 and 16 kDa extrinsic proteins of the oxygen evolving system

The isolation of everted thylakoid membranes made it possible to specifically release extrinsic proteins from the inner thylakoid surface where the water oxidation site is located (Murata and Miyao, 1985; Andersson and Åkerlund, 1987). Through such experiments, previously unknown proteins of 33, 23 and 16 kDa molecular weights were shown to be involved in the oxygen evolving reaction. A standard treatment to inhibit water oxidation and to release manganese from thylakoids is alkaline Tris-washing. When the Tris-mediated inactivation of oxygen evolution was performed by washing of everted thylakoids, not only manganese was released but also four polypeptides (Table 2). These polypeptides were designated the 33, 23, 16 and 10 kDa proteins according to their apparent molecular weights. This inactivation was irreversible, i.e. no reconstitution was obtained when the released proteins were added back to the washed thylakoid membranes.

The first reversible inactivation of oxygen evolution was accomplished by washing everted thylakoids with high concentrations of NaCl. This treatment released only the 23 and 16 kDa proteins while the manganese, the 33 and 10 kDa proteins remained membrane bound. Readdition of the 23 kDa protein resulted in a partial restoration of the lost activity. Several washing procedures have now been developed (Table 2) that allow reconstitution of oxygen evolution not only with the 23 kDa protein but also with the 16 and 33 kDa proteins. Of particular interest was the introduction of 1M $CaCl_2$ for washing of everted thylakoids. This treatment released only the 33, 23 and 16 kDa proteins while the manganese ions remained membrane-bound. However, the manganese became more unstable with time and accessible to foreign reductants. After $CaCl_2$ treatment, the 33 kDa protein can restore some of the lost oxygen evolving activity alhough maximal restoration required the other two proteins.

A crucial observation was that in everted thylakoids deprived of the 33, 23 and 16 kDa proteins oxygen evolution could be restored by the addition of 5-10 mM $CaCl_2$ or high concentrations of NaCl. These experiements show clearly that none of the three proteins is obligatory for in vitro catalysis of water oxidation. Rather, they excert regulatory roles required for stabilization but not ligation of the manganese cluster and regulation of the binding or sequestering of the calcium and chloride ions around the site of catalysis (Fig. 1). Since the water oxidation can proceed in the absence of the three proteins, possible binding sites for Ca^{2+}/Cl^- must reside on integral membrane proteins. The three proteins may therefore regulate the affinity of these putative ion binding sites or increase the local concentration of the two ions around the binding sites.

Analyses of the primary sequences of the 33, 23 and 16 kDa proteins
(Jansen et al., 1987; Tyagi et al., 1987) have not suggested any
high affinity sites for calcium and/or chloride ions. However, preliminary results suggest that the three proteins undergo calcium dependent conformational changes (B. Andersson, J. Bristulf and S.
Forsén, unpublished). When applied to hydrophobic chromatography the
three proteins all showed the same variations in the elution pattern
in response to the presence of calcium or EGTA typical for calmodulin and other calcium binding proteins. To unravel the molecular
mechanism by which the three proteins influence the calcium and chloride requirement in PS II certainly needs further experimentation.

The 33, 23 and 16 kDa proteins can be isolated by a number of
different techniques (Andersson and Åkerlund, 1987). None of the
isolated proteins, which all are perfectly water soluble, contain
any redox ligands in agreement with their non-catalytical role in
the water oxidation process. They are all composed of a high number
of charged amino acid residues, in particular arginine and lysine.
The primary amino sequences of the three proteins, deduced from nucleotide sequencing of cDNA clones (Jansen et al., 1987; Tyagi et al.,
1987), predict no membrane spans. This is consistent with the biochemical data suggesting that the three proteins are extrinsic membrane proteins bound to the inner thylakoid surface through electrostatic and/or hydrogen bonding. The amino acid sequences contain
putative amphipathic β-sheets and α-helices that are potential candidates for subunit/subunit interaction in such a binding to the
membrane. The 16 kDa protein appears to be bound to the 23 kDa protein, which in turn is anchored to the membrane by an integral membrane protein (Fig. 1). This is also the case for the 33 kDa protein. Immunoprecipitation nearest neighbour analysis suggest, that
the 33 and 23 kDa proteins are in close proximity to two proteins of
10 and 22 kDa apparent molecular weight. As will be discussed below
it has been experimentally verified that at least the 10 kDa protein
is essential for binding of the 23 kDa protein. Moreover, the increased binding of site-directed D_1-protein antibodies to everted thylakoids after Tris-washing suggests a close association between the
extrinsic proteins and the PS II reaction center (Sayer et al.,
1986). This association may be particularly tight for the 33 kDa
protein considering its presence in cyanobacterial thylakoids and
oxygen evolving PS II core preparations where the 23, 16 and 10 kDa
proteins are absent.

The integral 10 and 22 kDa proteins

When monospecific antibodies against the extrinsic 33 and 23
kDa proteins were added to partially solubilized PSII particles the
antibodies did not only precipitate the antigenic proteins but also
proteins of 10, 22 and 24 kDa (Ljungberg et al., 1984). Several li-
nes of evidence suggest that the 10 and 22 kDa proteins are components of PS II. Both proteins have been isolated after solubilization of thylakoid membranes followed by ion-exchange chromatography
in the presence of detergents (Ljungberg et al., 1986). Immunological analysis reveals that both proteins are enriched in subthylakoid
fractions, derived from the appressed thylakoid regions, while depleted in stroma lamellae thylakoids. Furthermore, the 22 kDa protein
is found in certain oxygen evolving PS II core preparations.
(Ghanotakis and Yocum, 1986). The 10 kDa protein is identical to the
10 kDa protein which is partially released from everted thylakoids

Table 2. Release of proteins and manganese from everted thylakoid
          membranes

Washing media	protein (kDa)					Mn
	33	23	16	22	10	
0.8 M Tris pH 8.5	+	+	+	-	(+)	+
1 M CaCl$_2$	+	+	+	-	-	-
1 M NaCl	-	+	+	-	-	-
1 M NaCl/0,06% TX-100	-	+	+	(+)	+	-
3 M NaSCN	+	+	+	+	(+)	+

together with the three extrinsic proteins and manganese upon Tris-
washing. Both the 10 and 22 kDa proteins can be released from everted thylakoids by washing with 1 M NaCl/0.06 % Triton X-100 or 3 M
NaSCN (Table 2). The NaCl/Triton treatment released as much as 88%
of the 10 kDa polypeptide and 68% of the 22 kDa protein in addition
to a complete loss of the 23 and 16 kDa proteins. Strikingly, the
33 kDa protein and manganese remained membrane bound after this
treatment. The release of the 10 and 22 kDa proteins, which resulted
in a nearly complete loss of oxygen evolving activity, required the
combination of salt and detergent. Addition of NaCl alone removed
only the 23 and 16 kDa proteins and 0.06% Triton X-100 alone did not
release any proteins. The lost oxygen evolving activity after the
NaCl/Triton X-100 treatment could be partly restored by CaCl$_2$, indicating that none of the two proteins are catalytic subunits of PS
II. The readdition of the 23 kDa protein could not restore the lost
activity which is the case after normal NaCl washings. This inability was caused by the failure of the 23 kDa protein to rebind to the
everted thylakoids after the NaCl/Triton X-100 treatment. The loss
of 23 kDa rebinding capacity (90%) correlated more closely with the
removal of the 10 kDa protein (88%) compared to the loss of the 22 kDa
protein (68%). This suggests that the 10 kDa protein provides the
binding sites for the 23 kDa protein at the inner thylakoid surface.
Moreover, the NaCl/Triton X-100 washing experiments suggest that the
33 kDa protein is not bound to the membrane through the 10 and 22
kDa proteins but more likely by the reaction center polypeptides as
discussed above.
     The NaCl/Triton X-100 treatment of everted thylakoids causes a
significant change in the properties of the acceptor side of PS II
(Henrysson et al., 1987). The inhibition of the PS II activity was
more pronounced when using an acceptor taking electrons from the
Q_B-site than when using an acceptor taking electrons from the Q_A-
site. Moreover, the NaCl/Triton treatment decreased the sensitivity
of the PS II activity to DCMU. Similar results have been obtained
when comparing PS II core preparations containing or lacking the 22
kDa protein (Ghanotakis and Yocum, 1986). The favored explanation is
that removal of the 22 kDa protein, rather than the 10 kDa protein,
is responsible for the changes in properties of the PS II acceptor
functions. It can therefore be speculated that the 22 kDa protein
may fulfill a function similar to the H-subunit of the reaction cen-

ter of photosynthetic purple bacteria. Determination of the amino
acid sequence of the 22 kDa protein will possibly give further support to this hypothesis. The sequencing of a full-sized cDNA-clone of
the 22 kDa protein is currently in progress.

The amino acid composition of both the 10 and 22 kDa proteins
has a fairly high proportion of hydrophobic amino acids. During Triton
X-114 fractionation, in which integral membrane proteins partition
into the detergent phase while extrinsic proteins partition into the
aqueous phase, both proteins were found in the detergent phase. The
amino acid sequence of the 10 kDa protein has been deduced from
nucleotide sequencing (Lautner et al., 1988). The predicted folding
pattern for the mature 10 kDa protein indicates one hydrophobic segment of ca 20 amino acid residues that is sufficiently long and
possesses sufficient hydrophobic moment to span the lipid bilayer in
an α-helical conformation (Fig. 2). This single transmembrane span
of the 10 kDa protein is flanked by a large N-terminal and a short
C-terminal hydrophilic domain of 72 and 7 amino acid residues, respectively. Observations such as accessibility of the 10 kDa protein
to proteases, its release from everted thylakoids by the various
washings procedures, and its role as docking protein for the extrinsic 23 kDa protein suggest that the large N-terminal domain is located towards the thylakoid lumen.

Interestingly, the 10 kDa protein exhibits a high degree of
homology (>84%) to a recently published sequence determined for an
unidentified polypeptide from a potato cDNA (Eckes et al., 1986).
This high degree of conservation suggests that the 10 kDa protein is
crucial for the function of PS II and it may therefore play additional roles to the structural function outlined above.

Photosystem II polypeptides of low molecular weight

Isolated PS II core preparations contain several low-molecular-
weight polypeptides (Ljungberg et al., 1986). Apart from the 4 kDa
subunit of cytochrome b_{559} we do not yet know the function of any of
these polypeptides. At least one of these polypeptides (5 kDa) is
hydrophilic and encoded by the nuclear DNA.

Light-harvesting chlorophyll a/b proteins of photosystem II

Apart from the 51 and 44 kDa chlorophyll a binding proteins PS
II contains a highly complex and heterogenous assembly of chlorophyll a/b proteins (J.P. Thornber, this volume). The main component
is LHCII which contains approximately 30% of the thylakoid protein
mass and up to 50% of the total chlorophyll. In spinach LHCII is
composed of two major closely related polypeptides of 27 and 25 kDa
in addition to some polypeptides of low abundance. LHCII is arranged
into two subpopulations (Larsson et al., 1987). One subpopulation is
closely associated with the PS II core, and contains only the 27 kDa
polypeptide. The other subpopulation is more peripherally located
with respect to PS II and contains the 27 and 25 kDa proteins in
about the same relative amounts. The latter LHCII population can be
heavily phosphorylated leading to reversible detachments from PS II
and subsequent lateral migration from the appressed thylakoid regions to the PS I-rich stroma lamellae. This pool of LHCII is therefore
necessary for the regulation of light-harvesting capacity of PS II.
The protein components of the LHCII complex are encoded by multiple
genes in the nuclear DNA and imported into the chloroplast. LHCII

is an integral membrane complex and its polypeptides are normally
thought to have 3 membrane spans (Karlin-Neumann et al., 1985). Apart
from LHCII, photosystem II contains additional chlorophyll a/b proteins designated CP29, CP27 and CP24 (see J.P. Thornber, this volume).
    Antibodies raised against a 20 kDa polypeptide resolved by SDS-
PAGE of isolated PS II membranes picked up a cDNA clone encoding a
precursor protein of 261 amino acids (Table 1). The mature protein
contained 210 amino acids corresponding to a molecular weight close
to 23 kDa. Sequence analysis suggests that this 20 kDa protein is a
chlorophyll a/b carrying protein (Steppuhn et al., in preparation).
The 20 kDa apoprotein appears phylogenetically more distant than
the various members of the LHCII gene family. The overall homology
with LHCII polypeptides is only 40% but there are clearly conserved
domains (60-80%) which are predominantly located in the two predicted transmembrane segments and their flanking regions. Remarkably,
the two regions in the 20 kDa protein can only be aligned with the
first and third predicted transmembrane α-helix of the LHCII apo-
polypeptides (Fig. 2). If correct the surprising consequence would
be that almost identical helices might be arranged in either an an-
tiparallel way, as in the 20 kDa protein, or parallel, as in the
LHCII proteins. Alternatively, the predicted models for these polypeptides need to be revised. Preliminary biochemical analyses suggest that the 20 kDa apopolypeptide may belong to the CP 24 chlorophyll a/b protein (M. Spangfort, personal communication).

ISOLATION OF RECOMBINANT DNAs FOR NUCLEAR-ENCODED PHOTOSYSTEM II
POLYPEPTIDES; GENE DOSAGE DETERMINATION

    As mentioned above, the understanding of PS II organization and
function have relied on a combination on biochemistry and molecular
genetics. The isolation of nuclear-encoded genes for PS II polypeptides has been fairly slow, probably due to the enormous complexity
of the genome. The identification of both plastid and nuclear genes
depends initially upon the preparation of active PS II fractions, the
production of antisera against individual subunits, and the immunological detection of specific polypeptides produced by cell-free
translations that are programmed with either poly A+- or plastid RNA.
In contrast to plastid-encoded proteins, all nuclear-encoded proteins
studied so far are decoded from polyadenylated RNA as precursor forms
with transit sequences of greatly varying sizes (Table 1).
    Hybrid selection translation and DNA-programmed coupled trans-
cription/translation using recombinant plastid DNA fragments of known
map position have been used to identify most plastome-encoded genes
and to locate them on the spinach plastid chromosome (summarized in
Herrmann et al., 1985). Clone selection for PSII polypeptides that
originate from nuclear genes followed a different strategy. It was
based on a two step procedure. The first step involved cloning of
cDNAs derived from polyadenylated RNA of spinach seedlings illuminated for various periods after etiolation. This was done in the expression vector lambda gt11 (Huynh et al., 1985), followed by immunological screening of the cDNA-library (Tittgen et al., 1985). This
approach inherently selects against "full-size" clones since it depends on "in-frame" fusions with the vector's lacZ gene. Recombinant phage that encode entire precursor polypeptides were obtained
in a second screening cycle utilizing a 5'-terminal probe of an im-
munologically selected insert for plaque screening by hybridization.
This 5'-primed DNA fragment was recloned in transcriptional expression vectors such as pSP64 or Bluescript M13+, and labelled to high

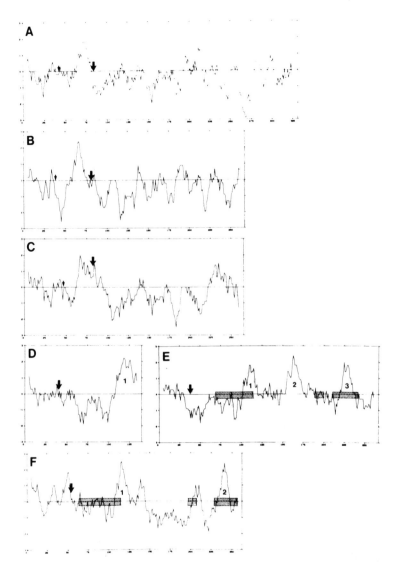

Fig. 2. Hydropathy profiles of the extrinsic 33 kDa (A), 23 kDa (B)
and 16 kDa (C) polypeptides of the water oxidation system,
the 10 kDa polypeptide (D), a member of the LHC II apoproteins (E) and of the 20 kDa chlorophyll a/b apoprotein (F).
The profiles were calculated according to (Kyte and Doolittle,
1982) using an 11-point moving interval. The large and small
arrows indicate the terminal and determined or predicted intermediate processing sites. Those for the LHC apoproteins
are tentative. Putative transmembrane spans are marked by
arabic numbers. Conserved regions between the LHC apoproteins
are boxed and hatched. The numbers on the Y-axis represent
the relative hydrophobicity while the numbers on the X-axis
represent intervalls of 25 amino acids.

specific activity by nick translation (Rigby et al., 1977), or transcription with SP6, T3 or T7 RNA polymerase.
      To date, the application of this strategy has allowed the iden-
tification of recombinant cDNA clones for 26 nuclear-encoded components, including 8 components associated with PS II. The identity of
the cDNA inserts was ensured in three ways, (i) serologically, (ii)
by combination of nucleotide sequence analyses and determination of
at least the N-terminal amino acid sequences of the mature proteins
by gas phase sequencing, and (iii) by in organello import of the precursor polypeptides made in vitro in cell-free translation assays.
In each instance consistent results were obtained.
      We have used these cDNA clones to evaluate structure and bio-
genesis of the PSII proteins. In addition, we have determined the
number of genes per haploid chromosome set by hybridization to
Southern blots of restricted DNA from isolated spinach nuclei. This
study has shown that, apart from LHCII, all proteins including the
20 kDa chlorophyll a/b apoprotein are products of single-copy genes
(A. Schneiderbauer, personal communication).

IV. BIOGENESIS OF PHOTOSYSTEM II

Gene expression and assembly

      The genes for PS II, which are distributed among two spatially
separated compartments within the cell, is expressed in an organ-
specific way, and hence the biogenesis of this complex is dependent
on two different sources of protein synthesis. These facts raise
intriguing questions concerning the biogenesis of thylakoid membrane
complexes in general, and PS II in particular. The availability of
gene probes and an appropriate collection of antisera have prompted
us to investigate the expression of genes during the development of
the photosynthetic apparatus during greening of etiolated spinach.
Four approaches were used that complement each other: dot blot, S1
analysis, and Northern and Western transfer techniques (in collaboration with P. Westhoff). Comparison of these data has established
several findings. First of all, biogenesis and function of PS II are
multilevel processes and are regulated at various interrelated stages (Herrmann et al., 1985). Secondly, functional PS I and PS II on
the one hand and the cytochrome b/f complex and ATP synthase on the
other accumulate differently in response to light vs. dark/light regimes. Thirdly, at least four mechanisms, (i) endonucleolytic cleavage at discrete sequence elements, (ii) attack to specific 3' ends,
(iii) processing at 5' termini, and (iv) excision of introns, regulate light-independently at a posttranscriptional level to process the
generally polycistronic transcripts in the plastid. These mechanisms
contribute to the differential accumulation of individual RNA species
that are processing products from generally polycistronic transcription units (Westhoff and Herrmann, 1988). Finally, the principal step
in photocontrol of plastome-encoded genes is translational rather
than transcriptional or posttranscriptional (Herrmann et al., 1985).
On the other hand, transcriptional light-regulation has been described for nuclear genes (see below).
      These analysis have also shown the remarkable fact that during
greening the various PSII polypeptides do not appear simultaneously.
Moreover their order of appearance is independent of the coding compartment and the presumed location within the complex and of vastly
different gene-dosages in nucleus (see above) and organelle (up to
5000 copies per cell). For instance, cytochrome b_{559} is structurally

closely associated with the D_1- and D_2-reaction center proteins. However, the cytochrome is present in the dark and does not increase substantially during greening, although, the D_1-, D_2-proteins are detectable only after 2-3 hours of illumination. This is also the case for the 51 and 44 kDa chlorophyll a proteins. Similarly, among the nuclear encoded components little increase in the extrinsic 33, 23 and 16 kDa proteins is observed during the dark/light transition. However, the integral 10 kDa protein and the chlorophyll a/b proteins, including the 20 kDa subunit, become detectable only in the light.

This pattern of appearance during greening may explain variations in the stoichiometric relation between the three extrinsic proteins and the PS II reaction center. Values in the range of 1-3 have been presented (Murata and Miyao, 1985; Andersson and Åkerlund, 1987). Probably only one copy of each extrinsic protein is associated with the PS II core. Under low light conditions the three proteins are synthesized in excess of available binding sites and are therefore present in an unbound form in the thylakoid lumen.

Surprisingly, all these PSII genes are constitutively expressed and all the steady state mRNA levels increase substantially (20- to 40-fold) regardless of the mode of appearance and accumulation of the polypeptides during development from etioplast to chloroplast. These discordant changes in mRNA and polypeptide levels without noticeable protein turn-over, except for the D_1- and D_2-proteins, indicate that light control in the nucleo/cytosolic compartment cannot be exclusively transcriptional. This is also suggested by the obvious uncoupling of transcription and translation in some instances. Although the analyses are not yet complete, it is expected that equivalent regulatory mechanisms that control plastid RNA stability during dark/ light transitions (Herrmann et al., 1985; Westhoff and Herrmann, 1988) are also operational for nuclear genes. To date, there is no rational explanation to why the individual complexes or subunits of the individual complexes are expressed so differently.

Transit sequences, protein import and sorting

A central aspect in the biogenesis of PS II is the posttranslational import of the nuclear-encoded precursor polypeptides from the cytosol into the chloroplast, and their subsequent assembly with the plastid-encoded components into functional PSII complexes. The amino acid sequence preceding the determined N-terminal residue of the mature proteins serve as transit sequences (Fig. 2), which are removed during or after import and bear topogenic signals for the intracellular sorting and possibly also for routing of the polypeptides to their final destination within the organelle. All nuclear-encoded components of PS II traverse the two organelle envelopes but their interaction with the thylakoids, the third membrane, is different. Some proteins are integrated (20 kDa protein, LHC II) or anchored (10 kDa protein) in the lipid matrix of the thylakoids, while others (33, 23 and 16 kDa proteins) pass this membrane completely.

Yet another complication in the assembly of PSII is its location within the appressed thylakoid region. It is likely that the new proteins first interact with the exposed stroma thylakoids and then diffuse along the membrane into the appressed membrane regions. The molecular mechanism for such a lateral trafficing of PS II proteins during biogenesis and degradation is an interesting area for future research.

Substantial information has now been obtained concerning the

transit peptides of the nuclear encoded PSII proteins. As previously
shown for plastocyanin, the transit sequences of the extrinsic luminal polypeptides and that of the 20 kDa protein are functionally divided and are processed in at least two steps (Tyagi et al., 1987;
Jansen et al., 1987 and unpublished data). Two endopeptidases, located in the stroma and thylakoid membrane respectively, have been described (Robinson and Ellis, 1984). They appear to be involved in the
cleavage of at least some of the components mentioned above. Intermediate processing products have been detected for the 33 kDa, 23 kDa,
16 kDa and 20 kDa polypeptides. Comparison of the transit sequences
for these proteins show that they possess a remarkably variable primary structure (<10% conserved residues). Similar to other transit
sequences, they possess a positively charged N-terminal region abundant in hydroxylated amino acid residues. These proteins obey the
-1/-3 rule for amino acids with small side chains preceding the terminal cleavage site (von Heijne, 1986), but any attempts to deduce
other primary sequence elements that could be of potential strategic
importance have failed. However, secondary structure predictions
strongly suggest that the transit sequences of the nuclear encoded
PS II polypeptides fall into three different classes (Fig. 2). The
transit sequences of the 33, 23 and 16 kDa proteins, and remarkably
also of the 20 kDa protein, are relatively long (Table 1). They all
have a typical hydrophobic domain preceding the terminal cleavage
site. The intermediate cleavage site is characterized by an amphipathic β-sheet (which in the case of the 23 kDa protein is sided but
not amphipathic). This secondary structure prediction also applies
to the transit sequence of plastocyanin. At least in two instances
the β-sheet structure has now been experimentally shown to provide
the intermediate cleavage site for the thylakoid protease (in collaboration with D. Bartling and C. Robinson).

The transit sequences of the 10 kDa polypeptide and the LHCII
polypeptides deviate substantially both from each other and from the
transit segments of the luminal proteins. They are relatively short
and lack the characteristic secondary elements, specifically the
hydrophobic domain, outlined above. The transit peptide of the 10
kDa protein exhibits remote similarity to transit sequences for the
Rieske Fe/S polypeptide of the cytochrome b/f complex and to the
delta subunit of the ATP synthase from spinach chloroplast (Steppuhn
et al., 1987; Herrmans et al., 1988). The elucidation of the molecular significance of these variations in the transit sequences is an
interesting area for future research.

REFERENCES

Alt, J., Morris, J., Westhoff, P. & Herrmann, R.G. (1984) Curr.
  Genet. 8, 597-606.
Anderson, J.M. & Andersson, B. (1982) Trends Biochem. Sci. 7, 288-292.
Andersson, B. & Åkerlund, H.-E. (1987) in Topics in Photosynthesis
  (Barber. J., ed.), vol. 8, pp. 379-420, Elsevier, Amsterdam.
Barber, J., Chapman, D.J. & Telfer, A. (1987) FEBS Lett. 220, 67-73.
Eckes, P., Rosahl, S., Schell, J. & Willmitzer, L. (1986) Mol. Gen.
  Genet. 205, 14-22.
Ghanotakis, D.F. & Yocum, C.F. (1986) FEBS Lett. 197, 244-248.
von Heijne, G. (1986) Nucleic Acids Res. 14, 4683-4690.
Henrysson, T., Ljungberg, U., Franzén, L.-G., Andersson, B. à Åker-
  lund, H.-E. (1987) in Progress in Photosynthesis Research (Biggins,
  J. ed.), vol. 2, pp. 125-128.

Hermans, J., Rother, Ch., Bichler, J., Steppuhn, J. & Herrmann, R.G. (1988) Plant Mol. Biol., 10, 323-330.

Herrmann, R.G., Westhoff, P., Alt, J., Tittgen, J. and Nelson, N. (1985) in Molecular Form and Function of the Plant Genome (v. Vloten-Doting, L., Groot, G. and Hall, T., eds.), pp. 233-256, Plenum Publ. Corp.

Herrmann, R.G., Alt, J., Schiller, B., Widger, W.R. and Cramer, W.A. (1984) FEBS Lett. 176, 239-244.

Huynh, T.V. Young, R.A., Davis, R.W. (1985) in DNA cloning: a practical approach (Glover DM, ed.), vol. 1, pp. 49-88, IRL-Press, Oxford, Washington, DC.

Jansen, Th., Rother, Ch., Steppuhn, J., Reinke, H., Beyreuther, K., Jansson, Ch., Andersson, B. and Herrmann, R.G. (1987) FEBS Lett. 216, 234-240.

Karlin-Neumann, G.A., Kohorn, B.D., Thornber, J.P. & Tobin, E.M. (1985) J. Mol. Appl. Genet. 3, 45-61.

Kyte, J. & Doolittle, R.F. (1982) J. Mol. Biol. 157, 105-132.

Larsson, U.K., Sundby, C. & Andersson, B. (1987) Biochim. Biophys. Acta 894, 59-68.

Lautner, A., Klein, R., Ljungberg, U., Bartling, D., Andersson, B., Reinke, H., Beyreuther, K. and Herrmann, R.G. (1987) J. Biol. Chem. in press.

Ljungberg, U., Åkerlund, H.-E. & Andersson, B. (1986) Eur. J. Biochem. 158, 477-482.

Ljungberg, U., Åkerlund, H.-E., Larsson, C. & Andersson, B. (1984) Biochim. Biophys. Acta 767, 145-152.

Ljungberg, U., Henrysson, T., Rochester, C.P., Åkerlund, H.-E. & Andersson, B. (1986) Biochim. Biophys. Acta 849, 112-120.

Michel, H. & Deisenhofer, J. (1986) in Encyclopaedia of Plant Physiology (Staehlein, L.A. and Arntzen, C.J., eds.) vol. 18, pp. 371-381, Springer Verlag, Heidelberg.

Murata, N. & Miyao, M. (1985) Trends Biochem. Sci. 10, 122-124.

Nanba, O. & Satoh, K. (1987) Proc. Natl. Acad. Sci. USA, 84, 109-112.

Rigby, P.W.J., Dieckmann, M., Rhodes, C. & Berg, P. (1977) J. Mol. Biol. 113, 237-251.

Robinson, C. & Ellis, R.J. (1984) Eur. J. Biochem. 142, 337-342.

Sayre, R.T., Andersson, B. & Bograd, L. (1986) Cell, 47, 601-608.

Steppuhn, J., Hermans, J., Jansen, Th., Vater, J., Hauska, G. & Herrmann, R.G. (1987) Molec. Gen. Genet. 310, 171-177.

Tittgen, J., Hermans, J., Steppuhn, J., Jansen, Th., Jansson, Ch., Andersson, B., Nechushtai, R., Nelson, N. & Herrmann, R.G. (1986) Molec. Gen. Genet. 204, 258-265.

Trebst, A. (1986) Z. Naturforsch, 41C, 240-245.

Tyagi, A., Hermans, J., Steppuhn, J., Jansson, Ch., Vater, J. & Herrmann, R.G. (1987) Molec. Gen. Genet. 207, 288-293.

Westhoff, P. & Herrmann, R.G. (1988) Eur. J. Biochem. 177, 551-564.

Westhoff, P., Farchaus, J.W. & Herrmann, R.G. (1986) Curr. Genet. 11, 165-169.

Lipid topology and the role of lipids in plant membrane structure

W. Patrick WILLIAMS.

Biochemistry Department, King's College, Campden Hill, London W8 7AH, United Kingdom.

SYNOPSIS

Aqueous dispersions of pure membrane lipids organise themselves into three basic configurations; micelles, bilayers or inverted micelles depending on the balance of forces existing between the molecules of lipid and their aqueous environment. The relevance of such topological arrangements to the structural organisation of plant biomembranes is explored.

1. INTRODUCTION

Current ideas on membrane structure tend to be centred around the fluid-mosaic membrane model of Singer & Nicolson (1972). This model in turn owes a great deal to earlier work emphasising the importance of chain-length and saturation on the phase properties of membrane lipids; particularly their role in determining the temperature of gel-to-liquid crystal transitions in lipid bilayers (see review of Chapman & Benga 1984). Such studies provided a major insight into the factors determining membrane fluidity. Recently a great deal of attention has been focussed on the topological properties of lipids and the existence of lipids that tend to form non-lamellar phases. This review is aimed at exploring the possible relevance of such "non-bilayer" phase behaviour to the structure and function of plant membranes.

2. LIPID PHASE BEHAVIOUR

2.1. Single Lipids

Membrane lipids dispersed in water aggregate to form one of three main types of structure. The structure adopted by a given lipid is partly determined by the overall hydrophilic-hydrophobic balance of the molecules and partly by its molecular geometry. A summary of the structures normally adopted by the main classes of membrane lipids is presented in Fig. 1. Single chain lipids such as

---

ABBREVIATIONS: DGDG digalactosyldiacylglycerol; MGDG, monogalactosyldiacylglycerol; NBL, non-bilayer lipid; PC phosphatidylcholine; PE, phosphatidylethanolamine; PG, phosphatidylglycerol; SQDG sulphoquinovosyldiacylglycerol.

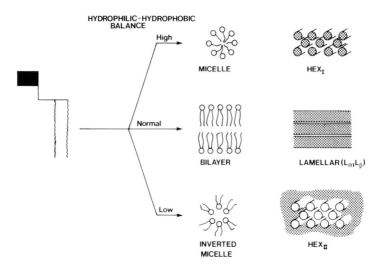

Fig. 1. Dependence of lipid phase behaviour on hydrophilic-hydrophobic balance.

lysolecithin tend to form dispersions of spherical or cylindrical micelles. Lipids of this type are not commonly found in membranes. The majority of lipid classes found in biological membranes form bilayers on dispersion in water. A significant proportion of membrane lipids, however, form structures based on inverted micelles. It is this group that is normally referred to as the non-bilayer forming lipids. In the case of plant membranes, MGDG and PE fall into the non-bilayer group while the remaining lipids including DGDG, SQDG, PG and PC are in the bilayer forming group (see reviews of Quinn & Williams 1983,1985).

The main phases encountered in studies on purified lipids are the liquid-crystal and the gel lamellar phases, denoted L_α and L_β respectively, and the inverted hexagonal phase Hex-II. All membrane lipids form lamellar phases in the gel state. The L_β phase, in which the hydrocarbon chains of the lipids are packed on a regular hexagonal lattice within the plane of the bilayer, is the simplest of such phases. A variety of other gel phases exist in which the chains exhibit other packing geometries but discussion of these is beyond the scope of this review.

The different phases formed by lipids are the results of the tendency of the lipids to maximise the contact between water and their polar headgroups while at the same time minimising contact between water and their hydrocarbon chains. The equilibrium structure adopted by the lipid is governed by the relative strengths of the different interactions between neighbouring lipid molecules, between the lipid molecules and water and steric constraints imposed by the geometry of the individual lipids. A number of theoretical treatments of the factors involved in the self-assembly of lipid aggregates have appeared in the literature (Tanford, 1973; Israelachvili et al., 1976,1980). Of these treatments, the analyses of Israelachvili and his co-workers, which gave rise to the "shape model" of lipid interaction, have had the most impact on membrane biologists.

The shape model is based on the calculation of a packing factor (p) which is defined in terms of the optimal surface area per lipid

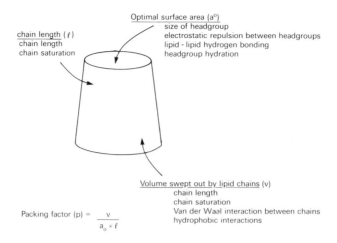

Packing factor (p) = $\dfrac{v}{a_o \times \ell}$

Fig. 2. Relationship between the lipid packing factor (p) and different molecular shape parameters.

----

molecule (a_o), the length of the lipid molecule (l) and the volume swept out by its hydrocarbon tails (v). The relationship between these variables, together with a summary of the main factors determining their value, are shown in Fig. 2. Classification of lipids in terms of their p values, as illustrated in Fig. 3, provides a direct insight into the type of aggregates that might be formed in a given aqueous dispersion. The non-lamellar structures described in this figure, it should be noted, are found only in the liquid-crystal state. As pointed out above, all membrane lipids form lamellar phases in the gel state. The concept of the packing factor is extremely useful in providing a simple qualitative understanding of the potential effects of such factors as temperature, lipid chain

----

Packing Parameter	Critical packing shape	Structures formed
<1/3	Cone 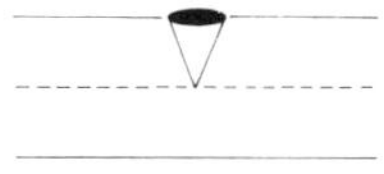	Spherical micelles 
1/2-1	Truncated cone 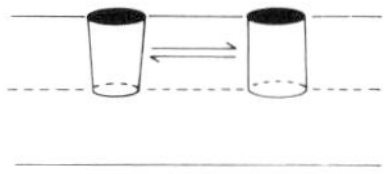	vesicles 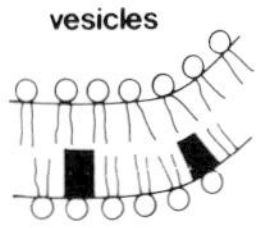
>1	Inverted truncated cone 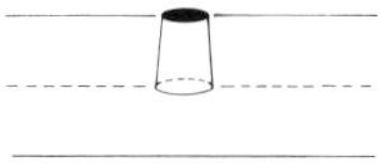	Inverted micelles 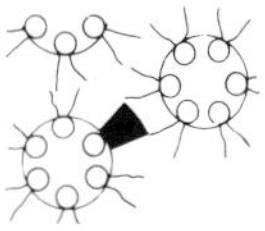

Fig. 3. Classification of lipids in terms of their characteristic packing factors (adapted from Israelachvili et al., 1980).

length and saturation, pH and ionic strength etc., on the phase
behaviour of a given lipid. Raising the temperature, for example,
will tend to increase the value of v, and hence of p, as the kinetic
energy of the hydrocarbon chains increases. This tends to stabilise
non-bilayer phases with respect to bilayer phases. Conversely,
increasing repulsion between headgroups as a result of changing the
pH of a dispersion of acidic lipids, for example, might be expected
to increase a_o and thus stabilise bilayer structures.

It must be stressed that the value of p is ultimately determined
by the overall thermodynamic balance of the aggregate and the
limitations of molecular geometry and that these are influenced by a
wide range of factors. Simple qualitative predictions of the type
cited above while often useful can, therefore, also sometimes be
misleading. This is particularly the case when attempts are made to
extrapolate the model to more complex systems. The temptation to
think of the basic shapes predicted for individual lipids as building
blocks that can be combined together to build up an overall picture
of membrane organisation must be especially strongly resisted.

The shape model with its concept of a variable packing factor
emphasises the fact that there is a continuum between lipids that
form bilayer and non-bilayer structures in the liquid-crystal phase.
Some lipids such as PE can form lamellar (L_a) or inverted hexagonal
(Hex-II) structures depending on the conditions. They tend to form
L_a phases on first entering the liquid-crystal phase but enter
Hex-II phases at higher temperatures. The temperature of the L_a-to-
Hex-II transition, as might be anticipated, is extremely sensitive to
such factors as hydrocarbon chain length, headgroup hydration, pH and
ionic strength (Harlos & Eibl 1981; Seddon et al., 1983). Other
lipids, such as MGDG form the Hex-II phase direct from the gel phase
with no intermediate L_a phase (Shipley et al., 1973; Sen et al.,
1981a).

At first sight, the organisation of lamellar and inverted
hexagonal phases appear to be very different. The rearrangement
involved in going from the L_a to Hex-II phase is, however,
relatively small. This is reflected in the low enthalphy changes
associated with L_a-to-Hex II transitions as opposed to L_a-to-
L_β transitions. In the case of dipalmitoyl phosphatidylethanolamine,
for example, the molar enthalpy value for the gel to liquid-
crystal transition is 33.1 $kJ.mol^{-1}$ while that for the bilayer to
non-bilayer transition is only 1.3 $kJ.mol^{-1}$ (Seddon et al., 1983).
The precise mechanism of bilayer/non-bilayer transitions is still
disputed but probably involves an initial fusion between adjacent
lipid surfaces to form first spherical and then cylindrical inverted
micelles (see Siegel 1986).

2.2 Lipid mixtures

The phase behaviour of even relatively simple lipid mixtures is
often complex. Phillips et al., (1970), using thermal measurements,
demonstrated that the crystallisation of binary mixtures of PC of
different chain length and saturation differs considerably from that
of the individual species. If the components are of similar chain
length and saturation, co-crystallisation takes place at temperatures
intermediate to their normal melting points. If the difference
between the components is large, the transition range of the higher-
melting point component tends to broaden and to occur at a lower
temperature reflecting the disordering effect of the lower melting-
point component on its more ordered chains. The same general
principles apply to mixtures of bilayer forming lipids of different
classes.

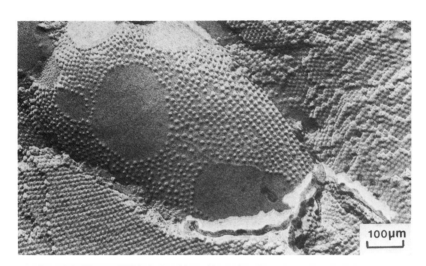

Fig. 4. Freeze-fracture electronmicrograph illustrating some of the different structures formed in aqueous dipersions of MGDG/DGDG (2:1). Data from Sen et al., 1982.

The situation with mixtures of bilayer and non-bilayer forming lipids is even more complex as it involves lipids which tend to form different liquid-crystalline phases. Mixtures containing relatively low proportions of non-bilayer forming lipids tend to stay in a bilayer configuration. A wide range of mixtures exists, however, containing intermediate proportions of bilayer and non-bilayer forming lipids which, on freeze fracture at least, show partial phase separation (Sen et al., 1981b, 1982a,b; Sprague & Staehelin 1984). Such mixtures are characterised by the presence of intermediate structures of the type shown in Fig. 4 consisting of inverted spherical or cylindrical micelles sandwiched within lipid bilayers and a variety of quasi-crystalline arrays.

It is not clear to what extent these structures are thermodynamically stable. Many, particularly those involving isolated spherical and cylindrical micelles, are clearly in a state of dynamic flux (Van Venetie & Verkleij, 1981). It is thus probable that many of the structures visualised by freeze-fracture reflect intermediate stages in conversions between different phases. Others resemble stable cubic phases of types reported for a variety of simple soap and lipid mixtures. There has been considerable debate regarding the relationship between the quasi-crystalline structures seen in freeze-fracture electronmicrographs of mixed dispersions of MGDG and DGDG and such phases (Hui & Boni, 1982; Williams et al., 1982; Brentel et al., 1985). It is, as emphasised by Rilfors et al., (1986), extremely difficult to distinguish between different cubic phases on the basis of freeze-fracture studies alone.

The low energy barriers existing between bilayer and inverted micellar structures means that the organisation of mixtures of bilayer and non-bilayer lipids are particularly susceptible to relatively minor changes in such factors as temperature, hydration, pH, ionic strength etc. In general the likely effect of a given perturbation can be predicted, or at least rationalised, by reference to the changes anticipated on the basis of the shape model. Increases in temperature and decreases in lipid hydration, for example, tend to stabilise non-bilayer structures with respect to

bilayer structures and hence increase the extent of phase
separation. The effect of other factors may be more subtle. In the
case of total polar lipid extracts of chloroplast membranes, we have
shown that lowering the pH of the dispersion or addition of metal
cations leads to extensive phase separation (Gounaris et al., 1983d).
Here, phase separation appears to be due to the neutralisation of
charges on the acidic bilayer forming lipids PG and SQDG rather than
direct effects on the neutral non-bilayer forming lipids.

2.3 Biomembranes

Membranes normally contain both bilayer and non-bilayer forming
lipids. The proportion of non-bilayer forming lipid varies with the
source of the membrane. Myelin, for example, contains little or no
non-bilayer forming lipids while the non-bilayer forming lipid MGDG
accounts for 40-50% of the membrane lipid of the photosynthetic
membranes of chloroplasts and cyanobacteria (Quinn & Williams, 1985).
Most membranes, including plant membranes, contain about 20-30% non-
bilayer lipid. Freeze-fracture studies performed on total polar
lipid extracts of chloroplasts show that extracts dispersed in media
used to isolate chloroplasts show extensive phase-separation (Quinn
et al., 1982). Structures of this type are, however, not normally
seen in chloroplasts under corresponding conditions. This strongly
suggests that lipid-lipid interactions are modified in the native
membranes in some way that tends to prevent phase-separation.

3. BIOLOGICAL RELEVANCE

The first question that needs to be asked in this context is
whether or not there is any real significance in the division of
lipids into bilayer and non-bilayer forming species. Membrane lipids
as emphasised in Section 2.1 show a continuum of physical
characteristics and the configuration adopted by any given lipid is
to a large extent dictated by its immediate environment. The fact
that a lipid forms a non-bilayer phase on dispersion in distilled
water does not necessarily imply that it will tend to take up such a
configuration in a biological membrane. Interaction with other
membrane components and exposure to cytoplasmic components may well
mean that it is thermodynamically as stable in a bilayer
configuration as any conventional bilayer-forming lipid.

Some indication of the possible importance of non-bilayer lipids
is given by the fact that certain micro-organisms appear to adjust
the proportions of bilayer and non-bilayer forming lipids in their
membranes in response to changes in their growth conditions.
Wieslander and his co-workers have shown that the mycoplasma
Acholeplasma laidlawii adjusts the relative proportions of bilayer
and non-bilayer forming lipids in its plasma mebrane to match changes
in its growth temperature (Wieslander et al., 1980a,b) and the
presence of foreign molecules such as hydrocarbons, alcohols and
detergents that can partition into its membrane (Wieslander et al.,
1986). The changes that occur appear to reflect an attempt to
minimise the tendency for non-bilayer lipid phase separation. The
pattern of changes seen in plant membranes is far more complex (see
Quinn & Williams, 1985 for discussion) but the possibility that
maintenance of such a balance is a driving force, even in higher
plants, cannot be ruled out at this stage.

Membranes containing high proportions of non-bilayer forming
lipids tend to share two common characteristics. They are often
highly convoluted and they tend to contain high proportions of

protein. This has led to suggestions that the non-bilayer lipids may either be concentrated in regions of high degrees of curvature (Murphy, 1983) or with the efficient sealing of membrane proteins into the lipid bilayer (Quinn et al., 1983, Williams et al., 1984).

Murphy (1983) calculated that approximately a third of the membrane lipids present in chloroplast membranes are located in the high curvature regions of stacked thylakoid membranes and suggested that the high proportion of MGDG in chloroplasts might be associated with the need to stabilise such structures. Analyses of the distribution of MGDG between the grana and stroma fractions of chloroplasts (Gounaris et al., 1983b, Murphy & Woodward, 1983), indicated that the proportion of MGDG present in the granal fraction was higher than in the stromal fraction. The fact that MGDG was, nevertheless, the predominate lipid in the stromal regions suggests, however, that membrane curvature is unlikely to be a major factor in the distribution of this lipid. There is some evidence to suggest that the non-bilayer character of MGDG plays a major role in the organisation of the pro-lamellar body formed in etioplasts. The pro-lamellar body is a complex three-dimensional network of membranes which, on exposure to light, develops into the thylakoid membrane system of the chloroplast (Gunning, 1965; Simpson, 1978). It contains the same lipids as the thylakoid membrane in rather similar proportions. Its protein content is, however much lower. The overall structure of the body, as pointed out by Larsson et al., (1980), closely resembles that of the bicontinuous cubic phase formed by aqueous dispersions of sunflower oil monoglyceride.

The proposal that non-bilayer lipids may be associated with the packaging of large intrinsic membrane protein complexes, such as the photosynthetic light-harvesting units, in plant membranes is largely based on studies on heat-stressed chloroplasts performed in our own laboratory (see Section 4.2). Evidence supporting this idea has also come from reconstitution studies in which the efficiency of energy transfer between chlorophyll-protein complexes (Siefermann-Harms et al., 1982) and the activity of the chloroplast coupling factor CF_1 (Pick et al., 1984, 1985,1987; Van Walraven et al., 1984) has been shown to be significantly enhanced by the presence of MGDG. A detailed discussion of the possible role of lipid-protein interactions in chloroplasts is provided in the article by Gounaris elsewhere in this volume.

The general idea that proteins can impose a bilayer structure on non-bilayer forming lipids has received considerable support from experiments involving model systems (see reviews of De Kruijff et al., 1985a,b). Taraschi et al., (1982), for example, have demonstrated the ability of glycophorin to stabilise the non-bilayer forming lipid dioleoyl-phosphatidylethanolamine into a vesicular configuration. They reported that each protein molecule could stabilise up to 200 molecules of lipid. Interestingly, addition of wheat agglutinin to the vesicles led to reversion of the lipid to the Hex-II state, presumably reflecting the destabilising effect of crosslinking the proteins into a smaller more compact grouping. Cytochrome c and apocytochrome c, in contrast, have been shown to destabilise the bilayer structure of mitochondrial membranes appearing to induce fusion between the inner and outer membranes (Van Venetie & Verkleij, 1982).

The nature and extent of non-bilayer lipid interaction with intrinsic membrane proteins, it must be stressed, is still unknown. The interactions are, however, almost certainly very weak. It is possible that the hydrocarbon chains of the non-bilayer lipids can sweep out a relatively larger volume than those of other lipids and this assists in stabilising interactions between the bulk lipid and the proteins. Another more attractive possibility is that the low hydration of the lipid headgroups allows a more efficient sealing of the protein into the bilayer.

Cullis, de Kruijff and Verkleij and their co-workers have suggested that non-bilayer forming lipids may play a functional as well as a structural role in biological membranes. They pointed out that the transitory formation of inverted lipid micelles could be involved in the transport of polar molecules across membranes and that they might also act as intermediates in membrane fusion processes (Cullis & de Kruijff, 1979; Cullis et al., 1979; Verkleij et al., 1979). Whilst little direct evidence has, as yet, appeared in the literature to support such views they should not be neglected (see Verkleij, 1984).

It is important to emphasise that with the possible exception of the pro-lamellar body, there is no evidence to suggest that the formation of stable non-bilayer lipid structures occurs in plant membranes occurs under normal conditions. There is however a great deal of evidence to suggest that such structures may be formed under conditions of environmental stress when the normal interactions between membrane components are disrupted. This evidence is discussed in the next section.

4.   LIPID PHASE BEHAVIOUR AND ENVIRONMENTAL STRESS

4.1   General Principles

The stability of a membrane is determined by the balance of interactions between its components, the lipids and proteins, and their aqueous environment. These interactions involve a variety of different factors including van der Waal interactions, hydrophobic interactions, hydrogen bond formation and steric considerations. The

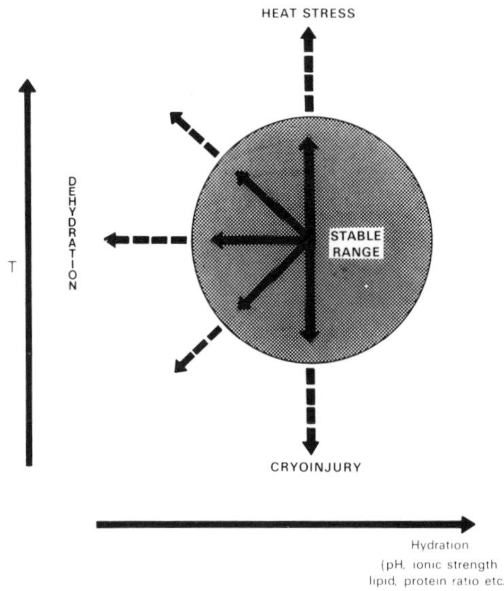

Fig. 5. Diagram illustrating the general concept of the dependence of membrane stability on temperature and lipid hydration.

dependence of the relative strengths of these interactions on such factors as temperature, hydration, pH, ionic strength and the presence or absence of divalent cations is very different. As such, the stability of the membrane necessarily varies with changes in such parameters. Any membrane can thus be considered to have only a limited range of stability. Outside this range, its components are likely to phase separate or if phase separation does not occur, the membrane is likely to become unacceptably fragile or leaky. The dependence of the stable range on temperature and hydration is represented diagrammatically in Fig. 5. These two parameters have been chosen as the main stresses encountered by plants are temperature stress and drought, either alone or in combination. In practice other factors such as changes in osmotic balance and ionic strength need to be taken into account but their inclusion would make the diagram unnecessarily complicated.

It is useful to consider the possible effects of temperature and hydration first on a single lipid, then on mixtures of lipids and then finally on a hypothetical membrane. A phase diagram for the non-bilayer lipid didodecyl phosphatidylethanolamine, taken from the work of Seddon et al., (1983), is shown in Fig. 6a. It shows that dehydration leads to a lowering of the characteristic temperature for phase transition between the L_α and Hex-II phases in the liquid crystalline state and a rather less marked raising of the temperature for the transition between the L_β and L_α states. Both of these changes can be rationalised in terms of the shape model. Removal of water is likely to reduce interaction of the lipid head groups with the aqueous phase and to increase any tendency for intermolecular hydrogen bonding between lipid headgroups. This tends to stabilise both the L_β and Hex-II phases with respect to the L_α phase. Yeagle & Sen (1986) have reported a similar increase in stabilisation of the Hex-II phase in PE in the presence of chaotropic ions which they attribute to an effective dehydration of the membrane surface.

Phase diagrams of mixtures of bilayer and non-bilayer lipids are less easy to find. They are also much more complex to interpret as the lipids are not restricted to L_α, L_β and Hex-II phases and can take up a variety of cubic phases. A simplified version of the phase diagram reported by Brentel et al., (1985) for a binary dispersion of MGDG and DGDG is presented in Fig. 6b. The overall form of the phase diagram is rather similar to that of a single non-bilayer forming

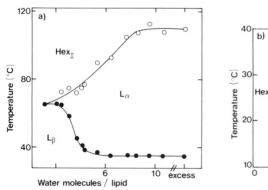
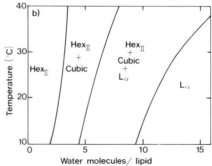

Fig. 6. Phase diagrams of (a) didoceyl phosphatidylethanolamine and (b) MGDG/DGDG (2:1 molar ratio). Data adapted from work of Seddon et al., (1982) and Brentel et al., (1985).

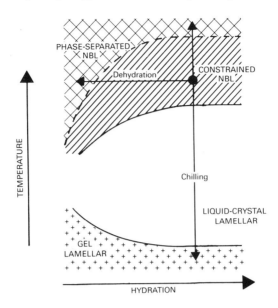

Fig. 7. Diagram illustrating the type of phase separations that might be expected for a natural membrane subjected to different forms of environmental stress.

lipid emphasising the fact that the phase behaviour of such mixtures tends to be dominated by the non-bilayer component (see Gounaris et al., (1983c) and Wieslander et al., (1986)).

Finally, a purely hypothetical diagram showing the sort of behaviour that might be expected for a natural membrane is presented in Fig. 7. One of the major differences between the natural membrane and a total polar lipid extract of such a membrane, as discussed above, is the tendency of other membrane components to constrain non-bilayer forming lipids within a bilayer configuration. This has been taken into account by dividing the non-bilayer region of the diagram into two sub-regions; one in which the non-bilayer lipids are constrained within the membrane and the other in which they are free of such constraints.

We can use this diagram to predict the possible effects of raising and lowering the temperature and of dehydration on the membrane. The extent to which such predictions are in agreement with observed effects is discussed below.

4.2   Thermal Stress

The effects of temperature can be divided into those associated with cooling and those associated with heating. In terms of cooling, a distinct division has to be made between cooling procedures that involve freezing and those that are limited to chilling temperatures above freezing. The effects of freezing damage, are more complicated as they involve a simultaneous dehydration of the membrane. This topic is, therefore, conveniently postponed to a later stage.

If we first consider chilling processes, a further distinction has to be made between chilling processes that involve the formation of gel phases and those that do not. The best studied example of the effects of chilling on a system involving gel-phase formation is that carried out on the thermophilic blue-green alga Anacystis nidulans. The formation of extensive phase-separated regions of gel-phase lipid at temperatures above $0°C$ of Anacystis has been studied by a number of groups using freeze-fracture (Armond & Staehelin, 1979; Brand et al., 1979; Furtado et al., 1979; Verwar et al., 1979; Ono & Murata, 1982), thermal techniques (Furtado et al., 1979; Ono et al., 1983; Mannock et al., 1985) and X-ray diffraction (Tsukamoto et al., 1980). It is well established that the interface between gel and liquid-crystal region in membranes tend to be particularly leaky (Haest et al., 1972). Measurements showing a close correlation between the extent of phase-separation of the lipids of the plasma membrane of Anacystis to the rate of leakage of K^+ ions from such cells (Ono & Murata, 1981a,b) provide convincing evidence for that this is the basis of chilling damage in these cells. There is, however, little or no evidence to suggest that similar large-scale phase separation takes place in the membranes of higher plants (see reviews of Quinn & Williams,(1978,1983) and Williams & Quinn, (1987)).

Murata and his co-workers (Murata 1982,1983; Murata et al., 1983) have pointed out the existence of a correlation betweeen the presence of highly saturated PG species in chloroplasts and chilling-sensitivity in a wide range of plants. The normal gel-to-liquid crystal phase transition temperature of such lipids, which typically represent approximately 5-8% of the total membrane lipid, is about $35°C$ (Murata & Yamaya 1984) but this might be expected to be much lower in the natural membrane for the reasons outlined in Section 2.2 (see also Raison & Wright (1983) and Raison & Orr (1986)). The question as to whether or not these lipids do phase separate in the chloroplast membranes of chilling-sensitive plants and whether such phase separation leads to chilling damage is still a matter of debate. Norman et al., (1986) and Raison & Orr (1987) have recently demonstrated that chilling-insensitive populations of species which are normally chilling-sensitive retain their relatively high contents of saturated PG species. This suggests that the possession of such species may not be as important to chilling sensitivity as originally thought.

An alternative explanation of such damage might be that the lowered temperature leads to a reduced ability of the non-bilayer forming lipids to seal key intrinsic membrane proteins into cell membranes and that this in turn leads to a leakage at protein/lipid interfaces akin to that occurring at gel/liquid crystal interfaces in Anacystis.

The effects of heat stress in plants have been widely studied (see reviews of Berry & Bjorkman (1980) and Quinn & Williams (1985)). The present discussion is, however, limited to the effects of such stress on membrane structure. Reference to Fig. 7, suggests that heat-stress might be expected to encourage the phase separation of non-bilayer forming lipids from plant membranes. There is good evidence to indicate that this happens in the case of chloroplasts at least. We have shown that an irreversible phase separation of non-bilayer occurs in chloroplasts Vicia faba (broad bean) subjected to mild heat-stress ($40-50°C$) (Gounaris et al., 1983a,1984; Williams et al., 1984). The phase-separated lipid normally takes the form of inverted cylindrical micelles, approximately 8-10 nm in diameter. Exposure to higher temperatures, leads to more extensive re-arrangements involving a vesiculation of the thylakoid membranes. We have, however, found no evidence of the formation of spherical inverted micelles of the type commonly seen in lipid mixtures.

There is a great deal of evidence in the literature based on measurements of fluorescence changes and electron transport studies

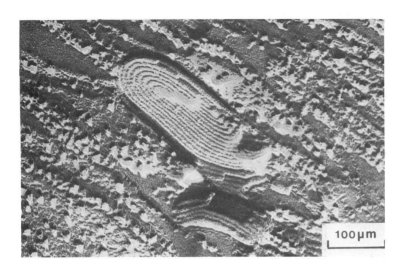

100μm

Fig. 8. Electronmicrograph illustrating the effect of low pH (pH 4.5) on the organisation of the thylakoid membranes of broad bean chloroplasts.
----

indicating that a major re-organisation of the PSII light-harvesting apparatus takes place over the same temperature range as we observe the phase separation of non-bilayer lipids (Quinn & Williams 1985). This has led us to suggest that the non-bilayer forming lipids may play a role in the stabilisation of PSII (Williams et al., 1984). This view is supported by studies carried out in our laboratory indicating that a physical dissociation of the antennae complexes from the core particles of PSII occurs at about 40-50°C (Gounaris et al., 1984). We have also reported evidence suggesting that the cytochrome f/b$_6$ particle, another major protein complex of the chloroplast membrane, becomes increasingly accessible to artificial electron donors over this temperature range (Thomas et al., 1986a). Interestingly, catalytic hydrogenation of the chloroplast membranes, which has the effect of increasing lipid saturation and hence tending to stabilise bilayer phases with respect to non-bilayer phases appears to lead to a shift in these changes to higher temperatures (Thomas et al., 1986b).

The phase-separation of non-bilayer structures, as pointed out in Section 2.2, probably involves a fusion between closely apposed lipid bilayers. As such the stacked thylakoid membrane would seem to be ideally suited for this purpose. It is probably significant that reports of phase-separations of this type in biological membranes are almost entirely restricted to stacked membrane structures such as the chloroplast and the outer segment of the visual rod (Corless & Costello, 1981). We have found it impossible to induce phase-separation in unstacked chloroplasts. A secondary effect of heat-stress is to bring about a destacking of chloroplast membranes (Armond et al., 1980; Gounaris et al., 1984). Phase separation is also not normally seen in such destacked regions. In practice, the easiest way to bring about non-bilayer phase-separation in chloroplasts is to reduce the pH below about 4.5 (Thomas et al., 1985). This is probably because it involves a destabilisation of the membrane without a concommitant tendency for thylakoid destacking. A typical example of such a phase separation is illustrated in Fig. 8.

4.5 Membrane Dehydration

The idea that lipids of dehydrated plant membranes might form a Hex-II phase was first put forward by Simons 1974. Dessicated seeds tend to be extremely leaky in the first stages of imbibing water. This led Simons (1974, 1978) to suggest that dry cellular membranes may be incapable of acting as effective permeability barriers as a result of their lipids undergoing a L_α-to-Hex II phase transition. It is easy to see by reference to Fig. 7 that such a transition might be possible in principle at least.

Tovio-Kinnucon & Stushinoff (1981) have reported that the lipid of the spherical lipid bodies lying close to the cell walls of the cotyledons of air-dried lettuce seeds appears to be amorphous when viewed by freeze-fracture electronmicroscopy but to take on a crystalline appearence at above 20% water content. They interpreted these changes as evidence for the formation of Hex-II phase lipid in the partially hydrated seeds. Other groups, however, have failed to find any evidence for Hex-II formation in seeds (Thomson & Platt-Aloia, 1982; Bliss et al., 1984; Vigil et al., 1984). McKershie & Stinson (1980) isolated the phospholipid fraction of dessication-tolerant and dessication-sensitive seeds of Lotus corniculatus (birdsfoot trefoil) and examined the two fractions at different degrees of hydration by X-ray diffraction. They were also unable to find evidence for Hex-II formation, even at the 5% hydration level.

A number of groups have investigated the possibility that non-bilayer structures are formed in pollen grains. Priestley & de Kruijff (1982) reported that the membrane lipids of the pollen grains of Typha latifolia (bulrush) are arranged in a bilayer configuration at hydration levels at least as low as 11% using ^{31}P-n.m.r. techniques. Below this level, the n.m.r. signals were too broad to distinguish the line-shapes characterising lipid organisation. Kroh & Knuiman (1985) reported the existence of hexagonally packed intramembranous particles in the plasma membranes of tobacco pollen tubes. These were not, however, thought to be of lipidic origin. Platt-Alloia et al., (1986) also failed to find any evidence for the occurrence of Hex-II in dried pollen.

Exposure of plant tissue to freezing temperatures leads to severe local deydration of membrane surfaces associated with the concentration of salts into the non-frozen fraction of the cell water. Steponkus and his co-workers have shown that this can, under appropriate conditions, lead to the phase separation of non-bilayer forming lipids. They found that protoplasts isolated from rye seedlings undergo a dramatic shrinkage associated with the loss of water to form extracellular ice on exposure to freezing conditions. This shrinkage was accompanied by an irreversible endocytotic vesiculation of the plasma membrane resulting in the formation of intracellular vesicles that lie directly under the plasma membrane surface (Dowgert & Steponkus, 1984; Gordon-Kamm & Steponkus, 1984a). Freeze-fracture studies revealed the presence of extensive regions of phase-separated non-bilayer lipid in the protoplast membranes resembling those seen in heat-stressed chloroplasts (Gordon-Kamm & Steponkus, 1984b). Protoplasts isolated from seedlings that had been subjected to chill-hardening showed an entirely different pattern of behaviour. No internal vesiculation was seen on shrinkage and non-bilayer lipid structures were not observed in the protoplast membrane. Protoplasts from hardened seedlings were characterised by exocytotic extrusions with osmophilic cores thought to contain lipid material deleted from their plasma membranes. Gordon-Kamm & Steponkus (1984b) suggest that the presence of closely apposed

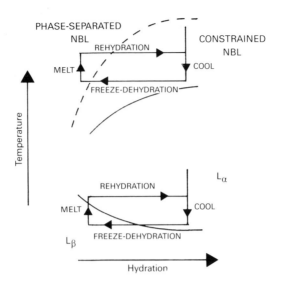

Fig. 9. Diagram illustrating the effects of freeze-dehydration on the phase behaviour of membranes and isolated lipids.

membranes of the type found in the non-acclimated, but not in acclimated, protoplasts might be a requirement for non-bilayer lipid phase separation. Significant changes in the lipid composition are observed on hardening (Lynch & Steponkus, 1987) but it is not fully clear whether this or other factors determine the very different behaviour of acclimated and non-acclimated protoplasts (Dowgert, et al., 1987).

Caffrey (1987) has reported a similar freeze-dehydration driven phase transition in dispersions of dioleoylphosphatidylserine. In this case, however, the phase behaviour of the lipid was followed by time-resolved X-ray diffraction and the transition was between the L_α and L_β phases. Caffrey found that on initial cooling the dispersions entered a super-cooled state in which the lipid persisted in an expanded L_α phase down to about -18°C. The dispersions then began to freeze bringing about a local dehydration of the membrane surfaces and the formation first of a more condensed L_α phase and then a L_β phase. The gel-to liquid crystal transition temperature of the dehydrated lipid was about -10°C, appreciably higher than that of the corresponding transition in the fully hydrated state. This as pointed out by Caffrey, can be readily understood if the lipid undergoes a freeze-dehydration cycle of the type illustrated in the lower part of Fig. 9. The non-bilayer lipid phase separation reported by Gordon-Kamm and Steponkus can be similarly explained as illustrated in the upper part of the same figure.

Pearce and his co-workers have reported an extensive study of the effects of membrane dehydration in wheat seedlings brought about by extracellular ice formation (Pearce, 1985a and Pearce & Willinson, 1985) and progressively more severe drought conditions at non-freezing temperatures (Pearce 1985b,c). They report the existence of extensive particle-free patches under both sets of conditions. In the case of seedlings exposed to slow drought, a ridged sub-structure was also sometimes seen (Pearce 1985c). The fact that intra-membranous particles were also seen in association with such structures suggested, however, that they were unlikely to be associated with Hex-II formation. Pearce and his collaborators also take pains to point out that the particle-free areas that they observe may not correspond to gel-phase lipid separations of the type seen in chilled micro-organisms.

The precise relevance of lipid phase separations to freezing damage in plants, as of the phase separations seen in heated chloroplasts to heat stress damage, is still a matter of considerable debate. The experimental observations reviewed here do, however, serve to underline the potential importance of studies of lipid topology to plant membrane structure and function.

REFERENCES

Armond, P.A & Staehelin, L.A. (1979) Proc. Natl. Acad. Sci. U.S., 76, 1901-1905.
Armond, P.A., Bjorkman, O. & Staehelin, L.A. (1980) Biochim. Biophys. Acta, 601, 433-442.
Bliss, R.D., Platt-Aloia, K.A. & Thomson, W.W. (1984) Plant Cell Environ. 7, 601-606.
Brand, J.J., Kirchanski, S.J. & Ramirez-Mitchell, R. (1979) Planta, 145, 63-68.
Brentel, I., Selstam, E. & Lindblom, G. (1985) Biochim. Biophys. Acta, 812, 816-826.
Caffrey, M. (1987) Biochim. Biophys. Acta 896, 123-127.
Chapman, D. & Benga, G. (1984). In "Biological Membranes Vol.5", (Chapman, D., ed.), pp. 1-56, Academic Press, London.
Corless, J.M. & Costello, M.J. (1981) Exp. Eye. Res 32, 217-228.
Cullis, P.R. & de Kruijff, B. (1979). Biochim. Biophys. Acta, 559, 399-420.
Cullis, P.R., de Kruijff, B., Hope, M.J., Nayer, R. & Schmid, S.L. (1980). Can. J. Biochem. 52, 1091-1100.
De Kruijff, B., Cullis, P.R., Verkleij, A.J., Hope, M.J., Van Echteld, C.J.A., Taraschi, T.F., Van Hoogevest, P., Killan, J.A., Rietveld, A. & Van der Steen, A.T.M. (1985). In "Progress in Protein-Lipid Interactions" (Watts, A. & De Pont, J.J.H.H.M., eds.) pp. 89-142, Elsevier, Amsterdam.
De Kruijff, B., Cullis, P.R., Verkleij, A.J., Hope, M.J., Van Echteld, C.J.A. & Taraschi, T.F. (1985). In "The Enzymes of Biological Membranes" (Martinosi, A., ed.) pp. 131-204, Plenum Press, New York.
Dowgert, M.F. & Steponkus, P.L. Plant Physiol., 1139-1151.
Dowgert, M.F., Woolfe, J. & Steponkus, P.L. (1987) Plant Physiol., 83, 1001-1007
Furtado, D., Williams, W.P., Brain, A.P.R. & Quinn, P.J. (1979) Biochim. Biophys. Acta, 555, 352-357.
Gordon-Kamm, W.J. & Steponkus, P.L. (1984a) Protoplasma, 123, 161-173.
Gordon-Kamm, W.J. & Steponkus, P.L. (1984b) Proc. Natl. Acad. Sci. U.S., 81, 6373-6377.
Gounaris, K., Brain, A.P.R., Quinn, P.J. & Williams, W.P. (1983a) FEBS Lett., 153, 47-52.
Gounaris, K., Sundby, C., & Andersson, B. & Barber, J. (1983b). FEBS Lett. 156, 170-174.
Gounaris, K., Mannock, D.A., Sen, A., Brain, A.P.R., Williams, W.P. & Quinn, P.J. (1983c) Biochim. Biophys. Acta, 732, 229-242.
Gounaris, K., Sen, A., Brain, A.P.R., Quinn, P.J. & Williams, W.P. (1983d) Biochim. Biophys. Acta, 728, 129-139.

Gounaris, K., Brain, A.P.R., Quinn, P.J. & Williams, W.P. (1984) Biochim. Biophys. Acta, 766, 198-208.
Gunning, B.E.S. (1965) Protoplasma, 60, 111-130.
Haest, C.W.M., De Grier, J, Van Es, G.A., Verkleij, A.J. & Van Deenen, L.L.M. (1972) Biochim. Biophys. Acta, 288, 43-53.
Harlos, K. & Eibl, H. (1981). Biochemistry 20, 2888-2892.
Hui, S.W. & Boni, L.T. (1982) Nature, 296, 175.
Israelachvili, J.N., Mitchell, D.J. & Ninham, B.W. (1976). J. Chem. Soc. Faraday Trans. II 72, 1525-1568.
Israelachvili, J.N., Marcelja, S. & Horn, R.G. (1980). Q. Rev. Biophys. 13, 121-200.
Kroh, M. & Knuiman, B. (1985) Planta 166, 287-299.
Larsson, K., Fontell, K. & Krog, N. (1980) Chem. Phys. Lipids, 27, 321-328.
Lynch, D.V. & Steponkus, P.L. (1987) Plant Physiol., 83, 761-767.
McKersie, B.D. & Stinson, R.H. (1980). Plant Physiol. 66, 316-320.
Murata, N. (1982) In "Effects of Stress on Photosynthesis" (Marcelle, R., Clijsters, H. & Van Prouke, M., eds.) pp. 285-293, Martinus Nijhoff, Dr. W. Junk Publishers, The Hague.
Murata, N. (1983) Plant Cell Physiol., 24, 81-86.
Murata, N. & Yamaya, J. (1984) Plant Physiol., 74, 1016-1024.
Murata, N., Sato, N., Takahashi, N. & Hamazaki, Y. (1983) Plant Cell Physiol., 23, 1071-1079.
Murphy, D.J. (1983). FEBS Lett. 150, 19-26.
Murphy, D.J. & Woodrow, I.E. (1983). Biochim. Biophys. Acta 725, 104-112.
Norman, H. A., McMillan, C. & Thompson, G.A. (198-) Plant cell Physiol., 25, 1437-1444.
Ono, T-A. & Murata, N. (1981a) Plant Physiol., 67, 176-181.
Ono, T-A. & Murata, N. (1981b) Plant Physiol., 67, 182-187.
Ono, T-A. & Murata, N. (1982) Plant Physiol., 69, 125-129.
Ono, T-A., Murata, N. & Fujita, T. (1983) Plant Cell Physiol. 24, 635-639.
Orr, G.R. & Raison, J.K. (1987) Plant Physiol., 84, 88-92.
Pearce, R.S. (1985a) J. Exp. Bot., 36, 369-381.
Pearce, R.S. (1985b) J. Exp. Bot., 36, 1209-1221.
Pearce, R.S. (1985c) Planta, 166, 1-14.
Pearce, R.S. & Willinson, J.H.M. (1985) Planta, 163, 304-316.
Phillips, M.C., Ladbrooke, B.D. & Chapman, D. (1970). Biochim. Biophys. Acta 196, 35-44.
Pick, U., Gounaris, K., Admon, A. & Barber, J. (1984) Biochim. Biophys. Acta., 765, 12-20.
Pick, U., Gounaris, K., Weiss, M. & Barber, J. (1985) Biochim. Biophys. Acta., 808, 415-420.
Pick, U., Weiss, M., Gounaris, K. & Barber, J. (1987) Biochim. Biophys. Acta., 891, 28-39.
Platt-Alloia, K.A., Lord, E.M., DeMason, D.A. & Thomson, W.W. (1986) Planta, 168, 291-298.
Priestly, D.A. & De Kruijff, B. (1982). Plant Physiol. 70, 1075-1078.
Quinn, P.J. & Williams, W.P. (1978) Prog. Biophys. Mol. Biol., 34, 109-173.
Quinn, P.J. & Williams, W.P. (1983) Biochim. Biophys. Acta, 737, 223-266.
Quinn, P.J. & Williams, W.P. (1985) In "Photosynthetic Mechanisms & the Environment" (Barber, J. & Baker, N.R., eds.) pp.1-47. Elsevier, Amsterdam.
Quinn, P.J., Gounaris, K., Sen, A. & Williams, W.P. (1982) In "Biochemistry & Metabolism of Plant Lipids" (Wintermans, J.F.G.M. & Kuiper, P.J.C. eds.) pp. 327-330. Elsevier, Amsterdam.

Raison, J.K. & Wright, L.C. (1983) Biochim. Biophys. Acta 731, 69-78.
Raison, J.K. & Orr, G.R. (1986) Plant Physiol. 81, 807-811.
Rilfors, L., Lindblom, G., Wiesl&er, A. & Christiansson, A. (1984). Biomembranes, 12, 205-245.
Rilfors, L., Eriksson, P-O., Arvidson, G. & Lindblom, G. (1986). Biochemistry 25, 7702-7711.
Rivas, E. & Luzzati, V. (1969). J. Mol. Biol. 41, 261-275.
Seddon, J.M., Cevc, G. & Marsh, D. (1983) Biochemistry, 22, 1280-1289.
Sen, A., Williams, W.P. & Quinn, P.J. (1981a) Biochim. Biophys. Acta, 663, 380-389.
Sen, A., Williams, W.P., Brain, A.P.R., Dickens, M.J. & Quinn, P.J. (1981b) Nature, 293, 488-490.
Sen, A., Williams, W.P., Brain, A.P.R. & Quinn, P.J. (1982a) Biochim. Biophys. Acta, 685, 297-306.
Sen, A., Brain, A.P.R., Quinn, P.J. & Williams, W.P. (1982b) Biochim. Biophys. Acta, 686, 215-224.
Sen, A., Mannock, D.A., Collins, D.J., Quinn, P.J. & Williams, W.P. (1983) Proc. Roy. Soc. Lond., B218, 349-364.
Shipley, G.G., Green, J.P. & Nicols, B.W. (1973) Biochim. Biophys. Acta 311, 531-544.
Siefermann-Harms, D., Ross, J.W., Kaneshiro, K.H. & Kamamoto, H.Y. (1982) FEBS Lett., 149, 191-196.
Siegel, D. (1986) Chem. Phys. Lipids, 42, 279-301.
Simon, E.W. (1978) In "Dry Biological Systems", (J.H. Crowe & J.S. Clegg eds.), Academic Press, New York, pp. 205-224.
Simon, E.W. (1974) New Phytol., 73, 377-420.
Simpson, D. (1978) Carlsberg Res, Comm. 43, 145-170.
Singer, S.J. & Nicolson, G.L. (1972) Science, 175, 720-731.
Sprague, S.G. & Staehelin, L.A. (1984) Biochim. Biophys. Acta, 777, 306-322.
Tanford, C. (1973). In "The hydrophobic effect: formation of micelles & biological membranes". Wiley, New York.
Taraschi, T.F., van der Steen, A.T.M., de Kruijff, B., Tellier, C. & Verkleij, A.J. (1982). Biochemistry 21, 5756-5764.
Thomas, P.G., Brain, A.P.R., Quinn, P.J. & Williams, W.P. (1985) FEBS Lett., 183, 161-166.
Thomas, P.G., Dominy, P.J., Vigh, L., Mansourian, A.R., Quinn, P.J. & Williams, W.P. (1986a) Biochim. Biophys. Acta, 849, 131-140.
Thomas, P.G., Quinn, P.J. & Williams, W.P. (1986b) Planta, 167, 133-139.
Thomson, W.W. & Platt-Alloia, K.A. (1978) Stain Technol., 57, 327-334.
Tovio-Kinnucan, M.A. & Stushnoff, C. (1981). Cryobiol. 18, 72-79.
Tsukamoto, Y., Yeki, T., Mitsui, T., Ono, T-A. & Murata, N. (1980) Biochim. Biophys. Acta, 602, 673-675.
Van Venetie, R. & Verkleij, A.J. (1981) Biochim. Biophys. Acta 645, 262-269
Van Venetie, R. & Verkleij, A.J. (1982) Biochim. Biophys. Acta, 692, 397-405.
Van Walraven, H.S., Koppenaal, E., Marvin,H.J.P., Hagendoorn, M.J.M & Kraayenhof, R. (1984) Eur. J. Biochem. 144, 563-569.
Verkleij, A.J. (1985) Biochim. Biophys. Acta, 779, 43-63.
Verkleij, A.J., Mombers, C., Geritsen, W.J., Leunissen-Bijvelt, L. & Cullis, P.R. (1979). Biochim. Biophys. Acta, 555, 358-361.
Verwer, W., Ververgaert, P.J.J.T., Leunissen-Bijvelt, J. & Verkleij, A.J. (1979) Biochim. Biophys. Acta, 504, 231-234.
Vigil, E.L., Steere, R.L., Wergin, W.P. & Christiansen, M.N. (1984) Am. J. Bot., 71, 601-606.

Wieslander, A., Christiansson, A., Rilfors, L., Khan, A., Johansson, L.B. & Lindblom, G. (1980a). FEBS Lett., 124, 273-278.
Wieslander, A., Christiansson, A., Rilfors, L. & Lindblom, G. (1980b).   Biochemistry 19, 3650-3655.
Wieslander, A., Rilfors, L. & Lindblom, G. (1986). Biochemistry, 25, 7511-7517.
Williams, W.P. & Quinn, P.J. (1987) J. Bioenerg. Biomemb., 19, 605-624.
Williams, W.P., Sen, A. & Quinn, P.J. (1982) Biochem. Soc. Trans., 10, 335-338.
Williams, W.P. Sen, A., Brain, A.P.R. & Quinn, P.J. (1982) Nature, 296, 175-176.
Williams, W.P., Gounaris, K. & Quinn, P.J. (1984) In "Advances In Photosynthesis Research Vol. III" (Cybesma, C., ed) pp. 123-130. Nijhoff/Junk, The Hague.
Yeagle, P.L. & Sen, A. (1986) Biochemistry, 25, 7518-7522.

Use of yeast lipid-synthesis mutants in establishing membrane function

Katharine D. ATKINSON

Department of Biology, University of California, Riverside, California 92521 USA

SYNOPSIS

Baker's yeast Saccaromyces cerevisiae provides a useful eukaryotic system for genetic and molecular investigations that may not be feasible in higher organisms. Lipid biosynthetic mutants provide in vivo access to cellular mechanisms for regulating lipid synthesis and functional requirements for specific lipid species. Yeast sphingolipid mutants have been identified by virtue of their unexpected ability to provide precursors for nitrogenous phospholipid synthesis. The sphingolipid mutants and corresponding normal cloned genes exhibit unorthodox genetic interactions that suggest complex functional interplay within cell membranes. The mutants provide test material for examining the role of inositol-sphingolipids in secretory membrane function and sphingolipid precursors in plasma membrane C-Kinase function.

INTRODUCTION

Genetic mutants of Baker's yeast, Saccharomyces cerevisiae, have proven useful in studying basic eukaryotic biosynthetic pathways, metabolic regulatory strategies, and biological roles of given cellular molecules. Many of the membrane lipids can be manipulated by combined genetic and molecular approaches that are particularly convenient in yeast. Fatty acid, sterol, and phospholipid biosynthetic mutants, reviewed by Henry (1982 & Henry et al., 1984) have been used primarily to investigate cross-pathway regulation and, to a lesser extent, membrane functions in the context of lipid abnormalities.

The fundamental problem in examining membrane function in the yeast phospholipid mutants is that the mutants create far-reaching, unexpected alterations in other branches of lipid synthesis, and eventually compromise overall cellular growth and viability. In a given mutant, it may be difficult to attribute altered membrane function to a specific lipid deficiency, rather than to secondary perturbations in lipid metabolism or even to overall cessation of growth or irreversible cell death.

Our own work on periplasmic enzyme secretion demonstrates the problem. Net plasma membrane growth is inositol-dependent, but secretion can be uncoupled, depending on the carbon source sustaining growth. In glucose-grown cells, secretion is inositol-dependent, failing coordinately with membrane growth, and leading to intracellular accumulation of secretory enzymes. In sucrose-grown cells, secretion proceeds unabated until general metabolism halts, despite identical patterns of interrupted phospholipid synthesis (Atkinson & Ramirez, 1984). A simple conclusion regarding inositol-lipid requirements for secretion cannot be drawn. Perhaps the metabolism of phosphatidylinositol, rather than its initial synthesis, will be found to differ in cells grown on glucose versus sucrose. However, it would be desirable to have genetic mutants, that make such a metabolic requirement accessible to direct and specific manipulation, to confirm its specific role in the secretory process.

The inositol-requiring mutants that we have now do not offer us an unambiguous answer to the question: Are inositol-lipids required for the secretory process? The mutants showed that the question was naively formulated, and that we need to know more about inositol-lipid metabolism to formulate a better question. The laboratory's current investigations focus on mutants in the sphingolipid biosynthetic pathway, where phosphatidylinositol is metabolized as a substrate.

SPHINGOLIPID MUTANTS

The yeast sphingolipids have been identified in Robert Lester's laboratory. A mutant that requires a sphingosine long-chain base has been isolated (Wells & Lester, 1983). The mutant does not merely cease growing when deprived of a sphingolipid precursor, but dies, reminiscent of inositol-less death. Unfortunately, it is difficult to flush out the long-chain base supplement, making it impossible to cleanly initiate starvation and compare with inositol starvation. My laboratory inadvertently obtained sphingolipid biosynthetic mutants that probably cover all steps in the pathway beyond the initial step altered in Lester's mutant. None of the yeast sphingolipid mutants have been rigorously matched with specific enzymes, but each exhibits distorted sphingolipid biosynthesis in vivo, compatible with the following identities.

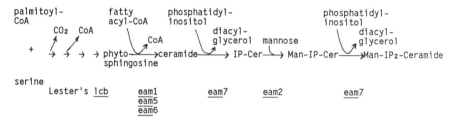

All of my laboratory's eam mutants show incomplete reductions in sphingolipid synthesis, beyond the initial serine-decarboxylation reaction. From the lethal nature of Lester's lcb mutant, we would expect that complete blocks at the later biosynthetic steps would be lethal, and beyond our ability to rescue by dietary supplementation. We can make complete mutants, as will be described later.

The eam mutants that we have were detected by virtue of an unexpected effect on nitrogenous phospholipid synthesis. Each of the eam mutants accumulates sphingolipid precursors that, if degraded, would yield all of the starting materials except for serine. The decarboxylated form of serine released is ethanolamine, a novel molecule that normal yeast do not make in a free form. The eam mutants make ethanolamine available to the nitrogenous phospholipid biosynthetic pathway. They were detected as spontaneous mutants that relieve the need for feeding ethanolamine to yeast chol

mutants, which cannot make phosphatidylethanolamine unless provided
with ethanolamine (Atkinson, 1983 & 1984).

UNORTHODOX GENETIC COMPLEMENTATION

The eam sphingolipid mutants demonstrate a novel genetic phenomenon. All of the eam mutants are recessive, and fall into a single
complementation group. However, the mutants map to at least five
distinct chromosomal locations:

```
                   cdc7      trp1
Chromosome———————+————————o——L———————+—————————
    #4           |   ---4---°|°—12——-|
                 eam2              eam7

                   cdc6            inol
Chromosome  ————————+————————————————+———————————-/ /——o
    #10     | -25-| ---15---| ---9-| ---12—-|
            eam6            eam1           eam5
```

The genetic maps show recombinational distance (centi-Morgans) between the indicated genetic loci. The complementation pattern, when
different mutants are placed together in a diploid, gives the false
indication that the mutants are all defective in the same gene. For

instance, eam1 and eam2 are each recessive, meaning that $\frac{eam1}{+}$ and
$\frac{eam2}{+}$ are each normal diploids that cannot make enough endogenous

ethanolamine available to sustain phosphatidylethanolamine synthesis
and cell growth. If we combine two independently-isolated eam1
mutants in a diploid, that diploid $\frac{eam1}{eam1}$ is mutant: both allelic

copies of the same gene are mutant. The unexpected development was
that diploids combining two different eam mutants, like $\frac{eam1}{+}$ $\frac{+}{eam2}$,

were mutant. Unlike the usual expectation, the non-allelic mutants
fail to complement (Atkinson, 1985). Several other laboratories
have discovered yeast or Drosophila mutants that fail to complement,
and are struggling, along with us, to understand this exceptional
genetic interaction. Apart from general genetics, our concern is:
What can the interaction between the eam mutants tell us about lipid
biosynthesis or membrane function?

FUNCTIONAL INTERACTIONS WITHIN MEMBRANES

The eam mutants either directly encode or indirectly regulate
sphingolipid biosynthetic reactions, all known to be localized
within yeast cellular membranes (Becker & Lester, 1980). It is
possible that genetic complementation failure among the eam mutants
reflects multiple-molecule interactions typical of membrane components. We have developed three general models to explain complementation failure, and molecular genetic tests that distinguish
these possibilities.

Direct multimeric protein interactions

If the protein products of the eam genes form a complex with
direct protein-protein interactions necessary for function, comple-

mentation failure could be caused by mutant proteins "poisoning" normal proteins. This explanation serves for complementation failure between recessive alpha-tubulin and beta-tubulin mutants in Drosophila melanogaster (Raff & Fuller, 1984 & Fuller, 1985). In a single-gene heterozygote, a^-/A^+ B^+/B^+, half the alpha-beta protein pairings are mutant, but the half that are normal sustain a normal phenotype. In a double heterozygote, a^-/A^+ B^+/b^-, only one-fourth of the alpha-beta protein pairings are fully normal, and a mutant phenotype results. The genetic test for this model involves null alleles, or mutants which make no protein at all. A double-null heterozygote, a°/A^+ B^+/b°, makes only normal proteins, at a half-normal level that sustains a normal phenotype.

If we clone the normal EAM$^+$ genes, use the clones to construct complete null mutants, and return these mutants to living cells, we can perform the critical test of the "alpha-beta" model. We expect some or all of the completely-defective eam mutants to be lethal. That problem can be circumvented by working with diploid strains and by mating potentially-inviable haploid spores directly with each other before a lethal condition develops.

Critical isozyme levels

If the genetic test with null alleles fails to support the "alpha-beta" model, others will be needed. One possibility is suggested by the similar in vivo biochemical alterations of the three chromosome #10 mutants, eam1, eam5, and eam6. These genes could encode identical isozymes that are all needed to adequately sustain ceramide synthesis. The haploid mutants demonstrate that two-thirds normal levels are not adequate. In diploids, with a total of six alleles, any two mutants alleles would reduce total isozyme levels to the inadequate two-thirds level. Null alleles should not behave any differently than defective-protein alleles in contributing to a quantitatively adequate level of total activity.

The crude biochemical information already available, suggests that the "isozymes" model might explain why eam1, 5 and 6 mutants fail to complement, but will not serve to explain complementation failure between eam1 and eam2 which have demonstrably different in vivo defects.

Crippled pathway

A third type of model for complementation failure does not involve any direct relationship or interaction between the eam proteins, other than their common participation in a single linear pathway. Any lipid product or intermediate, beyond the early serine-decarboxylation reaction in the sphingolipid pathway, will yield ethanolamine when degraded. Two different accumulating precursors can yield just as much ethanolamine as a double-accumulation of the same precursor. The detectably-mutant haploid strains indicate that eam1 mutants accumulate enough phytosphingosine to generate ethanolamine to sustain phosphatidylethanolamine synthesis, and eam2 mutants accumulate enough inositolphosphorylceramine. Each mutant alone in a heterozygous diploid would make exactly half-enough ethanolamine; together they would make enough, from two independently-generated sources. Again, null-allele mutants may have an exaggerated mutant phenotype, but certainly not an ability to restore a normal phenotype.

THE CLONED EAM+ GENES

Two of the normal EAM+ genes have been cloned, in preparation of the null alleles needed to test complementation failure models. The clones are just as peculiar as the in vivo eam mutants. One cloned gene "EAM X" complements any of the three eam mutants on chromosome #10 (roman numeral X). The other clone "EAM IV" complements both the mutants on chromosome #4 (IV). The cloned genes are carried in a shuttle vector, YCp50, that has a selectable yeast gene, URA3+, a yeast DNA replication origin, and a yeast centromere that maintains the construct at low or one-copy per cell. There are many reports of cloned genes that can complement the "wrong" mutants when the clone is carried in a high-copy vector and over-expressed. The EAM+ clones are the first to complement non-allelic mutations when engineered for normal low-level expression.

EXCLUSION OF PROTEIN-INTERACTION MODEL

Without any further information from the cloned genes, we can begin to rule out some possibilities amongst the models for eam mutant complementation failure. Only one variation of the "alpha-beta" model is compatible with non-allelic molecular complementation. An unstable alpha protein could be more effectively recruited into stable alpha-beta partnerships by raising the level of beta protein production. Excess beta should not be able to rescue stable but functionally defective alpha proteins. We now have only one allele each for the eam2 and eam7 genes, but have many for eam1, 5 and 6. The EAM X clone is able to complement all of the eam1, eam5, and eam6 mutants. We consider it impossible that every mutant in five genes has the same molecular feature. We expect the null allele tests to eventually confirm that the "alpha-beta" model is excluded.

Ruling out the "alpha-beta" model for eam mutant complementation failure is a major disappointment for other laboratories with mutants that fail to complement (discussed, but not yet in print). Based on the Drosophila tubulin mutants, complementation failure is thought to unequivocably indicate direct protein-protein interactions. As for my laboratory's primary interest in membrane function, ruling out the "alpha-beta" model indicates that genetic cooperation between steps in a membrane-embedded pathway may not be adequately explained by molecular paradigms developed for the soluble cytoplasmic phase of biology.

RESTRICTED ISOZYME MODEL APPLICABILITY

Non-allelic molecular complementation is perfectly compatible with the "isozymes" model. A haploid mutant strain with only one gene of an isozyme pair or trio defective, would be fully restored to normal levels of isozyme production by a cloned copy of any of the genes within the group. If the "isozyme" model applies to the eam genes, it must be restricted and cannot explain the entire phenomenon of eam complementation failure. The three genes on chromosome #10 could be an isozyme family, but the two genes on chromosome #4 must be, at least, a different isozyme family. Neither of the two EAM+ clones complements mutants on different chromosomes. The "isozyme" model is restricted to explaining complementation failure within each gene group. Another model must be invoked to explain complementation failure between eam1 and eam2, which cannot belong

to the same isozyme group. As already noted, distinct sphingolipid
disturbances evident in these two mutants support the idea that they
cannot encode isozymes. We expect that the "crippled pathway" model
will be required for a full explanation of eam mutant complementation failure. For the "crippled pathway" model, non-allelic molecular complementation is compatible with what we know about pathway
dynamics. Boosting one reaction in a pathway could drain or push
intermediates through a "bottle-neck" elsewhere in the pathway.

DNA SEQUENCE HOMOLOGY

     The three eam genes on chromosome #10 have demonstratable
sequence homology that lends support to applicability of the
"isozyme" model. Possible homology between the two chromosome #4
genes is still under investigation. Given only the knowledge of
which genetic mutants the EAM+ clones complement, we were unable to
assign a clear identity to either clone. In yeast, unlike other
eukaryotic systems, cloned segments of DNA recombine and integrate
into the natural genome by mechanisms that require sequence homology. We attempted to identify the EAM+ clones by asking which
genetic loci provided sequence homology for integrative recombination. Each cloned EAM+ segment was removed from the YCp50 vector, and transferred to a different plasmid vector, YIp5, that could
not sustain its own replication in yeast. The resultant constructs
must integrate into a replication-sustaining chromosome to be maintained. Just before introducing these constructs to yeast cells,
the EAM+ segment was cut open with a restriction enzyme to create
free DNA ends that provoke recombination.
     Fifteen independent integrants of the EAMX clone were located by
traditional fine-structure meiotic segregation analysis, with the
surprising result that integration occurred with equal frequency at
each of the three chromosome #10 eam gene locations. There is some
sequence homology between eam1, eam5, and eam6. We do not know how
much homology is required to support yeast recombination, since a
puzzle like this has not been seen before. Two integrants of the
EAMIV clone both map to the eam2 gene, and other integrants are
being examined for possible integration at eam7.
     Recombinational integration has failed to identify the EAMX
clone. It has, however, uncovered a homologous gene family, and set
the stage for obtaining identified clones from the eam1, eam5 and
eam6 genes. We now have genetically mapped strains with the unknown
clone integrated adjacent to each of the three genes, in the
following configuration:

There are two unique sites on the integrated DNA, between EAMX and
the vector's Ampicillin-resistance gene, that can be cut with a
restriction enzyme. Proceeding to the right on the diagram, the
next cut made by the same enzyme will be in or beyond the resident
chromosomal gene, shown as eam6 here. Ligation, and recovery of
ampicillin-resistant plasmids, will "loop out" part or all of the
resident gene, whose identity in each instance is known. With all
three genes in hand, we can compare them and ask: how homologous
are they at the positions involved in integrative recombination?
how homologous are the proteins they encode? how similar are their

regulatory sequences? We can also construct deletions of the coding regions, leaving flanking segments, and replace endogenous genes by gene conversion, to engineer the null alleles needed to complete testing of the complementation failure models.

RETURN TO SECRETION STUDIES

In the long run, we expect to obtain null mutants of eam2 and eam7, probably rescued by cloned normal genes that can be turned off to initiate defects in utilization of phosphatidylinositol as a sphingolipid substrate. This material should permit us to ask if secretion depends on inositol-sphingolipid biosynthesis, and other questions that may arise regarding the need for these lipids in membrane function.

SPHINGOLIPID REGULATION OF C-KINASE

The eam mutants and cloned genes provide the necessary material for an in vivo test of the newly developed idea that sphingolipids are natural cellular regulators of C-Kinase activity (Hannun et al., 1986). After stimulation, mediated by polyphosphoinositide lipids, sphingolipids are thought to inhibit C-Kinase, returning it to a responsive ground-state. Yeast have C-Kinase that is similar enough to the mammalian enzyme to be inhibited by antibody (F. O. Fields & J. Thorner, personal communication, manuscript in preparation). If sphingolipids dampen C-Kinase activity in vivo, the eam mutants, and the cloned EAM+ genes engineered for over-expression, should exhibit unusual C-Kinase activity. The full array of mutants and clones should indicate which specific sphingolipids, if any, control C-Kinase.

LIPIDS AND MEMBRANE FUNCTION

The laboratory's entire research program relies on the premise that the different lipid species each play a significant role in supporting normal membrane functions. On a very crude level, the premise is justified: most of the lipid mutants are lethal if not circumvented by nutritional supplementation. Identifying specific lethal events attached to each lipid will be a difficult task. For the most part, I suspect that identification of potentially-important membrane functions will come from the fields of biochemistry and cell biology, not from genetics. Genetic approaches to in vivo manipulation can provide suitable test material for evaluating theory and establishing the molecular requirements for viable membrane function.

REFERENCES

Atkinson, K. D. (1983) in Biosynthesis and Function of Plant Lipids (Thomson, W. W., Mud, B. & Gibbs, M., eds.), pp. 229-249, Amer. Soc. Plant Physiol., Rockville MD.
Atkinson, K. D. (1984) Genetics 108:533-543.
Atkinson, K. D. (1985) Genetics 111:1-6.
Atkinson, K. D. & Ramirez, R. M. (1984) J. Bacteriol. 160:80-86.
Becker, G. W. & Lester, R. L. (1980) J. Bacteriol. 142:747-754.

Fuller, M. T. (1985) in Gametogenesis and The Early Embryo (Gall, J.
  G., ed.), pp. 19-42, Alan R. Liss, New York.
Hannun, Y. A., Loomis, C. R., Merrill, A. H. & Bell, R. M. (1986) J.
  Biol. Chem. 261:12604-12609.
Henry, S. A. (1982) in Molecular Biology of the Yeast Saccaromyces
  (Strathern, J. N., Jones, E. W. & Broach, J. R., eds.), vol. 1, pp.
  101-158, Cold Spring Harbor Laboratory, New York.
Henry, S. A., Klig, L. S. & Loewy, B. S. (1984) Annu. Rev. Genet.
  18:207-231.
Raff, E. C. & Fuller, M. T. (1984) in Molecular Biology of the
  Cytoskeleton (Borisy, G. G., Cleveland, D. W. & Murphy, D. B.,
  eds.), pp. 293-304, Cold Spring Harbor Laboratory, New York.
Wells, G. B. & Lester, R. L. (1983) J. Biol. Chem. 258:10200-10203.

Surface electrical charges and their role in membrane function

James BARBER

AFRC Photosynthesis Research Group, Department of Pure & Applied Biology, Imperial College, London SW7 2BB

SYNOPSIS

A brief description of electrical double layer theory due to surface charges is given and its implication discussed in terms of the organisation and dynamics of the chloroplast thylakoid membrane.

INTRODUCTION

Biological membranes have at their surfaces electrically charged groups which may be either positive or negative. Taken together there is usually an excess of negative charge at physiological pH. This net negative charge may be derived from lipids, such as phospholipids or sulpholipids, or from protein residues such as the carboxyl groups of aspartic and glutamic acid. The precise value of the surface charge density will be a characteristic of a particular membrane being dependent on its composition and organisation. It is also important to note that the charge density and its polarity will be sensitive to the pH of the surrounding medium. Therefore depending on the relative concentrations of the different charged species and on their pKs, there will be a pH value at which the surface negativity is reduced to zero and the membrane is electroneutral (isoelectric). There will also be pH ranges where the surface becomes positively charged. Particle electrophoresis, and other techniques, have been widely used to investigate the surface electrical properties of a wide range of biological systems with the view to identifying the chemical nature of the exposed charged groups and their densities. These types of studies have proved important for understanding a wide range of biophysical, biochemical and cellular processes. For example, surface charge governs the concentration of charged solutes at the membrane/liquid interface; affects transmembrane electrical gradients; regulates pH induced phenomena via their pK values; governs the binding of extrinsic proteins; modifies ionic conductivity etc (Barber, 1980a, 1982a). In this paper I wish to focus on one aspect of surface electrical charges which seem to be very important for regulating the organisation and function of the thylakoid membrane of chloroplasts. This particular aspect addresses the question of interaction between adjacent membrane surfaces and the distribution of protein complexes within the membrane. Although the topic is dealt with in terms of the thylakoid membrane system, the basic concepts which I will present almost certainly have implications for other biological membrane phenomena ranging from cell fusion and formation of tight and gap junctions to membrane reorganisation in response to hormone and lectin binding (Barber, 1982b).

The financial support of the AFRC and SERC is gratefully acknowledged.

THE THYLAKOID MEMBRANE AND ITS SURFACE CHARGE DENSITY

With higher plant chloroplasts the thylakoid membrane appears as a complex folded structure giving rise to stacked and unstacked regions. Such an organisation means that some of its outer surface is exposed to the aqueous phase of the chloroplast stroma while the rest is tightly appressed. This membrane system contains the pigments and redox active components which are responsible for capturing light and converting it into chemical potential energy. The stabilized chemical products are ATP and NADPH and the by-product is molecular oxygen derived from the oxidation of water.

Isolated thylakoid membranes have been subjected to particle electrophoresis and a typical electrophoretic profile, as a function of pH, is shown in Fig.1.

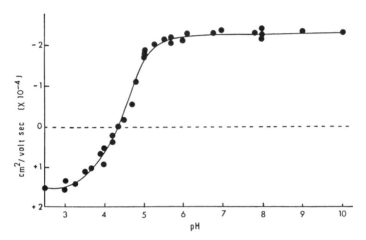

Fig. 1. Electrophoretic mobility of isolated pea thylakoid membranes as a function of pH (see Nakatani et al., 1978).

As can be seen, isolated thylakoids possess net negative surface charge at neutral pH and have an isoelectric point of about pH 4.3. The origin of this net charge seems to be the carboxyl group of glutamic and aspartic acid residues of exposed segments of integral membrane proteins with little or no contributions from the head groups of acidic lipids (Nakatani et al., 1978). Below pH 4.3 the surface becomes positively charged, the majority of which has been assigned to the guanidine group of exposed arginine residues (Nakatani & Barber, 1980). The data in Fig.1 can be used to estimate the surface charge density in Cm^{-2}. This is accomplished by combining the equations of Smoluchowski (Equ.1) and Gouy-Chapman (Equ.2).

$$u = \frac{\varepsilon_r \varepsilon_o \xi}{\eta} \qquad \ldots\ldots\ldots\ldots\text{Equ. I}$$

where u is the electrophoretic mobility, ε_r is the relative permittivity of the solution, ε_o is the permittivity of a vacuum having a value of 8.854×10^{-12} C^2 J^{-1} m^{-1}, η is the dynamic viscosity of the suspending medium and ξ is the zeta-potential. This latter quantity is the electrical potential at the hydro-dynamic plane of shear which will be at a small distance from the membrane surface. In order to calculate the surface charge density it is necessary to assume that the zeta-potential is equal to the electrical potential at the surface of the membrane (ψ_o). The relationship between ψ_o and the surface charge density σ is given by combining the Poisson, Boltzmann and Gauss equations in the way first described by Gouy and Chapman (see Barber, 1980a) to give:

$$\sigma = \pm \; [2\varepsilon_r \varepsilon_o RT \; \Sigma_i \, C_{i\alpha} \; (\exp(-Z_i \, F\psi_o / RT)-1)]^{1/2} \qquad \ldots\ldots\ldots \text{Equ.2}$$

where R is the gas constant, T is the absolute temperature, Z_i is charge carried by ion i, F is the Faraday and $C_{i\alpha}$ is the concentration of ion i in the bulk medium. Note that when the suspension medium contains a simple Z-Z electrolyte such as KCl or $MgSO_4$, Equ.2 reduces to

$$\sigma = 2A(C_{i\alpha})^{1/2} \; \sinh \; (Z_i \, F\psi_o / 2RT) \qquad \ldots\ldots\ldots \text{Equ.3}$$

where $A = (2RT\varepsilon_r \varepsilon_o)^{1/2}$

Numerical substitution into Equ.3 assuming T = 298 K gives

$$\sigma = 0.1174 \, (C_{i\alpha})^{1/2} \; \sinh \; (Z_i \, F\psi_o / 51.7) \qquad \ldots\ldots\ldots \text{Equ.4}$$

where σ is in Cm^{-2}, $C_{i\alpha}$ is in moles per dm^3 and ψ_o is in mV.

Using these equations and the data of Fig.1 the value of σ calculated for pea thylakoids was $-0.012 \; Cm^{-2}$ at neutral pH. Because the zeta-potential is assumed to be ψ_o the technique tends to underestimate the value of σ and this may account for higher values estimated by other techniques (Barber, 1982b; Chow & Barber, 1980). A number of calculations have favoured a value of $-0.025 \; Cm^{-2}$ but it must be remembered that this is only an average and takes no account of asymmetric distribution of charge along the membrane plane.

SPACE CHARGE AND ELECTROSTATIC SCREENING

The existence of net negative charge at the membrane surface requires that an equal amount of counterions are in the aqueous layer immediately adjacent to the surface. In order to attract these counterions an electrical potential is created by the surface charge. This potential ψ_x extends out from the surface decreasing as distance x increases. At x = 0 the ψ_x value becomes ψ_o and is related to σ and the bulk salt concentration $\Sigma_i C_{i\alpha}$ by Equ.2. At various distances x from the surface the electrical profile is given by integrating the combination of the Poisson and Boltzmann relationships (1) to give:

$$\frac{d\psi_x}{dx} = \pm \; \frac{(2RT)^{1/2}}{\varepsilon_r \varepsilon_o} \; [\Sigma_i C_{i\alpha} (\exp(-Z_i \, F\psi_x / RT)-1)]^{1/2} \qquad \ldots\ldots\ldots \text{Equ.5}$$

For a Z-Z electrolyte Equ.5 can be integrated to give

$$\kappa x = \ln \, [\tanh(Z_i \, F\psi_o / 4RT)] - \ln \, [\tanh(Z_i \, F\psi_x / 4RT)] \qquad \ldots\ldots\ldots \text{Equ.6}$$

$$\kappa^2 = \frac{2Z^2 \, F^2 \, C_{i\alpha}}{RT\varepsilon_o \varepsilon_r}$$

This equation can be used to calculate the profile of ψ_x as a function of x. This electrical potential profile will give rise to a diffuse layer of counterions, the concentration of which can be calculated for any distance x using the Boltzmann equation:

$$C_{ix} = C_{i\alpha} \; \exp \, (-Z_i \, F\psi_x / RT) \qquad \ldots\ldots\ldots \text{Equ.7}$$

Thus the way in which ions distribute themselves within the diffuse layer will vary depending on their valency and concentration in the bulk phase so the situation can be rather complex when both mono- and divalent cations are present (Barber, 1980a; Rubin & Barber, 1980). This fact has important implications with regard to electrostatic screening which is dependent on the space charge density in the diffuse layer. The space charge density ρ_x is related to the concentration of ions in the following way:

$$\rho_x = \Sigma_i Z_i FC_{ix} \qquad \qquad \text{.......... Equ.8}$$

The integrated space charge density ρ_x' is therefore given by

$$\rho_x' = \int_0^X \rho_x \, dx \qquad \qquad \text{.......... Equ.9}$$

Electrostatic screening becomes important when considering the forces of inter-actions between the adjacent macroscopic surfaces, such as two membrane surfaces. The electrical charges on the membrane surfaces will results in a coulombic repulsion between them. This repulsion will be in opposition to long range attractive forces due to electrodynamic events giving rise to van der Waals interactions. The magnitude of this attractive force will be dependent on the physical size and the dielectric properties of the macroscopic system and can be estimated using formulations derived from the Lifshitz theory (Barber, 1980a; Israelachvili 1973). Unlike the van der Waals attraction, the coulombic repulsive force will vary depending on the degree of electrostatic screening within the diffuse layer. Thus under a particular condition a balance is struck between van der Waals attraction and coulombic repulsion. Indeed it is this concept that forms the basis of the classical theory which describes the stability of lipo-phobic colloids known as the Derjaguin-Landau-Verwey-Overbeek (DLVO) theory (Overbeek, 1978).

Clearly the degree of electrostatic screening will be governed by the integrated space charge density ρ_x' and it is therefore expected that it is this parameter which will dictate the long-range spatial relationships between two electrically charged surfaces. An expression which relates the coulombic repulsive force F_r between two adjacent membrane surfaces and the space charge density ρ_x^f is given by the following (see Rubin & Barber, 1980).

$$F_r = (P_i - P_e) - 1/2[(2\varepsilon_o \varepsilon_i - \varepsilon_o) \{E_x/(1-\rho_x')\}_I^2$$
$$- (2\varepsilon_o \varepsilon_i - \varepsilon_o) \{E_x/(1-\rho_x')\}_{II}^2] \qquad \qquad \text{.......... Equ.10}$$

where (P_i, P_e) and $(\varepsilon_i, \varepsilon_e)$ are the Kelvin pressures and dielectric constants, respectively, in the electrolyte medium in contact with the internal (i) and external (e) faces of the membrane surfaces. The quantities $\{E_x/(1-\rho_x')\}_I$ and $\{E_x/(1-\rho_x')\}_{II}$ are the electrical fields E_i and E_e, respectively, acting at the internal and external faces of the membrane surfaces. Subscript I denotes the interplate region bounded by the midplane and the internal faces of the two membranes. Subscript II denotes the outer plate region bounded by the external face of each membrane and a plane in the bulk solution corresponding to a zero electric field, and E_x is defined as $E_x = -d\psi_x/dx$. Derivation of Equation 10 is given in Rubin & Barber, 1980.

THYLAKOID MEMBRANE ORGANISATION AND CHARGE SCREENING

Observations

When unwashed isolated thylakoids are suspended in low salt containing media (e.g. 100 mM sorbitol, 1 mM Hepes, brought to pH 7.5 with KOH), they maintain their normal configuration of stacked and unstacked regions and have a high F_m level of chlorophyll fluorescence (F_m is the level of fluorescence when all photosystem two (PS2) traps are closed, e.g. by illuminating with bright light in the presence of the herbicide 3-(3,4-dichlorophenyl)-1,1-dimethylurea (DCMU)). When low levels of monovalent cations (1 to 30 mM) are introduced into the suspension, the membranes totally unstack and the F_m level is lowered (Gross & Hess, 1974; Gross & Prasher, 1974; Barber et al., 1977) (see Fig.2).

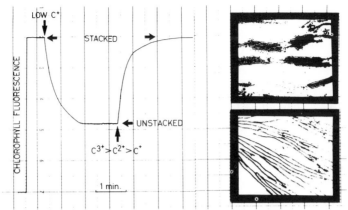

Fig. 2. Shows chlorophyll fluorescence and structural changes which occur when cations are added to unwashed, DCMU treated thylakoid membranes suspended in a medium essentially cation free (e.g. 100 mM sorbitol, 1 mM Hepes, brought to pH 7.5 with KOH).

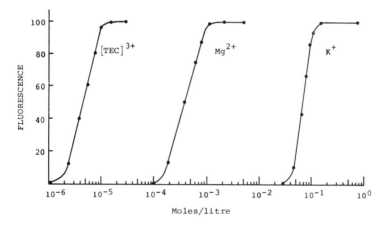

Fig. 3. Concentration requirements for the cation induced increase in room temperature chlorophyll fluorescence yield from isolated pea thylakoids showing the differential effect of the Tris (ethylenediamine)-cobaltic cation (TEC)$^{3+}$, Mg^{2+} and K$^+$. All salts were added as their chlorides (Barber & Searle, 1978).

From this condition the restacking and associated increase in F_m can be accomplished by introducing high levels of monovalent (>100 mM) or low levels of divalent (>5 mM) or low levels of trivalent cations (>0.1 mM) (see Fig.3). The relative order of effectiveness of different cations ($C^{3+} > C^{2+} > C^+$) in causing re-stacking and the fluorescence increase (Barber & Searle, 1979) indicates that the two processes are governed by electrostatics, a conclusion reinforced by the lack of dependency of the two phenomena on chemical species within a valency group or on osmotic potential changes (Mills & Barber, 1978). The dependency on valency could, however, imply that either ψ_o or $\rho_x^{'}$ is the main controlling factor but there are two observations which suggest that the latter parameter underlies the salt induced effects. Firstly the antagonistic action between low C^+ and high C^+ can be explained theoretically in terms of $\rho_x^{'}$ if at the beginning of the experiment described above the thylakoid membrane had divalent cations within its diffuse layer. Indeed analyses of unwashed isolated thylakoids and of intact chloroplasts indicates that Mg^{2+} is the dominating cation which screens the membrane surface charge (Nakatani et al., 1979). In low salt media these divalent cations are drawn tightly into the diffuse layer due to the large surface potential. Therefore, as can be calculated from the Poisson-Boltzmann equations (Equs.7 to 9 and the integrated form of Equ.5), the value of ρ_x' is relatively high close to the membrane surface (Rubin & Barber, 1980). When monovalent cations are introduced, the effect is to delocalize the space charge within diffuse layers so that the ρ_x' value near the membrane surface decreases. Clearly according to Equ.10 the effect of this is that there is an increase in the coulombic repulsion. However, as the monovalent cation level is increased to high levels the ρ_x' value increases with a consequent reduction in F_r. Fig. 4 shows such changes in F_r calculated for three background levels of divalent. This figure also shows the results of a chlorophyll fluorescence experiment carried out under ionic conditions similar to those assumed in the calculation. Clearly the experimental and theoretical curves share similar characteristics.

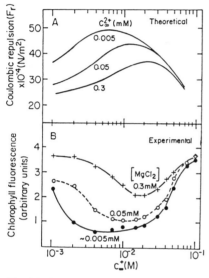

Fig. 4 A. Double layer repulsive force at constant surface potential as a function of monovalent cation concentration for three background divalent cation concs.(C_∞^{2+}) at a membrane separation distance of 2.0 nm. The curves have been calculated using Equ.10 taking the surface charge density at infinite membrane separation as -2.5×10^{-2} Cm^{-2}.
B. Chlorophyll fluorescence from isolated thylakoids treated with 10 μM DCMU and suspended in mixtures of mono- and di-valent cations C_∞^{2+} = Mg^{2+} and C_∞^+ = K^+, as indicated. Further details in Rubin et al., 1981.

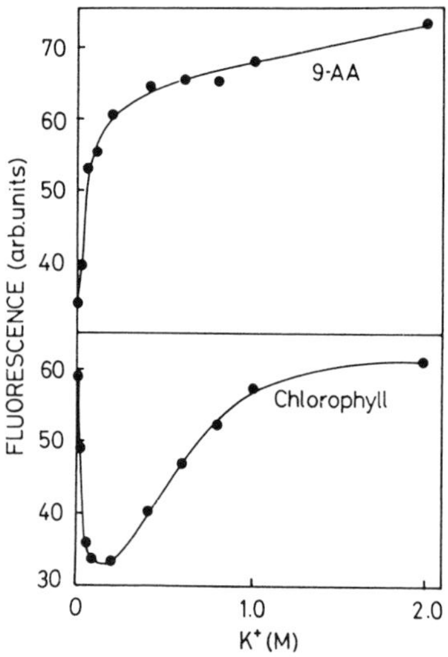

Fig. 5. Simultaneous measurements of chlorophyll and 9-aminoacridine fluorescence yields using isolated pea thylakoid membranes as a function of external levels of KCl (Barber & Searle, 1979).

A second, and more direct indication that salt induced phenomena are controlled by ρ_x' and not ψ_o comes from simultaneously measuring the fluorescence from chlorophyll and 9-aminoacridine (9-AA). It has been shown that 9-AA, which at neutral pH is a monovalent cation, becomes non-fluorescent when it is attracted into the diffuse layer due to a negative ψ_o (Chow & Barber, 1980). Reduction of ψ_o by addition of cations leads to a displacement of the dye molecule from the diffuse layer and a concomitant increase in its fluorescence yield. Such an increase can be seen in Fig.5 when isolated thylakoids were subjected to increasing levels of K^+. On the other hand, simultaneous measurements of chlorophyll fluorescence from the same sample, using a different detector, clearly shows the 'dip' phenomenon and a lack of correlation between the two fluorescence signals.

Explanation

How can changes in ρ_x^i bring about concomitant changes in thylakoid membrane stacking and yield of chlorophyll fluorescence? There are two important characteristics which suggest the coming together of adjacent thylakoid membrane surfaces on the addition of electrolytes is not simply due to the screening of fixed surface electrical charges as visualized in the DLVO theory for explaining colloid aggregation (Overbeek, 1978). Firstly, not all the membrane surface become appressed (i.e. stacked and unstacked regions form) indicative of heterogeneity in the surface properties. Secondly, where appression does occur the distance between adjacent surfaces is very short, being less than 4 nm. If the simplest form of the DLVO theory was applicable then, because of the relatively large average surface charge density, the expected inter-membrane distance would probably be greater than 10 nm. In fact, to account for a 4 nm inter-membrane distance requires a very strong van der Waals attractive interaction with a low electrostatic repulsion (Sculley et al., 1980; Rubin et al., 1981). The attractive force is increased if the protein to lipid ratio is high and if a substantial amount extends well beyond the bilayer. Such conditions are probably found in the appressed regions (Ford et al., 1982). The requirement for low electrostatic repulsion can be met if the surface charge density is low in regions where membrane appression occurs. Duniec et al. (1981) proposed that this was accomplished by ion binding while I (Barber, 1980b) gave an alternative explanation in which it was suggested that electrically charged components laterally migrate away from the regions where membrane appression occurs (see Fig.6). In this way domains with low surface charge would be formed with weak electrical double-layer repulsion.

Fig. 6.

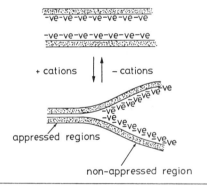

Unstacked with net negative charges randomised.

Lateral charge displacement with net negative charge localized in unstacked regions.

Because the formation of appressed and non-appressed membrane regions is paralleled by an increase in chlorophyll fluorescence yield, it was suggested that the lateral movement of charge was due to changes in spatial separations between chlorophyll protein complexes within the membrane. At room temperature it is known that chlorophyll fluorescence comes from a combination of two pigment-protein complexes, light harvesting chlorophyll a/b (LHC-2) complex and photosystem two (PS2) complex. The other pigment-protein complex of the thylakoids is photosystem one (PS1) which at room temperature is virtually non-fluorescent. When LHC-2–PS2 is close to PS1, energy transfer occurs to PS1 so that the yield of fluorescence is low. However, separation of the complexes will lead to an increase in fluorescence. As seen in Fig.7, the postulate was made (Barber, 1980b) that when the membranes restacked on the addition of cations, the PS1 complexes migrate to the unstacked regions while LHC-2–PS2 form the domains which give rise to close membrane interaction. Thus the model assumes that PS1 together with the other complexes in the unstacked regions (the ATP synthetase CF_o-CF_1 and a proportion of the cytochrome b_6-f) carry significant levels of net electrical charge on their exposed surfaces (steric factors may also be important, especially in the case of CF_o-CF_1). On the other hand the model would predict that the combination of

LHC-2 and PS2 would give rise to a low surface charge. Using data of Davies et al. (1986), Telfer (1987) recently calculated that the surface charge density of the appressed and non-appressed regions of lettuce thylakoids is -0.006 and -0.057 Cm^{-2}, respectively.

The charge displacement model presented above has many observations to support it. It has been shown that the restacking of thylakoids requires a fluid membrane (Barber et al., 1980) and the postulated movements of complexes has been observed by freeze-fracture (e.g. Staehelin, 1976). The lateral distribution of complexes as indicated in Fig.7 is now well accepted due to the pioneering work of Andersson & Anderson (1980) and Albertsson et al. (1982) using a phase-separation technique. The stacking of thylakoids by increasing ρ_x^s contrasts with the more trivial stacking induced by neutralizing the net surface charge on the membrane by strong ionic binding or by lowering the pH to the isoelectric point of 4.3 (Barber, 1980b). In this case more extensive stacking occurs which is independent of lipid fluidity and which maintains a randomisation of complexes (Barber et al., 1980). As would be expected there is no characteristic increase in chlorophyll fluorescence.

CONSEQUENCES

The above considerations indicate that the organisation of the thylakoid membrane and the spatial relationships between various intrinsic protein complexes are governed, at the first approximation by long range forces of attraction and repulsion. The important consequence of this is that we can interpret lateral separation of different types of complexes into the appressed and non-appressed regions in terms of simple electrostatic theory. We can therefore understand how perturbation of the electrical properties of this system will give rise to changes in membrane structure. Varying the ionic condition of the bathing medium is an extreme example of this. Another possibility is to alter the nature or extent of the electrical charge on the surface of a particular complex. It is this latter possibility which seems to underlie the molecular mechanism by which protein phosphorylation can regulate energy distribution between PS1 and PS2 via the phenomenon known as State transitions (Barber, 1983,1986; Bennett, 1983; Horton, 1983, Allen, 1983). The phosphorylation occurs at the threonyl residues close to N-terminus of the LHC-2 polypeptide. This phosphorylation is catalyzed by a membrane-bound kinase activated when PS2 is excited more than PS1 so that the electron transport system which functionally links the two photosystems (in particular, the plastoquinone pool) becomes over reduced. When the kinase is not activated, for example in the dark under oxidising conditions, or in excess PS1 light, membrane-bound phosphatase brings about the dephosphorylation of LHC-2. The phosphorylation/dephosphorylation processes have been shown to give rise to changes in energy distribution between PS2 and PS1 (Allen et al., 1981; Horton & Black, 1980). Indeed, this regulatory process, under limiting light intensities, gives rise to a better balance of energy distribution between PS1 and PS2 and since these photosystems must cooperate together to pass electrons from H_2O to NADP, then a higher photosynthetic efficiency is obtained with a particular lighting condition (Canaani et al., 1984).

How can phosphorylation of LHC-2 regulate energy distribution between PS2 and PS1? Bearing in mind the surface electrical properties of the thylakoid as described above, it would be predicted that the phosphorylation/dephosphorylation of LHC-2 would alter its surface charge properties and thus dictate its preference to partition into the appressed and non-appressed regions of the membrane (Barber, 1982a; Barber, 1986). It seems that only a proportion of the LHC-2 are involved in this process and are therefore mobile. In the non-phosphorylated condition, the mobile LHC-2 would be closely associated with PS2 complexes in the appressed regions while the introduction of negative charge onto its surface by phosphorylation would induce coulombic repulsive forces leading to lateral diffusion into the PS1 enriched non-appressed membranes (see Fig.7). In this way the mobile form of LHC-2 can act either as antenna for PS2 or PS1. There is considerable support for this model ranging

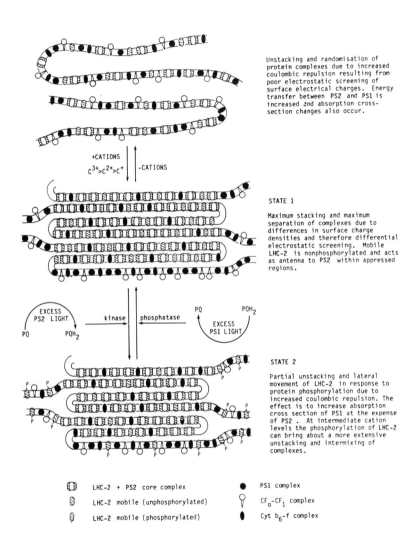

Unstacking and randomisation of
protein complexes due to increased
coulombic repulsion resulting from
poor electrostatic screening of
surface electrical charges. Energy
transfer between PS2 and PS1 is
increased and absorption cross-
section changes also occur.

+CATIONS

$c^{3+} > c^{2+} > c^{+}$ -CATIONS

STATE 1

Maximum stacking and maximum
separation of complexes due to
differences in surface charge
densities and therefore differential
electrostatic screening. Mobile
LHC-2 is nonphosphorylated and acts
as antenna to PS2 within appressed
regions.

EXCESS
PS2 LIGHT   kinase   phosphatase   PQ PQH_2

PQ PQH_2 EXCESS
                                              PS1 LIGHT

STATE 2

Partial unstacking and lateral
movement of LHC-2 in response to
protein phosphorylation due to
increased coulombic repulsion. The
effect is to increase absorption
cross section of PS1 at the expense
of PS2. At intermediate cation
levels the phosphorylation of LHC-2
can bring about a more extensive
unstacking and intermixing of
complexes.

LHC-2 + PS2 core complex   ● PS1 complex

LHC-2 mobile (unphosphorylated)   CF₀-CF₁ complex

LHC-2 mobile (phosphorylated)   Cyt b₆-f complex

Fig. 7

**Diagrammatic representation of changes in thylakoid membrane
organisation in response to protein phosphorylation and
modification of cation levels.**

from freeze-fracture electron microscopy (Kyle et al., 1983) and biochemical analyses (e.g. Chow et al., 1981; Andersson et al., 1982) to energy-transfer and quantum-yield studies (e.g. Farchaus et al., 1982; Telfer et al., 1986). However, as yet the quantitative details of this regulatory process are not firmly established. We do not know precisely how much of the LHC-2 becomes functionally linked to PS1. It is also unclear whether LHC-2 moves alone or whether some LHC-2–PS2 complexes also laterally diffuse into the non-appressed regions. If the latter occurs then there will be changes in energy transfer between PS2 and PS1 (spillover) as well as changes in the absorption cross section of the two photosystems. Indeed, with isolated thylakoids the degree of LHC-2 and LHC-2–PS2 diffusion in response to phosphorylation is, in agreement with the electrostatic model, sensitive to the level of cations in the suspension medium (Telfer et al., 1983; Telfer et al., 1984).

CONCLUSION AND FINAL COMMENT

The electrical charges on the surface of the chloroplast thylakoid membrane seems to dictate its conformational state. Any perturbation of the surface electrical properties therefore bring about a reorganisation of the membrane. The phosphorylation of LHC-2 seems to be a beautiful demonstration of how such perturbations are used in order to regulate and optimise the efficiency of photosynthesis. Other proteins of the thylakoid membrane can be reversibly phosphorylated but the precise reason why is as yet unclear. It could be that in a similar way to LHC-2 these phosphorylations bring about key changes in surface electrical properties of the exposed portions of the proteins and therefore effect their interaction with neighbouring complexes by changes in coulombic forces.

REFERENCES

Albertsson, P-A., Andersson, B., Larsson, C. & Akerlund, H-E. (1982) Meth. Biochem. Anal. 28, 115-150
Allen, J.F. (1983) Trends Biochem. Sci. 8, 369-373
Allen, J.F., Bennett, J., Steinback, K.E. & Arntzen, C.J. (1981) Nature 291, 21-25
Andersson, B. & Anderson, J.M. (1980) Biochim. Biophys. Acta 593, 427-440
Andersson, B., Akerlund, H-E., Jergil, B. & Larsson, C. (1982) FEBS Lett. 149, 181-185
Barber, J. (1980a) Biochim. Biophys. Acta 594, 253-308
Barber, J. (1980b) FEBS Lett. 118, 1-10
Barber, J. (1982a) Ann. Rev. Plant Physiol. 33, 261-295
Barber, J. (1982b) BioScience Rep. 2, 1-13
Barber, J. (1983) Photobiochem. Photobiophys. 5, 181-190
Barber, J. (1986) in Encyclopedia of Plant Physiology, New Series (Staehelin, L.A. & Arntzen, C.J. eds.), vol. 19, pp 653-664, Springer-Verlag, Berlin and New York.
Barber, J., Mills, J. & Love, A. (1977) FEBS Lett. 74, 174-181
Barber, J. & Searle, G.F.W. (1978) FEBS Lett. 92, 5-8
Barber, J. & Searle, G.F.W. (1979) FEBS Lett. 103, 241-245
Barber, J., Chow, W.S., Scoufflaire, C. & Lannoye, R. (1980) Biochim. Biophys. Acta 591, 92-103
Bennett, J. (1983) Biochem. J. 212, 1-13
Canaani, O., Barber, J. & Malkin, S. (1984) Proc. Natl. Acad. Sci. USA 81, 1614-1618
Chow, W.S. & Barber, J. (1980) J. Biochim. Biophys. Methods 3, 173-185
Chow, W.S., Telfer, A., Chapman, D.J. & Barber, J (1981) Biochim. Biophys. Acta 638, 60-68
Davies, E.C., Chow, W.S. & Jordan, B.J. (1986) Photosyn. Res. 9, 359-370
Duniec, J.T., Israelachivili, J.N., Ninham, B.W., Pashley, R.M. & Thorne, S.W. (1981) FEBS Lett. 129, 193-196

Farchaus, J.N., Widger, W.R., Cramer, W.A. & Dilley, R.A. (1982) Arch. Biochem. Biophys. 217, 362-367

Ford, R.C., Chapman, D.J., Barber, J., Pedersen, J.Z. & Cox, R.P. (1982) Biochim. Biophys. Acta 681, 145-151

Gross, E.L. & Hess, S.C. (1974) Biochim. Biophys. Acta 339, 334-346

Gross, E.L. & Prasher, S.H. (1974) Arch. Biophys. Biochim. 164, 460-468

Horton, P (1983) FEBS Lett. 152, 47-52

Horton, P. & Black, M.T. (1980) FEBS Lett. 119, 141-144

Israelachvili, J.N. (1973) Q. Rev. Biophys. 6, 341-387

Kyle, D.J., Staehelin, L.A. & Arntzen, C.J. (1983) Arch. Biochem. Biophys. 222, 527-541

Mills, J. & Barber, J. (1978) Biophys. J. 21, 257-272

Nakatani, H.Y. & Barber, J. (1980) Biochim. Biophys. Acta 591, 82-91

Nakatani, H.Y., Barber, J. & Forrester, J.A. (1978) Biochim. Biophys. Acta 504, 215-225

Nakatani, H.Y., Barber, J. & Minski, M.J. (1979) Biochim. Biophys. Acta 545, 24-35

Overbeek, J. Th. G. (1978) J. Colloid Interface Sci. 58, 408-422

Rubin, B.T. & Barber, J. (1980) Biochim. Biophys. Acta 592, 87-102

Rubin, B.T., Chow, W.S. & Barber, J. (1981) Biochim. Biophys. Acta 634, 174-190

Sculley, M.J., Duniec, J.T., Thorne, S.W., Chow, W.S. & Boardman, N.K. (1980) Arch. Biochem. Biophys. 201, 339-346

Staehelin, L.A. (1976) J. Cell Biol. 71, 136-158

Telfer, A. (1987) in Progress in Photosynthesis Research, Vol.2, pp 689-696 (Biggins, J. ed.), Pub. Martinus Nijhoff Publ.

Telfer, A., Hodges, M. & Barber, J. (1983) Biochim. Biophys. Acta 724, 167-175

Telfer, A., Hodges, M., Millner, P.A. & Barber, J. (1984) Biochim. Biophys. Acta 766, 554-562

Telfer, A., Whitelegge, J., Bottin, H. & Barber, J. (1986) J. Chem. Soc., Faraday Trans. 2 (Special Edition) 82, 2207-2215

Catabolic regulation of thylakoid membrane structure and function during senescence

Howard THOMAS

Plant and Cell Biology Department, AFRC Institute of Animal and Grassland Production, Welsh Plant Breeding Station, Plas Gogerddan, Aberystwyth, Dyfed, SY23 3EB, UK

SYNOPSIS

The D1 protein of Photosystem II (PSII), cytochrome f and LHCP-2 (the chlorophyll a/b binding protein of the PSII light-harvesting complex) are normally labile during foliar senescence. A mutation of a single Mendelian locus in Festuca pratensis results in greatly enhanced stability of these and other proteins of intrinsic thylakoid membrane complexes. Catabolism of LHCP-2 is coordinated with, and possibly regulated by, chlorophyll breakdown. Proteolysis of LHCP-2 is preceded by a protein synthesis-dependent removal of phytol from the prosthetic group, requires oxygen and is disabled in the Festuca mutant after the dephytylation step. The specificity of proteolysis may reside in the ability of degradative enzymes to identify altered, non-functional proteins as preferred substrates. The kinetic consequences of such a mechanism are evaluated and implications for the control of cellular turnover processes are considered.

INTRODUCTION

The rate of photosynthesis diminishes during leaf senescence. Although many components of the photosynthetic apparatus decline in function and quantity, it is generally believed that loss of thylakoid proteins, particularly those of PSII and the cytochrome b_6/f complex, is the rate-determining process (Holloway et al., 1983). It has been proposed that decreases in these components are a consequence of cessation of synthesis coupled with continued turnover (Woolhouse & Batt, 1976). A recent study of thylakoid protein metabolism in senescing bean leaves comes to essentially the same conclusion (Roberts et al., 1987). This model of the molecular control of senescence assumes that catabolic processes degrading protein during senescence are the same as those responsible for turnover during membrane assembly or at the steady-state in pre-senescent leaves. As discussed later, this contention is far from proven. A non-yellowing mutant of Festuca pratensis, in which thylakoid disassembly is impaired but which can

----

ABBREVIATIONS: PSI/II, Photosystem I/II; LHC-2, light-harvesting complex of PSII; LHCP-2, chlorophyll a/b binding protein of LHC-2; EM, electron microscopy; SDS, sodium dodecyl sulphate; Chl, chlorophyll; HPLC, high performance liquid chromatography; PC, polar Chl derivative; TLC, thin-layer chromatography

assemble functional chloroplasts in an apparently normal manner
(Thomas, 1987; Thomas & Hilditch, 1987), appears to provide
decisive evidence to the contrary. The first two sections of this
paper describe the behaviour of chloroplast membrane proteins and
pigments during senescence of normal and mutant leaf tissue. These
examples introduce a general discussion of the specificity of
degradative processes in metabolic turnover.

PROTEINS OF THYLAKOID MEMBRANE COMPLEXES

        Figure 1a presents a Western analysis using antibodies to the
D1 protein of Photosystem II (Nixon et al., 1986) applied to blots
of thylakoid proteins from normal yellowing (Rossa) and mutant
(Bf993) leaf tissue senesced in darkness for up to 5 days. The
enhanced stability of D1 in Bf993 confirms previous observations
based on radiolabelling and herbicide binding (Hilditch et al.,
1986). In addition to the major antigenic band at 32 kD, there is
a 24-25 kD component which we believe is identical with the product
of in vivo light-dependent proteolysis described by Greenberg et
al. (1986). This fragment clearly remains in the membrane and its
further degradation in the mutant is impaired during senescence.
This is additional evidence that D1 is turned over by two distinct
catabolic processes (Greenberg et al., 1986). Cytochrome f is also
rather less labile in Bf993 than in Rossa (Figure 1b), as
originally observed by Thomas (1983) in a study based on haem
peroxidase determination.
        Cytochrome f and D1 protein are both coded by chloroplast
DNA. Transcription of the chloroplast genome (and possibly even of
key nuclear genes for chloroplast components) may well be
restricted in anticipation of the onset of senescence - but the
behaviour, in the Festuca mutant, of the proteins described here
says that switching off synthesis is not enough. It is likely that
part or all of the degradation process operating in senescence is
specific to that phase of leaf development.

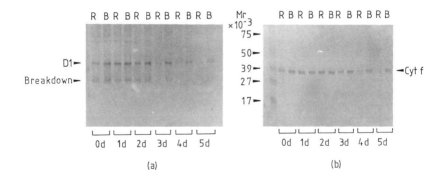

(a)   (b)

Fig. 1. (a) Western blot of D1 proteins from normal (R) and mutant
(B) Festuca pratensis leaf tissue senescing in darkness. Antibody
generously supplied by Peter Nixon, Imperial College. (b) Western
blot of cytochrome f from normal (R) and mutant (B) Festuca
pratensis leaf tissue senescing in darkness. Antibody to Sinapsis
cytochrome f the kind gift of John Gray, Cambridge University.

Hilditch (1986) quantified the behaviour of the major nuclear-encoded chlorophyll-binding intrinsic protein, LHCP-2, by enzyme-linked immunosorbent assay, and confirmed its abnormal stability in the mutant during foliar senescence. Qualitative alterations in LHCP-2 and its subcellular distribution in Bf993 and the wild-type comparison Rossa have been further characterized using Western blotting and immunogold electron microscopy (EM). Detached leaf tissue of Bf993, and of the yellowing genotype Rossa, was incubated in darkness for up to 6 days and total proteins extracted into buffered sodium dodecyl sulphate (SDS) solution. Figure 2a presents Western blots of LHCP-2 separated by SDS-polyacrylamide gel electrophoresis. Progressive loss of the main antigen band from Rossa is apparent, whereas, in the mutant, LHCP-2 persists to the end of the treatment. A series of what appear to be proteolytic fragments become more prominent in Bf993 as senescence proceeds. Although these may represent true in vivo intermediates of LHCP-2 breakdown accumulating as a consequence of impaired further catabolism in the mutant, we believe they are more likely to be extraction artefacts. Figure 2b is a similar sequence but here tissue was rapidly separated into soluble and membrane fractions before detergent solubilisation. The abundance of fragments is much reduced, although some degradation is still seen at day 6. There is no evidence of antigenic fragments in the soluble fraction, only intact LHCP-2 which had not fully pelleted during centrifugation. The persistence of full-length LHCP-2 polypeptide in the mutant throughout a period of intense proteolysis acting on stroma and extrinsic thylakoid proteins (Thomas & Hilditch, 1987) implies that LHCP-2 is protected from attack. We infer that the lesion in Bf993 concerns impaired accessibility of intrinsic membrane protein substrates, or absence of a specific proteinase.

Thomas (1977) previously described the ultrastructure of senescing mutant plastids. Stroma material is lost but thylakoids endure as loosely appressed, swirling membrane figures. Plastoglobuli are far less abundant than in normal plastids at a

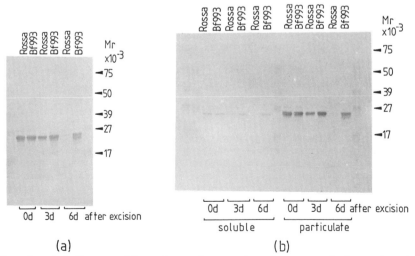

Fig. 2. (a) Western blot of total proteins from normal (Rossa) and mutant (Bf993) Festuca leaf tissue using rabbit antibody to Festuca LHCP-2 protein. (b) LHCP-2 Western blot of Festuca leaf proteins separated by centrifugation into soluble and particulate fractions.

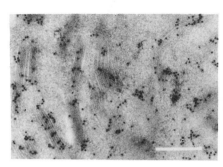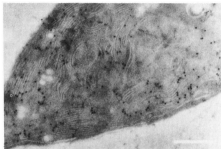

(a) Day 0   (b) Day 4

Fig. 3. (a) Immunogold labelling of LHCP-2 in a chloroplast of non-senescent Bf993 leaf tissue. (b) Immunogold labelling of LHCP-2 in a chloroplast of Bf993 after 4 days senescence. Bar = 500 nm.

comparable stage. These observations were confirmed in an EM study of the localisation of LHCP-2 using immunogold staining. Figure 3a is part of a chloroplast of non-senescent Bf993 showing that LHCP-2 is most abundant in the stacked regions of the thylakoid system. Non-senescent Rossa plastids are essentially identical in appearance. Figure 3b is a section of a 4 day senesced plastid of the mutant. With the loss of tightly-stacked grana there is a randomisation of particles, indicating dispersal of LHCP-2 throughout the membrane, a conclusion supported by freeze-fracture studies (Marti, 1984). Rossa plastids at the same stage displayed negligible labelling. These results imply that inaccessibility of LHCP-2 to proteolysis in the mutant is not a consequence of persisting lateral heterogeneity of its distribution in the membrane.

COORDINATION OF PROTEIN AND PIGMENT CATABOLISM

It is suggestive that the abnormally stable proteins of Bf993 chloroplasts have prosthetic groups (haem, chlorophyll). Experiments with mutants have established that synthesis and disassembly of chlorophyll-proteolipid complexes requires close coordination of pigment and protein metabolism (Bellemare et al., 1982; Thomas & Hilditch, 1987). The biosynthetic pathways of chlorophyll (Chl) and chloroplast polypeptides have been intensively studied and possible points of regulatory interaction have been identified; but the routes whereby pigments and proteins are degraded are obscure. Recently, progress has been made towards establishing the pathway of chlorophyll catabolism (Matile et al., 1987). I describe here the characterisation of early intermediates in the breakdown sequence and demonstrate that these catabolites are still associated with the apoproteins of the major pigment-proteolipid complex, the light-harvesting Chl-protein of PSII (LHC-2).

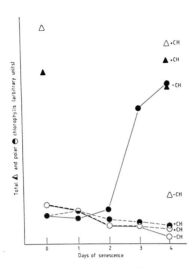

Fig. 4. Total chlorophyll (△ ▲) and polar chlorophyll derivatives (o •) in detached Festuca leaf tissue during dark senescence. Open symbols Rossa (normal genotype), closed Bf993 (mutant). ± CH = incubated in presence (+) or absence (-) of 1 mM cycloheximide.

        The total quantity of green pigments declines very little during dark incubation of leaf tissue from Bf993 (Thomas, 1987); but liquid chromatography (HPLC) analysis revealed some loss of Chls a and b and accumulation of more polar green pigments (PCs) during the early stages (Gut et al., 1987). PCs are a group of dephytylated derivatives of Chl a and Chl b (K. Bortlik, unpublished). Figure 4 shows the occurrence, in mutant and normal Festuca, of PCs separated by anion exchange chromatography. PCs increase to maximal levels at 4 days of senescence. Cycloheximide, at a concentration sufficient completely to inhibit yellowing in the normal genotype, prevents PC accumulation. Levels of PCs in normal leaf tissue are insignificant.
        In experiments with Rossa and with barley primary leaves, however, PC-like pigments are accumulated to high levels if tissue is incubated in darkness under an O_2-free atmosphere (D.Reutsch, unpublished). It is proposed, therefore, that PCs are normal intermediates in Chl catabolism, which begins with a protein synthesis-dependent dephytylation followed by oxidation leading ultimately to open-chain pyrrole derivatives (Matile et al., 1987). PCs accumulate in the mutant because the metabolic lesion blocks pigment-proteolipid degradation after the dephytylation step.
        PCs in 4 day senesced mutant leaves are associated with LHC-2. This was demonstrated by fractionating thylakoids using sucrose gradient centrifugation (Thomas et al., 1985) and separating pigments from gradient fractions by reverse phase TLC (Schoch et al., 1984; Figure 5a). PCs are visible as the most mobile green bands. Figure 5b is a reflectance scan of pigments from PSI compared with those from Mg^{++}-precipitated LHC-2. PCs are clearly associated with the latter complex. Thus separation of Chl from its associated proteins and its further catabolism occurs after the dephytylation step. The subsequent route of breakdown is

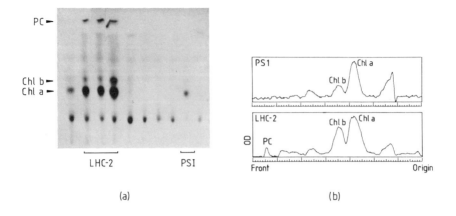

(a)   (b)

Fig. 5. (a) Pigments of Triton X-100-solubilised thylakoids from 4 day senesced mutant leaf tissue fractionated by sucrose gradient centrifugation. Pigments from gradient fractions were separated by reverse phase TLC. Positions of LHC-2 and PSI were identified by fluorescence, Chl a:b ratio and polypeptide composition of electrophoresed samples. (b) Pigments of PSI and Mg^{++}-precipitated LHC-2 from sucrose gradient fractionation of thylakoids.

a complex sequence of reactions and relocations of intermediates, since the next stable identifiable products, pink open-chain pyrroles, are found almost exclusively in the cell vacuole (Matile et al., 1988).

SPECIFICITY OF PROTEIN TURNOVER

During senescence, catabolic cellular processes predominate over synthetic or maintenance activities. The behaviour of thylakoid membrane proteins in the Festuca mutant is consistent with a high degree of specificity operating on the degradative side of the turnover equation. Whereas there is (so far as we know) a single mechanism of synthesis applicable, with essentially trivial variations, to all proteins wherever they are found, it is highly unlikely that a process of comparable generality exists to account for protein breakdown. Mature proteins carrying out their specific functions in specific cellular microenvironments (and often having undergone specific post-translational modifications to their structures) must present far too diverse a range of substrates for a single catabolic process to work on.

A given protein may be degraded by different routes depending on particular cellular conditions. Turnover is a central element in the homeostatic maintenance of equilibrium. For example, if the stoichiometric production of the separate components of a macromolecular complex is perturbed, there is often specific degradation of those subunits that are in excess to bring the proportions of the constituents into line (Bennett, 1983; Mishkind et al., 1985; Duysen et al., 1987). Indeed, it may be that post-translational fine tuning of this sort, rather than sophisticated coordinated modulations at the genetic level, is the principal means of regulating assembly and maintenance of thylakoid complexes and other multi-partite structures in plants. If developmental or environmental pressures necessitate a change to a new state outside current homeostatic limits, turnover is once more the agency whereby the protein complement of the cell is remodelled. Again, there are numerous examples of new proteins replacing old in response to an endogenous or environmental inductive cue (Davies, 1982). In the extreme case of a shift to cellular conditions beyond the capacity of the homeostatic or adaptive machinery, then turnover becomes catastrophic, either as part of a last-ditch survival effort or else as a symptom of declining viability (Davies, 1986).

What is not at all clear is the extent to which the same catabolic system might be responsible for all modes of protein turnover - homeostatic, adaptive, catastrophic. If there are separate systems (which seems likely), how are their activities coordinated? What determines which degradative pathway will operate to break down a given protein? And what are the biochemical properties of these catabolic sequences? Despite the high fidelity with which structures are made and functions performed in cells, some individual molecules in a protein population will become distinguishable from the rest, perhaps because they have been structurally damaged, displaced from their normal cellular location or stripped of some stabilizing factor. An important function of turnover would be to prevent such casualties of subcellular life from accumulating to chaotic levels. Kinetic analyses of turnover (Davies, 1980) have generally assumed that all individuals in the protein population are equally likely to be degraded. A simple model of synthesis, breakdown and alteration may be used to show what happens when only altered proteins are candidates for catabolism.

RANDOM AND BIASED CATABOLISM

'Random' here refers to lack of discrimination between altered and unaltered molecules on the part of the degradative process. Where only altered protein molecules are eligible for breakdown during turnover, the process is said to be 'biased'. Assuming synthesis to be zero-order and alteration and degradation to be first-order with respect to protein, the following equations describe the behaviour of the total population and number of altered molecules in the population during a small interval of time $\delta \underline{t}$.

Random case
$$\delta \underline{n} = \sigma \delta \underline{t} - \underline{n} \delta \underline{t} \tag{1}$$

Biased case
$$\delta \underline{n} = \sigma \delta \underline{t} - \beta \underline{a} \delta \underline{t} \tag{2}$$

Both cases
$$\delta \underline{a} = \alpha \delta \underline{t} (\underline{n} - \underline{a}) - \beta \underline{a} \delta \underline{t} \tag{3}$$

where σ, α and β are the rate constants for synthesis, alteration and degradation respectively, $\underline{n}$ is the total number of molecules in the population and $\underline{a}$ the number of altered molecules (Figure 6).

Figure 7 presents iterative solutions in which the changing magnitudes of $\underline{n}$ and $\underline{a}$ are mapped after successive small intervals of time. An efficient turnover system must be capable of regulating

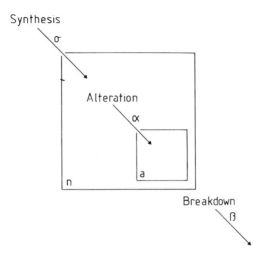

Fig. 6. Parameters defining the turnover kinetics of a population of protein molecules, a proportion of which have been structurally altered.

both the total size of the population and the proportion of altered molecules within it. In the random case (Figure 7a), control of n is readily accomplished by altering either the rate of synthesis or the rate of breakdown, whereupon n and a adjust themselves to new equilibrium values. Adjustment of the number of altered entities present, a, can be achieved by increasing or decreasing the overall turnover rate, the rates of synthesis and degradation being maintained in the same proportion in order that the value of n does not change. Considering now the biased model (Figure 7b): here the values of n and a are mutually related in a fashion unlike that seen for the random case. Introducing any variation in the number of altered individuals, a, provokes considerable changes in the total population size. The biased model is certainly capable of controlling a, but seems unable to do this independently of n.

Both models provide suitable control of the number of altered molecules within the population. The random model is able to do this while the total size (including altered members) of the set is constant. In contrast, in the biased model the value of a cannot be changed by simple modulation of parameters without provoking wild fluctuations in the total number n. The biased model would thus be suitable for controlling the number of altered molecules in a population where the number of unaltered members was of more importance than the total amount of protein present. The random system, on the other hand, could control the occurrence of altered proteins in a population of defined total size and is consistent with the notion of a general non-specific mechanism for proteolysis. The biased model is reminiscent of the one protein - one proteinase concept which, for several reasons, is not considered likely to be of widespread occurrence in cells (Ferreira and Davies, 1986).

These observations lead to the conclusion that degradation of the population as a whole, without discrimination between altered and unaltered molecules, is fully capable of controlling population size and the number of altered members within it. Specific

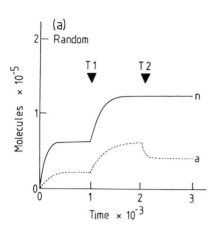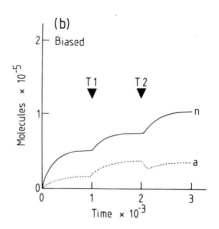

Fig. 7. (a) The 'random' model. From given initial values of n and a, equilibrium levels are reached, which may be altered by changing rates of synthesis or degradation separately or together as turnover rate. Initial parameters: $\sigma = 6 \times 10^2$; $\beta = 10^{-2}$; $\alpha = 5 \times 10^{-3}$. At time T1, β reduced to 5×10^{-3}; at time T2 σ increased to 1.2×10^3 and β increased to 10^{-2}. (b) The 'biased' model. Equilibrium values are again reached, but altering β affects n and a to an equal extent; attempting to change a via increased turnover affects n. Initial parameters: $\sigma = 2 \times 10^2$; $\beta = 10^{-2}$; $\alpha = 5 \times 10^{-3}$; at time T1; β reduced to 5×10^{-3}; at time T2 σ increased to 4×10^2 and β increased to 10^{-2}.

degradation of redundant molecules might occur in certain cases, sufficiently important to warrant their own complex infrastructure, but there is no necessity for this to happen in the general case. Where rates of damage to a protein are high, degradation targetted specifically at the damaged molecules would be better able to control them - but alone would provide poor control of the population size, so it might be that this protein would have two systems, one aimed only at damaged molecules, and one directed at the population as a whole, giving control and flexibility. For a protein where alteration is slow, rates of turnover are still essential to give control of population size, and here the random model would be appropriate.

How are these conclusions relevant to the question of thylakoid membrane protein turnover? Consider the D1 protein which, as discussed previously is known to be catabolized by two separate routes, one rapid and light-dependent, the other slower and light-independent (Hilditch et al., 1986). We speculate that the former pathway is a specific damage-related mechanism similar to the biased model discussed above, while the light-independent pathway is a general ('random') method which could control population size. Furthermore, it may now be justifiable to relate systems of the random type primarily to the adaptive mode of turnover, and to consider catabolic bias to be more significant in homeostasis at one extreme and catastrophe at the other. An additional corollary of this hypothesis is that quite separate metabolic pathways must exist to degrade proteins during adaptive

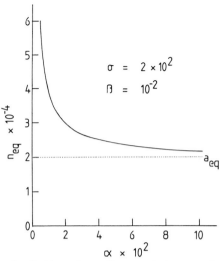

Fig. 8. Control of the size of a protein population at the steady state by biased catabolism acting through the degree of alteration. The curve describes the relation

$$\underline{n}_{eq} = \frac{\sigma(\alpha + \beta)}{\alpha\beta}$$

as compared with homeostatic and catastrophic adjustments, and that the altered locus in the <u>Festuca</u> mutant is part of coterie of genes concerned with regulating this adaptive pathway.

The foregoing discussion of the significance of turnover has concerned itself with what may be termed the 'classical' conception, namely the function of turnover in controlling levels of abnormal or unwanted proteins. But in recent years a different, essentially inverted, view has gained credence: turnover controlled by modification. Proteins destined for degradation by a biased catabolic system are envisaged as becoming 'tagged' in structure-modifying reactions, the rates and specificities of which define the turnover characteristics of those proteins (Dice, 1987). By varying the rate constant for alteration or tagging, the size of the total protein population at the steady state ($\underline{n}_{eq}$) may be varied over the range $a_{eq} \leqslant n_{eq} \leqslant \infty$ (Figure 8). Such a mechanism for regulating protein levels is intuitively satisfying and finds some experimental support (Stadtman, 1986); but the potential costs and benefits of this version of biased turnover have to be weighed up in similar terms to those discussed earlier.

ACKNOWLEDGEMENTS

I thank my collaborators Paul Hilditch, Emyr Davies, Linda Nock, Lyndon Rogers, Barry Thomas and Philippe Matile for allowing me to use some of their ideas and unpublished data. This work was supported by the Agricultural and Food Research Council, the Science and Engineering Research Council and the Swiss National Science Foundation.

REFERENCES

Bellemare, G., Bartlett, S.G. & Chua, N-H (1982)   J. Biol. Chem.
  257, 7762-7767
Bennett, J. (1983) Biochem. J. 212, 1-13
Davies, D.D. (1980) in   Advances in Botanical Research Vol. 8.
  (Woolhouse, H.W. ed.), pp. 66-126, Academic Press, London
Ferreira, R.B. & Davies, D.D. (1986) Planta 169, 278-288
Davies, D.D. (1982) in Encyclopaedia of Plant Physiology,
  (Boulter, D. & Parthier, B. eds.), New Series vol. 14A,   pp.
  189-228, Springer-Verlag, Berlin
Davies, K.J.A. (1986) J. Free Radicals Biol. Med. 2, 155-173
Dice, J.F. (1987) FASEB J. 1, 349-357
Duysen, M., Huckle, L., Mogen, K. & Freeman, Th. (1987)
  Photosynth. Res. 14, 159-170
Ferreira, R.B. & Davies, D.D. (1986) Planta 169, 278-288.
Greenberg, B.M., Gaba, V., Mattoo, A.K. & Edelman, M. (1986) EMBO
  J. 6, 2865-2869
Gut, H., Rutz, C., Matile, Ph. & Thomas, H. (1987) Physiol. Plant.
  70, 659-663
Hilditch, P. (1986) Plant Science 45, 95-99
Hilditch, P., Thomas, H., Rogers, L.J. (1986)   FEBS Letts. 208,
  313-316
Holloway, P.J., MacLean, D.J. & Scott, K.J. (1983) Plant Physiol.
  72, 795-801
Marti, R. (1984) Diplomarbeit. ETH, Zurich
Matile, Ph., Ginsburg, S., Schellenberg, M. & Thomas, H. (1987)
  J. Plant Physiol. 129, 219-228
Matile, Ph., Ginsberg, S., Schellenberg, M. & Thomas, H. (1988)
  Proc. Nat. Acad. Sci. USA, in the press
Mishkind, M.L., Jensen, K.H., Branagan, A.J., Plumley, F.G. &
  Schmidt, G.W. (1985)   in Current Topics in Plant
  Biochemistry & Physiology (Randall, D.D., Blevins, D.G. & Larson,
  R.L., eds.), vol. 4 pp. 34-50, University of Missouri-Columbia
Nixon, P.J., Dyer, T.A., Barber, J. & Hunter, C.N. (1986)   in
  Progress in Photosynthesis Research (Biggins, J., ed.), vol. III,
  pp. 779-782, Dordrecht: Nijhoff
Roberts, D.R., Thompson, J.E., Dumbroff, E.B., Gepstein, S. &
  Mattoo, A.K. (1987) Plant Mol. Biol. 9, 343-354
Schoch, S., Rüdiger, W., Lüthy, B. & Matile, Ph. (1984)   J. Plant
  Physiol. 115, 85-89
Stadtman, E.R. (1986) Trends Biochem. Sci. 11, 11-12
Thomas, H. (1977) Planta 137, 53-60
Thomas, H. (1983) Photosynthetica 17, 506-514
Thomas, H. (1987) Theor. Appl. Genet. 73, 551-555
Thomas, H. & Hilditch, P. (1987)   in Plant Senescence: Its
  Biochemistry and Physiology (Thomson, W.W., Nothnagel, E.A. &
  Huffaker, R.C., eds.), pp. 114-122, ASPP, Rockville
Thomas, H., Lüthy, B. & Matile, Ph. (1985) Planta 164, 400-405
Woolhouse, H.W. & Batt, T. (1976)   in Perspectives in
  Experimental Biology (Sunderland, N., ed.), vol. 2 - Botany, pp.
  163-175, Oxford, Pergamon.

Heat stress and membranes

Kurt A. SANTARIUS and Engelbert WEIS

Botanisches Institut, Universität Düsseldorf, Universitätsstrasse 1,
D-4000 Düsseldorf, Federal Republic of Germany

SYNOPSIS

Effects of heat stress on leaf cells, mainly the thermolabile
chloroplast membranes, are discussed. Changes in thylakoid structure
are related to reversible and irreversible heat-induced alterations in
photosynthetic reactions and to acclimation to high temperature.

INTRODUCTION

It is widely accepted that membranes play a key role in the response of cellular functions to variable stress conditions. As the
heat-tolerance limits of various plants and microorganisms differ in a
wide range, considerable differences in the thermostability of individual cellular membranes and their adaptability to supraoptimal temperature must exist. In the present review we shall concentrate on the
effect of high temperature on membranes of leaf tissue, i.e. cells of
higher plants containing chloroplasts and so far capable of photosynthesis.

Responses of leaf cells to heat are rather complex and, thus, in-
terpretations of experimental data are difficult. Mild heating of
whole leaves causes reversible membrane alterations, i.e. recovery
from injury can occur. In contrast, more severe heat stress leads to
irreversible membrane damage. To distinguish between reversible and
irreversible heat effects is rather complicated as both the extent of
damage and the time required for partial or complete recovery depend
on the severity of the heat stress. Membrane injury is largely determined by the absolute temperature, but also by the time of exposure to
thermal stress, i.e. cells can be killed either within minutes or even
seconds by heat shocks of extreme temperatures or during continuous
treatment for hours or days at moderate thermal conditions (Alexandrov, 1977; Levitt, 1980). Moreover, exposure of cells to supraoptimal, but non-lethal temperatures results in adaptive responses. Such
acclimation to high temperatures can be reflected in an increased
thermal stability of the membranes.

Functional and structural changes occurring in membranes of leaf
cells upon heat treatment *in vitro* and *in situ* have been extensively
investigated and were last reviewed by Berry & Björkman (1980) and
Quinn & Williams (1985). In spite of the experimental data considered
in these articles and added during the last few years, the molecular
mechanisms of reversible and irreversible heat damage, recovery from
transient thermal inhibitions and acclimation to high temperature are
still far from clear.

HEAT SENSITIVITY OF CELLULAR MEMBRANES AND THERMAL TOLERANCE LIMIT

It is well established from numerous studies with intact leaves that photosynthesis is particularly sensitive to high temperature (see review by Berry & Björkman, 1980). Inhibition of dark respiration in whole leaves and protoplasts does not occur until photosynthesis is almost completely inactivated (Björkman et al., 1980; Thebud & Santarius, 1982). The distinct difference in the heat sensitivity of these two metabolic processes *in situ* is confirmed with mitochondria and chloroplasts isolated from heat-treated spinach leaves (Thebud & Santarius, 1982). Further on, heat-induced leakage of solutes from leaf tissue, isolated mesophyll protoplasts and vacuoles is observed at temperatures far above those which cause complete inhibition of photosynthetic CO_2 fixation (Björkman et al., 1980; Badger et al., 1982; Thebud & Santarius, 1982; Weigel, 1983; Nelles, 1985) indicating that changes in the permeability of tonoplast and plasma-membrane cannot be responsible for heat injury to the photosynthetic apparatus. A comparable high thermostability is also shown for chloroplast envelope membranes (Krause & Santarius, 1975; Bauer & Senser, 1979). Hence, thermal inactivation of photosynthesis and related damage to leaves is not primarily caused by an unspecific disturbance of cellular compartmentation.

Soluble enzymes are supposed to be more resistant against thermal inactivation, compared to primary photosynthetic reactions associated with the thylakoid membrane (Berry & Björkman, 1980). A reversible depression of the ribulose-1,5-bisphosphate carboxylase/oxygenase activity has been observed in isolated chloroplasts (Weis, 1981a, b) and in intact plants (Weis & Berry, 1988), but this enzyme is known to be highly regulated and inhibition is caused by thermal perturbation of activation processes, while the enzyme protein itself is relatively heat stable. Also, the activity of enzymes which require photochemically generated reducing power for their activation exhibit a relative high sensitivity to heat, but this may be linked to the heat inhibition of the membrane-bound photosynthetic electron transport rather than to thermal inactivation of the enzymes themselves (Björkman et al., 1980; Badger et al., 1982). Therefore, in the following we shall focus to heat-induced changes in thylakoid membranes.

In intact tissue a thermally induced decline in photosynthesis can be reversible or irreversible. Mostly, the temperature optimum of photosynthesis is far below the heat tolerance limit of the plant. The heat tolerance limit indicates a characteristic level of thermal stress beyond which lethal damage takes place. Below that limit there is a wide range of supraoptimal temperatures, where recovery from moderate thermal perturbations of the photosynthetic apparatus can occur.

A heat-induced rise in the basal level of chlorophyll fluorescence, F_o, has frequently been used to determine the heat tolerance limit of intact leaves (Schreiber & Berry, 1977; Armond et al., 1978; Smillie, 1979; Björkman et al., 1980; Yordanov & Weis, 1984; Downton et al., 1984; Seemann et al., 1984, 1986). Detailed comparative determination of heat tolerance of a variety of vascular plant leaves carried out by Bilger et al. (1984, 1987) with the post-culture necrosis method and by measurement of heat-induced chlorophyll fluorescence changes clearly shows that the critical temperature for the onset of increase in basal fluorescence coincides with the critical temperature for irreversible damage of the leaf tissue. The mechanistic basis for this fluorescence effect will be discussed below.

A large number of mechanistic studies has been carried out with isolated membranes. But different to what has been found with intact plants, most thermal effects observed *in vitro* are virtually irreversible within the experiment, including those effects which would have been reversible *in situ*. Moreover, it has often been observed that the relative thermal stability of isolated membranes is drastically influenced by the composition of the suspending medium. Therefore, with respect to possible physiological implications, *in vitro* studies should be interpreted with caution.

THERMAL EFFECTS ON THYLAKOID MEMBRANE FUNCTIONS

Photosystem II

     A pronounced thermal sensitivity of the PS II-mediated electron transport was demonstrated in numerous experiments carried out with different plant materials *in vivo* and *in vitro* (see Berry & Björkman, 1980). It is now clear that this sensitivity can be explained by different mechanisms. Chlorophyll fluorescence has frequently been used as a non-intrusive approach to examine the different thermal effects at the pigment level (for a review see Krause & Weis, 1984).
     Upon severe thermal stress, the primary process of energy conservation, the photochemical charge separation at PS II reaction centres, has been shown to be inactivated (Döring et al., 1969). The substantial rise in the basal level of chlorophyll fluorescence, F_o, which occurs when isolated membranes (Schreiber & Armond, 1978; Armond et al., 1978) and intact leaves (Schreiber & Berry, 1977) are exposed to high temperatures, has been related to a direct thermal blockage of PS II reaction centres. Apparently, this inactivation is an irreversible process and may actually determine the heat tolerance limit of plants (see above).
     In contrast, moderate heat treatment which does not lead to inactivation of PS II reaction centres has been shown to cause a substantial decline in the rate of rise in fluorescence from the basal (F_o) to the maximum level (F_m) and, subsequently, to a progressive suppression of the variable part of fluorescence (Krause & Santarius, 1975; Pearcy et al., 1977; Santarius & Müller, 1979). Under such conditions, F_o is not affected. Substitution of electrons from the water-splitting reaction by use of hydroxylamine largely restores the induction of variable fluorescence in isolated thylakoid membranes (Weis, 1982b). This thermally induced decrease of variable fluorescence may be related to a limitation of electron donation to PS II reaction centres. Heat-induced inactivation of the water-splitting reaction and related inhibition of PS II-dependent electron transport has frequently been demonstrated in isolated systems and intact cells (for a review see, e.g., Quinn & Williams, 1985). It has been attributed to loss of manganese (Cheniae & Martin, 1970), chloride (Krishnan & Mohanty, 1984; Critchley & Chopra, 1988) and certain polypeptides (Yamamoto & Nishimura, 1983; see also Volger & Santarius, 1981) from the entire PS II complex. Chloride depletion accelerates heat-induced release of manganese and proteins from isolated PS II particles containing the oxygen evolution system (Nash et al., 1985). But no direct correlation seems to exist between release of manganese on heating and solubilization of the 33 kDa protein (Franzén & Andréasson, 1984) and between loss of protein and inactivation of oxygen evolution (Nash et al., 1985). While alkaline pH accelerates thermal inactivation, metal cations have a substantial stabilizing effect (Weis, 1982a). It is suggested that the thermal stability of the entire PS II complex partially decreases with the density of negative charges exposed to the membrane surface. A specific stabilizing effect of divalent cations on oxygen-evolving PS II particles and of non-oxygen-evolving PS II core complex against heat inactivation was recently shown by Inouê et al. (1987). Ionic effects may also account for light-induced stabilization described be-

low. Obviously, the relationship between binding of chloride and other
ionic effects and thermal inactivation needs further examination.

In intact plants heat-depression of electron donation to PS II
may not be an irreversible process. Complete recovery takes place upon
transfer of heat-stressed leaves to lower temperatures, even when the
photosynthetic electron transport or the variable chlorophyll fluorescence were previously largely depressed (Bauer & Senser, 1979; Weis et
al., 1986).

The reversible inactivation of the oxygen-evolving reaction may
be overlayed and, under very moderate stress, even preceded by specific alterations of the membrane that affect the quantum efficiency
rather than the reaction mechanism and the photochemical capacity of
PS II. Exposure to elevated temperatures causes a redistribution of
light energy absorbed by the pigment system in favour of PS I, at the
expense of PS II excitation (Bhardwaj & Singhal, 1981; Weis, 1983,
1984b, 1985; Sane et al., 1984; Havaux et al., 1987; Havaux & Lannoye,
1987). Under moderate stress this transition has been shown to be completely reversible with the two photoreactions remaining fully functional (Weis, 1984b, 1985; Havaux & Lannoye, 1987), while extreme perturbation of energy distribution observed upon progressive heating
(Havaux et al., 1987) may be irreversible. Thermal mobilization of
membrane protein complexes may generally facilitate migration of PS II
from the grana to the PS I-rich regions of thylakoid membranes, where
excitation energy can be 'spilled over' to PS I (see below). Light-dependent control of energy distribution as mediated by enzymatic phosphorylation of the light-harvesting chlorophyll a/b complex (Bennett
et al., 1980; Horton & Black, 1980) and its relationship to thermal
effects will be discussed in section LIGHT AND HEAT STRESS.

Moreover, in intact leaves this thermally induced increase in
spillover is accompanied by an increase in the ß-phase of the fluorescence induction curve (Weis, 1983). The ß-phase has been attributed to
a state of PS II characterized by a low quantum yield of photochemis-
try (Melis & Homann, 1978). It was shown with isolated thylakoid membranes that mild heat treatment actually causes a reversible conversion of PS II from the high-efficient α-state to the low-efficient ß-
state (Sundby et al., 1986; Andersson et al., 1987). In the temperature range where most of these changes occurred, very little inhibition of electron transport capacity was observed.

Evidently, the functional relationship between thermally induced
change in spillover and the conversion of PS II$_\alpha$ to PS II$_\beta$ and its
physiological significance needs further examination. Both, high
spillover and conversion of PS II$_\alpha$ to PS II$_\beta$ are supposed to reduce
the quantum efficiency rather than the capacity of total electron
transport. From studies of steady-state fluorescence and photosynthetic gas exchange with leaves of cotton plants at different temperatures
it has been concluded that reversible temperature-dependent quenching
of fluorescence occurring at moderately elevated temperatures (> 25°C)
is actually related to a substantial non-photochemical and non-radia-
tive dissipation of absorbed energy (Weis & Berry, 1988). Possibly,
this could be a protective mechanism to avoid photooxidative damage of
PS II at high light intensities which usually accompany heat stress in
the field (see section LIGHT AND HEAT STRESS). It has also been suggested that a thermally enhanced spillover of excitation energy from
PS II to PS I could stimulate a cyclic electron transport around PS I
and, hence, could cause additional proton pumping (Weis, 1985).

Photosystem I

In contrast to the pronounced heat-lability of the PS II complex,
PS I-driven electron transport is much more heat-stable. This was

shown in numerous experiments with thylakoid membranes either isolated
from heat-inactivated leaf tissue or subjected to heat treatment *in
vitro* (see review by Quinn & Williams, 1985). In intact leaves of *Pas-
siflora edulis* inhibition of PS I-mediated photooxidation of cyto-
chrom-554 was not observed until PS II activity was almost completely
inactivated (Smillie, 1979). When isolated thylakoid membranes became
incubated at elevated temperature, even a considerable stimulation of
the light-saturated PS I-mediated electron transport was frequently
observed (Santarius, 1975; Armond et al., 1978; Stidham et al., 1982;
Gounaris et al., 1983). Similarly, a significant enhancement in PS I
activity was observed in chloroplasts isolated from heat-treated wheat
leaves (Mohanty et al., 1987). The threshold temperature for PS I
stimulation in isolated thylakoids was very similar to that for heat-
induced inhibition of PS II activity (Santarius, 1975; Thomas et al.,
1984, 1986) indicating that the initial heat-induced increase in PS I
activity is supposed to be reversible. This thermal stimulation may
partially be related to the removal of a kinetic limitation of electron donation to PS I. It is suggested that heat stress causes a reorganization of the thylakoid membrane resulting in the exposure of new
electron-acceptor sites located within the cytochrom f/b_6 complex of
the electron transport chain (Thomas et al., 1986).

At non-physiological high temperatures (65-70°C) a second rise in
chlorophyll fluorescence emission became apparent in leaves of various
plants and isolated stroma thylakoids and has been explained by a
stimulation of fluorescence emission of PS I (Downton & Berry, 1982).
This interpretation, however, has been questioned by Mannan et al.
(1986).

Photophosphorylation

Photophosphorylation is another reaction of the thylakoid membrane which already becomes inhibited at moderate heat stress. The extensive experimental data accumulated in this field are rather contra-
dictory (see reviews by Berry & Björkman, 1980; Quinn & Williams,
1985). Studies with thylakoid membranes isolated from heated leaves of
Nerium oleander indicate that thermal inhibition of non-cyclic photophosphorylation was limited by inhibition of the whole-chain electron
transport (Badger et al., 1982). However, coupling of photophosphorylation to the electron transport also appears to be sensitive to heat.
While in a number of studies thermal uncoupling has been related to an
increase in proton leakage, there is also indication for heat-induced
inactivation of photophosphorylation without depressing light-dependent proton uptake (for literature see Quinn & Williams, 1985). After
moderate high temperature treatment of spinach leaves and isolated
chloroplasts the rate and the extent of light-scattering changes
(Weis, 1981a, c) and pH-dependent 'high energy' quenching of chlorophyll fluorescence (Laasch, 1987; Bilger et al., 1987) increased, indicating that the light-induced uptake of protons into the intrathylakoid space was not inhibited. Even stimulation of proton uptake has
been observed in isolated chloroplasts after heat stress that partially inactivated photophosphorylation (Krause & Santarius, 1975). A col-
lapse in the electrical component of the proton-motive force rather
than in the proton gradient has been suggested to be a primary cause
in thermal uncoupling of thylakoids (Emmett & Walker, 1973). But the
discussion about the involvement of electrical events in energy conservation is still controversial. The electrochromic absorbance change
at 515 nm (see next section) was found to be substantially more sensitive to heat than proton uptake and photophosphorylation indicating
that thermal suppression of electrical field gradients related to the
electrochromic reaction may not be related to uncoupling of photophos-

phorylation (Weis, 1981a).

Although little data are available on the effect of heat on the phosphorylating enzyme itself, it seems likely that this protein is not a primary site of heat damage. The chloroplast coupling factor both attached to the membrane and separated from it, was found to be more stable than photophosphorylation (Mukohata et al., 1973; Santarius, 1975), and heat treatment that inactivated thylakoid membranes did not lead to dissociation of the complex protein (Volger & Santarius, 1981). Evidently, it seems to be the coupling mechanism rather than the phosphorylating enzyme itself which is primarily affected by heat, but more information is necessary to understand the underlaying mechanism.

Electrochromic absorbance change at 515 nm

A light-driven electrochromic absorbance change peaking at 515 nm and related to electrical field gradients at the thylakoid membrane has been found to be particularly sensitive to mild heat stress (Santarius & Müller, 1979; Weis, 1981a, c, 1982b; Fork et al., 1985). In intact leaves substantial thermal depression of the pigment absorption at 515 nm caused by moderate thermal stress has been found to be completely reversible (Weis, 1981a, c; Weis et al., 1986). It is mainly the 'slow' component of this reaction which is affected by mild heat treatment. Since this component is produced by electrogenic reactions connected to the electron transport from plastoquinone to PS I (Velthuys, 1978), its depression may be related to very specific reversible perturbations in the vicinity of the cytochrom f/b_6 complex, but may not necessarily indicate thermal uncoupling or severe disruption of other thylakoid functions. Although heat suppression of the 515-nm absorbance change is proved to be one of the most sensitive test for sub-lethal thermal stress, the present data do not allow unambiguous interpretation of this effect.

THERMAL EFFECTS ON THYLAKOID MEMBRANE STRUCTURE

Heat inactivation of PS II-mediated electron transport and photophosphorylation of spinach thylakoids *in situ* and *in vitro* was not accompanied by chemical alterations in the membrane lipids: neither oxidation of unsaturated fatty acids nor hydrolysis of lipids and release of free fatty acids was observed under conditions leading to complete inactivation of PS II activity (Santarius, 1980). However, distinct extrinsic polypeptides became separated from isolated thylakoids and PS II particles upon heat stress (Volger & Santarius, 1981; Yamamoto & Nishimura, 1983; Franzén & Andréasson, 1984; Nash et al., 1985). This thermally induced protein release, which seems to occur also during heat inactivation *in vivo* (Volger & Santarius, 1981), is suggested to be a consequence of irreversible alterations in the membrane structure. The existence of differential thermal sensitivities of specific PS II proteins was recently suggested from heat-induced decrease in the affinity of the membranes for DCMU (Wiest, 1986).

Heat-induced structural changes in thylakoid membranes has been examined by freeze-fracture electron microscopy. Quantitative analysis of particle-size frequency and particle density of freeze-fracture micrographs from *Nerium oleander* leaves revealed that increase in membrane damage was accompanied by a progressive loss of lager particles from the exoplasmic fracture face (Armond et al., 1980). These particles are presumed to be the chlorophyll a/b light-harvesting complex and the core complex of PS II. Moreover, progressive decrease of PS II-mediated electron flow of chloroplasts isolated from broad beans and heated in the range from 35 to 45°C was accompanied by destacking of the grana thylakoids (Gounaris et al., 1983, 1984; Williams et al., 1984). Membrane adhesion in grana stacks is believed to be mediated *via* the chlorophyll a/b light-harvesting complexes which are normally attached to PS II and located in the appressed regions of the membranes. Heat-induced loss of grana stacking appears to be associated with a physical dissociation of the light-harvesting units from the core complex of PS II. Rearrangement of the particles within different membrane sites is reflected in a clustering of the antennae complexes of PS II into regions of membrane adhesion, whilst excluding the core complexes of PS II from these regions. - Increasing the incubation temperature above 45°C led to an extensive irreversible phase separation of non-bilayer-forming lipids producing three-dimensional aggregates of cylindrical inverted lipid micelles.

Additional evidence for changes in the lateral organization and a partial destacking of spinach chloroplast membranes at elevated temperatures was provided by subfractionation of thylakoids into stroma lamellae and vesicles representative of the non-appressed and appressed regions of the membranes, respectively. Sundby & Andersson (1985) suggested from changes in the distribution of chlorophyll-protein complexes between those particles a heat-induced lateral migration of the PS II core with a portion of its tightly bound light-harvesting antenna from the grana out into the PS I-rich stroma regions leaving mainly free light-harvesting units behind in the appressed membranes (see also Sundby et al., 1986; Andersson et al., 1987). Very recently, a characterization of the two different subpopulations of the light-harvesting chlorophyll a/b protein complex has been performed (Larsson et al., 1987). The structural rearrangements of the thylakoid membranes occurred at very moderate thermal stress (30-40°C) and were partially reversible upon lowering the temperature. The observation that recovery took place even in isolated chloroplast lamellae suggests that the reversibility of the heat-induced lateral movement of protein-pigment complexes within the membranes is based on a spontaneous, temperature-dependent reorganization of the thylakoid membrane structure rather than on a specific repair mechanism.

The general pattern of reversible structural alterations coincides well with suggestions on thermal-induced conversion of PS II$_\alpha$ to PS II$_\beta$ and changes of energy transfer between different pigment complexes, as obtained from fluorescence studies (see section 'Photosystem II'). Clustering of membrane proteins and phase separation of particular lipids may determine the level of thermal stress, beyond which structural reorganization of thylakoid membranes becomes irreversible. It is likely that irreversible damage of the photosynthetic apparatus, which is attributed to a functional disruption of PS II centres, is not manifest before the entire PS II complex is fully dissociated.

ACCLIMATION TO HIGH TEMPERATURES

The ability of plant cells to increase thermotolerance when grown for a longer period at elevated temperatures or subjected to a transient heat stress was described in numerous papers (see reviews by Alexandrov, 1977; Levitt, 1980; Berry & Björkman, 1980; Kappen, 1981). The capacity for heat hardening enables plants to adjust their heat tolerance to seasonal or even diurnal variations in the environmental temperature regime. At least two types of heat acclimation processes can be distinguished which seem to be based on different mechanisms: (1) long-term heat hardening generally occurring in the time range of at least several days, rather weeks or months at elevated growth temperatures, and (2) short-term heat acclimation which can be induced by

exposure of plants for a few hours to moderate high temperature or
even by a brief but severe heat shock. Both processes are supposed to
involve distinct alterations in the thylakoid membrane structure and,
definitely during long-term heat acclimation, also in other cellular
membranes.

Long-term acclimation

     Gradual acclimation of various plants from both hot and cool en-
vironments to higher growth temperatures is accompanied by a distinct
increase of the thermostability of cellular membranes, mainly the
chloroplast membranes. This is evident from data obtained with intact
leaves and isolated chloroplast membranes. Heat-induced changes of
chlorophyll fluorescence occurred in leaves of heat-adapted plants at
considerably higher temperature than was found in non-acclimated leaf
tissue (Schreiber & Berry, 1977; Pearcy et al., 1977; Armond et al.,
1978; Downton et al., 1984; Seemann et al., 1984, 1986). Similar differences in various photosynthetic activities of thylakoid membranes
were also found when the membranes were isolated from leaves of heat-
hardened and non-acclimated plants previously exposed to increasing
heat stress (Pearcy et al., 1977; Björkman et al., 1980; Badger et
al., 1982; Chetti & Nobel, 1987). When chloroplast membranes were isolated from leaves of different heat tolerance and subsequently subjected to heat treatment *in vitro*, thylakoids from heat-acclimated
leaves exhibited a significantly higher thermostability than membranes
isolated from non-acclimated tissue. Obviously, changes in the thermostability of chloroplast membranes occurring in the course of long-
term adaptation to high temperature were maintained during the isolation procedure. This was evident from the temperature response of
whole-chain electron transport of thylakoids isolated from *Larrea di-
varicata* (Armond et al., 1978) and *Nerium oleander* (Badger et al.,
1982) grown at different thermal regimes. Corresponding freeze-fracture electron microscopic observations of thylakoids from *Nerium oleander* revealed an enhanced thermostability of the entire PS II complex
in the samples from the heat-adapted plants (Armond et al., 1980).
Thylakoids isolated from heat-acclimated and non-hardened bean plants
exhibited a different pattern of membrane changes in response to high-
temperature stress: while the thermostability of membrane energization
and light-induced proton gradient formation was considerably enhanced
in heat-adapted thylakoids, the oxygen-evolving system was practically
unchanged (Yordanov et al., 1987). However, in this study the isolated
chloroplasts have been frozen in liquid nitrogen before thermal treatment *in vitro*. Although experimental data in this field are rare, it
is likely that alterations in the membranes themselves seem to play a
dominant role in acclimation of leaves to high temperature.
     There is wide agreement that changes in chemical composition and
physical properties of acyl lipids of cellular membranes take place
during long-term acclimation to high temperatures. Growth at elevated
temperatures generally results in decrease in the abundance of polyunsaturated fatty acids, which is associated with reduced membrane
fluidity (Levitt, 1980; Raison et al., 1980; Quinn & Williams, 1985).
According to a model based on a balance between hydrophobic and hydrophilic interactions among proteins and lipids it is suggested that
changes in the lipid viscosity may be related to loss of physiological
functions of membranes at high temperatures, i.e. the thermal stability of membranes is linked to the physical properties of their lipids
(Raison et al., 1980).
     Acclimation of the photosynthetic apparatus to high temperatures
has been shown to be accompanied by an increase in saturation of the
total polar lipids in leaves of *Atriplex lentiformis* (Pearcy, 1978)

and *Nerium oleander* (Raison et al., 1982a) and in thylakoid membranes
isolated from the latter species (Raison et al., 1982b) and from bean
plants (Süss & Yordanov, 1986). While the relative proportion of the
major classes of polar lipids in chloroplast membranes isolated from
Nerium oleander did not alter significantly during acclimation, the
ratio of diacylmonogalactoside to diacyldigalactoside decreased in
bean thylakoids with increasing growth temperature.

However, it is not yet prooved whether changes in lipid composition and saturation are primarily responsible for changes in the thermal stability of cellular membranes. A gradual decrease in the degree
of unsaturation of the fatty acids of thylakoid membrane lipids not
related to changes in thermostability was found when spinach plants
were subjected for 3 days to elevated temperature (Santarius & Müller,
1979). Moreover, the acquisition of high-temperature tolerance was not
accompanied by marked changes in overall fatty acid saturation and
composition of chlorenchyma membrane fractions for two desert succu-
lents (Kee & Nobel, 1985). Determination of the fluidity of membrane
lipids from *Nerium oleander* leaves by electron spin resonance studies
and by analysis of polarization of fluorescence from the probe, trans-
parinaric acid, indicated that the membrane viscosity varies with the
growth temperature but no lipid phase separation occurred in the range
of high temperatures which caused loss of functional integrity of the
membrane (Pike et al., 1979; Raison et al., 1980, 1982a, b). Similarly, neither in isolated pea thylakoids nor in separated stromal and
granal membrane fragments a break in the temperature dependence of polarization of DPH fluorescence was found up to about 60°C (Barber et
al., 1984). This indicates that no significant phase transition of its
bulk lipids took place at injurious temperatures. Thus, the role of
lipids in determining thermostability of chloroplast membranes is not
yet clear. Possibly, an increase in saturation of the membrane lipids
at elevated growth temperature may be important for maintaining an optimum membrane fluidity necessary for effective functional activities
at moderate temperatures. Nevertheless, it cannot ruled out that the
decrease in the actual fluidity of the membrane lipids during long-
term heat acclimation could shift the appearance of irreversible alterations such as complete destacking and bilayer/non-bilayer trans-
formations (Gounaris et al., 1983, 1984) to higher temperatures.

Long-term acclimation of plant cells to high temperature seems
not to be confined to changes in the chloroplast membranes. In leaves
of *Nerium oleander* the cell membrane exhibited a higher thermal stability in 45°C-grown than in 20°C-grown plants (Björkman et al., 1980;
Badger et al., 1982). However, in this case, it could not be excluded
that changes in the cellular environment of the membranes could contribute to variations in their thermostability. A stabilizing effect
of various low- and high-molecular-weight solutes such as carbohydrates and proteins on isolated thylakoid membranes towards heat
stress has been repeatedly reported (Santarius, 1973; Krause & Santarius, 1975; Volger & Santarius, 1981; Venediktov & Krivoshejeva, 1984;
Seemann et al., 1986; Yordanov et al., 1987). Similar to the appearance of heat-shock proteins upon brief exposure to high temperatures
(see next section), the synthesis of 'protective' compounds and their
accumulation in distinct cell compartments could enhance thermal stability. Actually, a correlation between thermotolerance of leaves and
their osmotic potential was found for a variety of plants indicating
that increase in the concentrations of some small molecules might also
confer increased heat stability (Hellmuth, 1971; Seemann et al.,
1986). Induction of heat-shock proteins by treating of desert succu-
lents at 50°C seems generally to require three or more days (Kee &
Nobel, 1986). Thus, the appearance of soluble cell constituents could
contribute to membrane stabilization in plants generally adapted to

elevated temperature in their environment. However, without doubt
changes within the membrane structure and, possibly, in its chemical
composition are primarily responsible for the increase of heat tolerance under these conditions. The nature of these alterations remains
obscure.

Short-term acclimation

    Reversible variations in the thermostability of a wide variety of
plant cells induced by short-time heat treatment have been frequently
reported (for literature see Alexandrov, 1977; Levitt, 1980; Kappen,
1981). Responses of leaf cells to fast variations in temperature which
seem to be involved in short-term heat hardening are (1) the synthesis
and accumulation of protective substances, mainly of heat-shock proteins, and/or (2) heat-induced alterations of cellular membranes.
    Exposure of whole spinach and bean plants and detached leaves for
few hours to elevated, but sub-lethal temperatures increased the heat
tolerance by several degrees C (Santarius & Müller, 1979; Weis, 1983,
1984a; Yordanov & Weis, 1984). In addition, the range of moderate
thermal stress causing reversible alterations of the photosynthetic
apparatus (see above) was shifted to higher temperatures. Analysis of
membrane lipids of thylakoids isolated from heat-hardened and non-acclimated spinach leaves revealed no temperature-induced changes in
lipid content and fatty acid composition (Santarius & Müller, 1979).
The enhanced thermostability of the thylakoid membranes related to
short-term acclimation was maintained in intact chloroplasts isolated
from adapted leaves, but, in contrast to long-term acclimation, was
completely abolished after liberation of the membranes from their
stromal environment. This points to alterations in the stroma that
stabilize the membranes. Sugar analysis exhibited a considerable decrease in the carbohydrate concentration in the chloroplast stroma during short-term heat acclimation showing that particular membrane
preservation by low-molecular-weight carbohydrates cannot be responsible for the observed increase in heat stability (Santarius & Müller,
1979). It has been proposed that alterations in the ionic milieu,
which considerably affect the heat sensitivity of isolated thylakoid
membranes (see next section) could account for changes of their thermal stability (Weis, 1982a). However, there is no direct experimental
evidence so far that ionic effects are actually involved in heat acclimation of cells.
    A great deal of experimental data became accumulated during the
last decade indicating marked changes in protein synthesis on transfer
of plants to supraoptimal temperatures (Kee & Nobel, 1986; Cooper &
Ho, 1987; Fabijanski et al., 1987; Heuss-LaRosa et al., 1987; Mansfield & Key, 1987; Necchi et al., 1987; Ougham, 1987; Valliammai et
al., 1987; and many others). Both, incubation of plant tissue for few
hours at elevated temperatures and exposure to brief but severe heat
stress, induces the synthesis and characteristic intracellular distribution of a limited number of heat-shock proteins. These changes
are reversible, i.e., when the tissue is returned to its original temperature regime, the synthesis of heat-shock proteins progressively
declines and reappearance of a normal pattern of protein synthesis occurs within few hours. Although the function of the heat-shock proteins are not yet clear, they could be involved in the reversible increase in thermostability of the tissue. For example, the pronounced
accumulation of these proteins in the chloroplasts (see, e.g., Vierling et al., 1986; Süss & Yordanov, 1986) could protect thermo-labile
thylakoid membranes against the deleterious effect of heat stress.
However, conclusive evidence which would support this idea is missing.
Interestingly, the potential for heat hardening of leaves of various

plants declines during aging and is almost absent in senescent tissue where the capacity for protein synthesis is low (Yordanov & Weis, 1984).

Short-time exposure of leaves to moderate heat stress which induces an increase in thermal resistance is also accompanied by changes of the structural organization and functional characteristics of the thylakoid pigment systems. In spinach leaves thermally induced alteration in the distribution of absorbed light energy in favour of PS I has been found under conditions which also led to a fast increase in thermotolerance (Weis, 1984a, b, 1985). The structural basis of such reversible temperature-induced changes in energy distribution and its possible physiological significance has been discussed above. However, it is not clear whether short-term acclimation to high temperature and heat-induced alterations in the structural arrangement and function of the thylakoid pigment complexes are based on the same mechanism or whether they are independent processes occurring under similar thermal conditions.

LIGHT AND HEAT STRESS

Interactions between light and high temperature stress have been repeatedly reported. Two opposing effects can be distinguished: light can increase membrane stability and, during severe thermal stress, excess radiation can cause photoinhibition.

A stabilizing effect of illumination on the heat tolerance of the photosynthetic apparatus has been described for leaves of cotton (Veselovskii et al., 1976), *Tidestromia oblongifolia*, two *Atriplex* species (Schreiber & Berry, 1977), *Tradescantia*, cucumber (Kislyuk, 1979) and spinach (Weis, 1983). Mechanistic investigations on isolated thylakoids have shown that the protective effect of light on the water-splitting complex is related to the light-induced pumping of protons across the membranes (Weis, 1982a, b, 1983). In isolated thylakoids the heat sensitivity of the oxygen-evolving complex largely depend on the pH and the ionic milieu: increasing of the pH in the range of 6 to 9 (Krause & Santarius, 1975; Weis, 1982a; Venediktov & Krivoshejeva, 1984) and reduction of the cation concentration below a critical limit (Weis, 1982a; Inoué et al., 1987) drastically decreased the heat stability of the oxygen-evolving complex. Illumination, however, causes increase in the proton concentration in the intrathylakoid space and, hence, stabilization of the water-splitting apparatus. The loss of the protective effect of light in the presence of uncouplers clearly shows that the light-induced proton gradient across the thylakoid membrane is responsible for the stabilizing effect of light (Weis, 1982b). Likewise, when the pH in the medium is low, the water-splitting system is much less labile and, then, light-induced acidification of the intrathylakoid space has no further stabilizing effect.

Since this light-dependent stabilization has been shown to depend on the absolute concentration of protons within the thylakoid space rather than on the level of light (Weis, 1982b), it may be connected to the energetic balance of plants in the field. From comparison between photosynthetic CO_2 fixation and 'energy'-dependent quenching of chlorophyll fluorescence in intact plants under steady-state conditions (Weis & Berry, 1987) it has been concluded that even at high light intensities, when a high rate of photosynthetic energy consumption is maintained, the proton gradient across the thylakoid membrane is rather low. However, when photosynthetic CO_2 fixation is limited and light is in excess, a high proton gradient is build-up and the related acidification of the intrathylakoid space possibly could cause stabilization of the PS II complex. Further work is required to support this idea. - Under conditions which do not lead to a considerable increase in the proton concentration within the thylakoid membranes, e.g. in uncoupled chloroplasts, light even destabilizes PS II during thermal stress (Ageeva, 1977; Weis, 1982b).

In the nature high leaf temperatures are normally accompanied by high light intensities. Unfortunately, interactive effects of excess light and heat stress are scarcely investigated. Ludlow & Björkman (1984) found that in intact attached leaves of the tropical pasture legume Siratro high leaf temperature exacerbates photoinhibitory damage, which was already observed in water-stressed plants with increasing photon flux density up to full sunlight values. The degree of injury was dependent upon the severity of heat stress and the light intensity. Gradual recovery of leaves from heat and photoinhibitory damage was observed after return of the plants to more moderate conditions. Similarly, transient photoinhibition in attached olive leaves has been recently shown upon exposure of plants to high temperature at high levels of radiation (Bongi & Long, 1987).

It appears possible that reversible effects of moderate thermal stress may even serve to protect the pigment system against light stress. As outlined in previous sections heat induces a reversible detachment of a distinct pool of chlorophyll a/b light-harvesting complexes from PS II and a lateral migration of the PS II core with a tightly bound subpopulation of the light-harvesting units into the stroma thylakoids (Sundby & Andersson, 1985). The related conversion of PS II$_\alpha$ to PS II$_\beta$ and change in energy distribution may be a protective mechanism to avoid overexcitation and subsequent photooxidative damage to PS II (Weis, 1985; Sundby et al., 1986; Andersson et al., 1987; Havaux & Lannoye, 1987).

Adaptive changes in the lateral organization of the thylakoid membranes can also occur in response to changes in the light regime. If PS II is overexcited relative to PS I, a reversible phosphorylation of the light-harvesting chlorophyll a/b complex and subsequent migration from the appressed regions into the PS I-rich non-appressed lamellae is induced (for reviews see Staehelin & Arntzen, 1983; Bennett, 1984). This may serve to re-establish a well balanced distribution of excitation energy between the two photosystems, necessary for an efficient use of the absorbed energy and for avoiding photodamage. Since both thermal stress and protein phosphorylation are supposed to mediate dissociation of the PS II complex, a relationship between these effects is expected. When spinach leaves were kept in 'state 1' (presumably the state where the light-harvesting complexes are dephosphorylated), mild thermal stress had only little effect on energy distribution (Weis, 1985) indicating that the thermally induced migration of pigment complexes was partially regulated by protein phosphorylation. Only at more severe heat stress, 'uncontrolled' migration might occur.

Protective mechanisms different from those described above are effective under light stress, when the total amount of absorbed energy exceeds that required by the carbon metabolism. Part of the excess energy can be dissipated at PS II by direct thermal de-excitation (Demmig & Björkman, 1987; Krause et al., 1988). However, little information is available on the relationship between this mechanism and heat stress.

Exposure of leaves and chloroplasts to heat stress led to an increased susceptibility of the chloroplast membranes to photooxidative bleaching of chlorophyll (Gounaris et al., 1983; Thomas et al., 1984; Williams et al., 1986). This photobleaching occurred in the presence of strong white light and resulted in cell death. In heat-stressed broad bean and pea chloroplasts the threshold temperature for chlorophyll bleaching coincided closely with that for thermal inhibition of PS II and was accompanied by destacking and disruption of chlorophyll-protein complexes of both photosystems. Obviously, severe heat stress which already causes damage to PS II and, hence, reduces the ability of the leaf to dissipate excitation energy through photochemical reactions will tend to increase photoinhibition and photooxidative bleaching of chlorophyll (for literature see Ludlow, 1987).

CONCLUSIONS

Exposure of plant tissue to heat stress causes structural alterations in cellular membranes, mainly the highly thermo-labile chloroplast thylakoids, which impaires several membrane properties: (1) reversible decline in the activity of various biochemical and biophysical membrane reactions, (2) reversible acclimation to a higher functional efficiency at elevated temperature, (3) reversible increase in heat tolerance, and (4) reversible adaptation to other stress factors accompanying heat stress, e.g. excess light. Gradual recovery from heat-induced alterations in membrane structure and functions can occur, while beyond a critical level of thermal stress, disruption of membrane structure is associated with a complete breakdown of its functional integrity. Heat-induced selective modifications in different membrane properties reflecting injury, acclimation and recovery processes are closely linked together and can occur simultaneously. They are strongly influenced by the severity of heat stress, mainly the absolute temperature and time of exposure to stress conditions. However, it is fairly unknown whether a brief severe heat shock or the accumulation of a heat dosage produced at less severe temperature stress leads to identical or different patterns of membrane alterations, i.e. whether membrane damage is caused by the same or different mechanisms. Moreover, it is not clear whether reversible changes in the membrane structure which precede the irreversible effects are related to the same structural events which, at more severe heat stress, determine the onset of irreversible membrane damage. Obviously, it is still difficult to relate the different damaging, acclimation and repair processes occurring at high temperatures to particular molecular mechanisms.

The pronounced susceptibility of PS II to high temperature signifies the key role of primary energy conservation for damage or survival of plants at severe heat stress. Damage to PS II stops the energy supply leading to a general breakdown of structural and functional integrity of the cells. The susceptibility of this pigment-protein complex to thermal stress may determine the overall heat tolerance of plant tissue. Hence, changes in the thylakoid structure that affects the stability of the PS II complex seem to be a key factor in adaptation to high temperatures.

The experimental data available on the effect of heat stress on cellular membranes are mainly derived from analyses of photosynthetic tissue. If in leaf cells thylakoid membranes are the primary site of irreversible heat injury, the question arises on differences in the heat sensitivity of particular membranes and soluble proteins in nonphotosynthetic tissue. Very little is known about reversible and irreversible heat-induced changes in cellular membranes beyond the chloroplast level.

REFERENCES

Ageeva, O.G. (1977) Photosynthetica 11, 1-4
Alexandrov, V.Ya. (1977) Cells, Molecules and Temperature, Ecological Studies, vol. 21, Springer, Berlin

Andersson, B., Sundby, C., Larsson, U.K., Mäenpää, P. & Melis, A.
(1987) in Progress in Photosynthesis Research (Biggins, J., ed.),
vol. 2, pp. 669-676, Martinus Nijhoff Publ., Dordrecht
Armond, P.A., Björkman, O. & Staehelin, L.A. (1980) Biochim. Biophys.
Acta 601, 433-442
Armond, P.A., Schreiber, U. & Björkman, O. (1978) Plant Physiol. 61,
411-415
Badger, M.R., Björkman, O. & Armond, P.A. (1982) Plant Cell Environ.
5, 85-99
Barber, J., Ford, R.C., Mitchell, R.A.C. & Millner, P.A. (1984) Planta
161, 375-380
Bauer, H. & Senser, M. (1979) Z. Pflanzenphysiol. 91, 359-369
Bennett, J. (1984) Physiol. Plant. 60, 583-590
Bennett, J., Steinback, K.E. & Arntzen, C.J. (1980) Proc. Natl. Acad.
Sci. USA 77, 5253-5257
Berry, J.A. & Björkman, O. (1980) Ann. Rev. Plant Physiol. 31, 491-543
Bhardwaj, R. & Singhal, G.S. (1981) in Photosynthesis V, Chloroplast
Development (Akoyunoglou, G., ed.), pp. 407-416, Balaban Internat.
Sci. Serv., Philadelphia, Pa.
Bilger, H.W., Schreiber, U. & Lange, O.L. (1984) Oecologia 63, 256-262
Bilger, H.W., Schreiber, U. & Lange, O.L. (1987) in Plant Response to
Stress - Functional Analysis in Mediterranean Ecosystems (Tenhunen,
J.D., Catarino, F.M., Lange, O.L. & Oechel, W.C., eds.), NATO Advan.
Sci. Inst. Ser., pp. 391-399, Springer, Berlin
Björkman, O., Badger, M.R. & Armond, P.A. (1980) in Adaptation of
Plants to Water and High Temperature Stress (Turner, N.C. & Kramer,
P.J., eds.), pp. 233-249, J. Wiley & Sons Lim., Chichester
Bongi, G. & Long, S.P. (1987) Plant Cell Environ. 10, 241-249
Cheniae, G.M. & Martin, I.F. (1970) Biochim. Biophys. Acta 197, 219-
239
Chetti, M.B. & Nobel, P.S. (1987) Plant Physiol. 84, 1063-1067
Cooper, P. & Ho, T.D. (1987) Plant Physiol. 84, 1197-1203
Critchley, C. & Chopra, R.K. (1988) Photosynth. Res. 15, 143-152
Demmig, B. & Björkman, O. (1987) Planta 171, 171-184
Döring, G., Renger, G., Vater, J. & Witt, H.T. (1969) Z. Naturforsch.
24b, 1139-1143
Downton, W.J.S. & Berry, J.A. (1982) Biochim. Biophys. Acta 679, 474-
478
Downton, W.J.S., Berry, J.A. & Seemann, J.R. (1984) Plant Physiol. 74,
786-790
Emmett, J.M. & Walker, D.A. (1973) Arch. Biochem. Biophys. 157, 106-
113
Fabijanski, S., Altosaar, I. & Arnison, P.G. (1987) J. Plant Physiol.
128, 29-38
Fork, D.C., Mohanty, P. & Hoshina, S. (1985) Physiol. Vég. 23, 511-521
Franzén, L.G. & Andréasson, L.E. (1984) Biochim. Biophys. Acta 765,
166-170
Gounaris, K., Brain, A.P.R., Quinn, P.J. & Williams, W.P. (1983) FEBS
Lett. 153, 47-52
Gounaris, K., Brain, A.P.R., Quinn, P.J. & Williams, W.P. (1984) Bio-
chim. Biophys. Acta 766, 198-208
Havaux, M., Canaani, O. & Malkin, S. (1987) Plant Cell Environ. 10,
677-683
Havaux, M. & Lannoye, R. (1987) Photosynth. Res. 14, 147-158
Hellmuth, E.O. (1971) J. Ecol. 59, 365-374
Heuss-LaRosa, K., Mayer, R.R. & Cherry, J.H. (1987) Plant Physiol. 85,
4-7
Horton, P. & Black, M.T. (1980) FEBS Lett. 119, 141-144
Inoué, H., Kitamura, T. & Noguchi, M. (1987) Physiol. Plant. 71, 441-
447

Kappen, L. (1981) in Physiological Plant Ecology I (Lange, O.L., No-
bel, P.S., Osmond, C.B. & Ziegler, H., eds.), Encyclopedia of Plant
Physiology, New Series, vol. 12 A, pp. 439-474, Springer, Berlin
Kee, S.C. & Nobel, P.S. (1985) Biochim. Biophys. Acta 820, 100-106
Kee, S.C. & Nobel, P.S. (1986) Plant Physiol. 80, 596-598
Kislyuk, I.M. (1979) Photosynthetica 13, 386-391
Krause, G.H., Laasch, H. & Weis, E. (1988) Plant Physiol. Biochem. 26,
in press
Krause, G.H. & Santarius, K.A. (1975) Planta 127, 285-299
Krause, G.H. & Weis, E. (1984) Photosynth. Res. 5, 139-157
Krishnan, M. & Mohanty, P. (1984) Photosynth. Res. 5, 185-198
Laasch, H. (1987) Planta 171, 220-226
Larsson, U.K., Sundby, C. & Andersson, B. (1987) Biochim. Biophys.
Acta 894, 59-68
Levitt, J. (1980) Responses of Plants to Environmental Stresses, vol.
1, Chilling, Freezing, and High Temperature Stresses, 2nd ed., Aca-
demic Press, New York
Ludlow, M.M. (1987) in Photoinhibition (Kyle, D.J., Osmond, C.B. &
Arntzen, C.J., eds.), pp. 89-109, Elsevier Sci. Publ. B.V., Amsterdam
Ludlow, M.M. & Björkman, O. (1984) Planta 161, 505-518
Mannan, R.M., Periyanan, S., Kulandaivelu, G. & Bose, S. (1986) Photo-
synth. Res. 8, 87-92
Mansfield, M.A. & Key, J.L. (1987) Plant Physiol. 84, 1007-1017
Melis, A. & Homann, P.H. (1978) Arch. Biochem. Biophys. 190, 523-530
Mohanty, N., Murthy, S.D.S. & Mohanty, P. (1987) Photosynth. Res. 14,
259-267
Mukohata, Y., Yagi, T., Higashida, M., Shinozaki, K. & Matsuno, A.
(1973) Plant Cell Physiol. 14, 111-118
Nash, D., Miyao, M. & Murata, N. (1985) Biochim. Biophys. Acta 807,
127-133
Necchi, A., Pogna, N.E. & Mapelli, S. (1987) Plant Physiol. 84, 1378-
1384
Nelles, A. (1985) Biochem. Physiol. Pflanzen 180, 473-475
Ougham, H.J. (1987) Physiol. Plant. 70, 479-484
Pearcy, R.W. (1978) Plant Physiol. 61, 484-486
Pearcy, R.W., Berry, J.A. & Fork, D.C. (1977) Plant Physiol. 59, 873-
878
Pike, C.S., Berry, J.A. & Raison, J.K. (1979) in Low Temperature
Stress in Crop Plants (Lyons, J.M., Graham, D. & Raison, J.K., eds.),
pp. 305-318, Academic Press, New York
Quinn, P.J. & Williams, W.P. (1985) in Photosynthetic Mechanisms and
the Environment (Barber, J. & Baker, N.R., eds.), pp. 1-47, Elsevier
Sci. Publ. B.V., Amsterdam
Raison, J.K., Berry, J.A., Armond, P.A. & Pike, C.S. (1980) in Adapta-
tion of Plants to Water and High Temperature Stress (Turner, N.C. &
Kramer, P.J., eds.), pp. 261-273, J. Wiley & Sons Lim., Chichester
Raison, J.K., Pike, C.S. & Berry, J.A. (1982a) Plant Physiol. 70, 215-
218
Raison, J.K., Roberts, J.K.M. & Berry, J.A. (1982b) Biochim. Biophys.
Acta 688, 218-228
Sane, P.V., Desai, T.S., Tatake, V.G. & Govindjee (1984) Photosyn-
thetica 18, 439-444
Santarius, K.A. (1973) Planta 113, 105-114
Santarius, K.A. (1975) J. Thermal Biology 1, 101-107
Santarius, K.A. (1980) Physiol. Plant. 49, 1-6
Santarius, K.A. & Müller, M. (1979) Planta 146, 529-538
Schreiber, U. & Armond, P.A. (1978) Biochim. Biophys. Acta 502, 138-
151
Schreiber, U. & Berry, J.A. (1977) Planta 136, 233-238

Seemann, J.R., Berry, J.A. & Downton, W.J.S. (1984) Plant Physiol. 75, 364-368
Seemann, J.R., Downton, W.J.S. & Berry, J.A. (1986) Plant Physiol. 80, 926-930
Smillie, R.M. (1979) Aust. J. Plant Physiol. 6, 121-133
Staehelin, L.A. & Arntzen, C.J. (1983) J. Cell Biol. 97, 1327-1337
Stidham, M.A., Uribe, E.G. & Williams, G.J. (1982) Plant Physiol. 69, 929-934
Süss, K.H. & Yordanov, I.T. (1986) Plant Physiol. 81, 192-199
Sundby, C. & Andersson, B. (1985) FEBS Lett. 191, 24-28
Sundby, C., Melis, A., Mäenpää, P. & Andersson, B. (1986) Biochim. Biophys. Acta 851, 475-483
Thebud, R. & Santarius, K.A. (1982) Plant Physiol. 70, 200-205
Thomas, P.G., Quinn, P.J. & Williams, W.P. (1984) in Advances in Photosynthesis Research (Sybesma, C., ed.), vol. 3, pp. 35-38, Martinus Nijhoff/Dr. W. Junk Publ., The Hague
Thomas, P.G., Quinn, P.J. & Williams, W.P. (1986) Planta 167, 133-139
Valliammai, T., Gnanam, A. & Dharmalingam, K. (1987) Plant Cell Physiol. 28, 975-985
Velthuys, B.R. (1978) Proc. Natl. Acad. Sci. USA 75, 6031-6034
Venediktov, P.S. & Krivosheieva, A.A. (1984) Planta 160, 200-203
Veselovskii, V.A., Leshinskaya, L.V., Markarova, E.N., Veselova, T.V. & Tarusov, B.N. (1976) Fiziol. Rast. 23, 467-472
Vierling, E., Mishkind, M.L., Schmidt, G.W. & Key, J.L. (1986) Proc. Natl. Acad. Sci. USA 83, 361-365
Volger, H. & Santarius, K.A. (1981) Physiol. Plant. 51, 195-200
Weigel, H.J. (1983) Planta 159, 398-403
Weis, E. (1981a) Planta 151, 33-39
Weis, E. (1981b) FEBS Lett. 129, 197-200
Weis, E. (1981c) Z. Pflanzenphysiol. 101, 169-178
Weis, E. (1982a) Planta 154, 41-47
Weis, E. (1982b) Plant Physiol. 70, 1530-1534
Weis, E. (1983) in Effects of Stress on Photosynthesis (Marcelle, R., Clijsters, H. & van Poucke, M., eds.), pp. 295-304, Martinus Nijhoff/Dr. W. Junk Publ., The Hague
Weis, E. (1984a) Plant Physiol. 74, 402-407
Weis, E. (1984b) in Advances in Photosynthesis Research (Sybesma, C., ed.), vol. 3, pp. 291-294, Martinus Nijhoff/Dr. W. Junk Publ., The Hague
Weis, E. (1985) Biochim. Biophys. Acta 807, 118-126
Weis, E. & Berry, J.A. (1987) Biochim. Biophys. Acta 894, 198-208
Weis, E. & Berry, J.A. (1988) in Plant and Temperature (Long, S., ed.), Proc. Soc. Applied Biology, in press
Weis, E., Wamper, D. & Santarius, K.A. (1986) Oecologia 69, 134-139
Wiest, S.C. (1986) Physiol. Plant. 66, 527-535
Williams, W.P., Gounaris, K. & Quinn, P.J. (1984) in Advances in Photosynthesis Research (Sybesma, C., ed.), vol. 3, pp. 123-130, Martinus Nijhoff/Dr. W. Junk Publ., The Hague
Williams, W.P., Sen, A. & Fork, D.C. (1986) Photosynth. Res. 10, 75-92
Yamamoto, Y. & Nishimura, M. (1983) in The Oxygen Evolving System of Photosynthesis (Inoué, Y., Crofts, A.R., Govindjee, Murata, N., Renger, G. & Satoh, K., eds.), pp. 229-238, Academic Press, Tokyo
Yordanov, I.T., Goltsev, V., Stoyanova, T. & Venediktov, P. (1987) Planta 170, 471-477
Yordanov, I.T. & Weis, E. (1984) Z. Pflanzenphysiol. 113, 383-393

Glycerolipid synthesis

John L. HARWOOD

Department of Biochemistry, University College, Cardiff CF1 1XL, U.K.

SYNOPSIS

The lipids of chloroplast thylakoids have been conserved remarkably in different plant species. Three glycosylglycerides and phosphatidylglycerol make up over 90% of the total acyl lipids. These components are distributed asymmetrically across the thylakoid membrane and are made by the cooperation of chloroplast-located synthesis with that outside the organelle. In particular, de novo fatty acid formation is concentrated in the chloroplasts while synthesis of polyunsaturated fatty acids utilises the participation of the endoplasmic reticulum. Similarly, formation of the major thylakoid lipids, the galactosylglycerides, uses substrates from both within and without chloroplasts. The details of these reactions, and their modification in certain plants species, will be discussed.

THE NATURE OF CHLOROPLAST LIPIDS

Four different chloroplast membranes can be identified – appressed and non-appressed thylakoids and the inner and outer envelope membranes. Of these, the two types of thylakoid membranes appear to have a rather similar lipid compositions in contrast to their protein contents (see Gounaris et al., 1986). That is, the chloroplast thylakoids show little, if any, lateral heterogeneity with regard to lipids. However, the lipid to protein ratio of stromal thylakoids appears to be significantly higher (about 4) than granal membranes where it is about 2.5 (Gounaris et al., 1986). The thylakoids from a wide variety of plants have been analysed and shown to contain four major lipids. These lipids are identical to those of cyanobacterial photosynthetic membranes. In contrast to other plant membranes, phosphatidylglycerol is the only significant phosphoglyceride and the majority of the total lipid is composed of three glycosylglycerides - monogalactosyldiacylglycerol (MGDG),

---

ABBREVIATIONS: ACP, acyl carrier protein; MGDG, monogalactosyl-diacylglycol; DGDG, digalactosyldiacylglycerol; SQDG, sulphoquino-vosyldiacylglycerol; PG, phosphatidylglycerol; PC, phosphatidyl-choline.

Acknowledgements: Experiments on lipid metabolism in the author's laboratory have been supported by the A.F.R.C. and S.E.R.C.

digalactosyldiacylglycerol (DGDG) and sulphoquinovosyldiacylglycerol
(SQDG).   The two galactolipids contain very high amounts of
α-linolenic acid (Harwood, 1980a; Joyard and Douce, 1987) which, in
consequence, is the major leaf fatty acid.   However, MGDG and DGDG
contain a number of fatty acyl constituents and, consequently, exist
as several molecular species.   The balance of these molecular forms
varies with plant species.   Some plants, such as pea, contain almost
only α-linolenate in MGDG and are, therefore, termed '18:3 plants'.
By contrast, other species, such as spinach, have some linolenate
replaced by cis 7,10,13-hexadecatrienoate and are known as '16:3-
plants' (Heinz, 1977).   When present, hexadecatrienoate is
exclusively esterified to the sn-2 position of MGDG.   It is not
found in DGDG or other lipid classes.   The absence of hexadecatrienoate from DGDG is particularly interesting in view of the
postulated metabolic connections between the two galactolipids (see
later).   Palmitic acid is a significant component of DGDG and occurs
to a small extent in MGDG; it may be esterified to either the sn-1
or sn-2 positions on the glycerol backbone, depending on the plant
species.

The combination of a C_{18} acid at the sn-1 position with a C_{16}
acid at the sn-2 position is characteristic of cyanobacteria.
Therefore, molecular species of galactolipids having this
combination are referred to as 'prokaryotic'.   A plant species
having MGDG molecules almost exclusively of the 'prokaryotic' type
is Anthriscus.   In contrast, pea which has MGDG with C_{18}
(α-linolenate) acids at both positions is referred to as having
'eukaryotic' lipids.   Other plants, such as spinach, have MGDG
molecular species of both 'prokaryotic' and 'eukaryotic' types
(Heinz, 1977; Joyard and Douce, 1987).   The naming of 'prokaryotic'
and 'eukaryotic' molecular species is a useful concept because it
has particular significance when one considers the synthesis of
galactosylglycerides during thylakoid formation (Joyard and Douce,
1987; Roughan, 1987; see later).

The two remaining major lipids of thylakoids are both negatively
charged at physiological pH's.   They both contain predominantly
'prokaryotic' molecular species.   In the case of SQDG the species
with linolenate at the sn-1 position and palmitate at the sn-2
position is most widespread (Harwood, 1980b).   Phosphatidylglycerol
(PG) is unique amongst lipids in containing trans-Δ3-hexadecanoate.
This unusual acid is exclusively located at the sn-2 position
(Harwood, 1980a).   Although the C_{16} acids are concentrated at the
sn-2 position, some plants may contain large amounts of palmitate at
the sn-1 position also.   This can lead to molecular species of PG
which have high gel/liquid phase transitions and the content of
'disaturated' PG has been correlated with freezing susceptibility
(see Murata, 1987).   Typical thylakoid lipid compositions for
representatives of 16:3- and 18:3-plants are shown in Table 1.

In contrast to thylakoid membranes, it has been known for some
time that isolated envelope membranes contain much higher amounts of
phosphoglycerides (e.g. Harwood, 1980a).   Recently, analyses of
inner and outer envelopes have been undertaken.   While the inner
membrane is relatively similar to the thylakoid in lipid
composition, the outer envelope resembles that of the endoplasmic
reticulum (Table 2).   For example, phosphatidylcholine (PC) the
major extra-chloroplastic lipid is almost exclusively confined to
the outer envelope in spinach (Dorne et al., 1985) and pea (Cline et
al., 1981).

Table 1.   Thylakoid lipid composition in 16:3 and 18:3 plants.

	Lipid (%)	Fatty acid composition (% total)					
		16:0	16:1	16:3	18:1	18:2	18:3
Barley	MGDG (42)	4	–	–	1	3	90
(18:3-plant)	DGDG (28)	9	2	–	3	7	78
	SQDG (12)	32	3	–	2	5	55
	PG (10)	18	27	–	2	11	38
Spinach	MGDG (39)	1	–	25	1	2	72
(16:3-plant)	DGDG (26)	4	–	2	2	2	87
	SQDG (11)	39	tr.	tr.	1	7	53
	PG (14)	11	32	–	2	4	47

   In view of the cooperation of extra-chloroplastic compartments in chloroplast lipid synthesis, the different compositions of the two envelope membranes raise intriguing questions as to the details of such formation.

FATTY ACID FORMATION

   This topic has been recently reviewed (Harwood, 1988a) and only a brief summary will be made here. Of course, the ultimate source of carbon for chloroplast fatty acid synthesis in most circumstances is photosynthesis. However, there is some doubt about whether chloroplastic pyruvate dehydrogenase has sufficient activity to provide all necessary actyl-CoA in every plant. Moreover, pyruvate must be provided by glycolysis and an uninterrupted pathway from 3-phosphoglycerate seems to be missing in many cases (Stitt and ap Rees, 1979). However, it is accepted generally that acetyl-CoA for fatty acid synthesis is generated mainly by plastid pyruvate dehydrogenase activity (Liedvogel, 1987; Harwood, 1988a).

Table 2.   Comparison of envelope membrane compositions with other membranes.

		Lipid composition (%)				
		MGDG	DGDG	SQDG	PG	PC
*Spinach	Outer envelope	17	29	6	10	32
	Inner envelope	49	30	5	8	6
	Thylakoid	52	26	7	10	5
+Pea	Outer envelope	6	33	3	6	44
	Inner envelope	45	31	2	7	10

References: * Block et al. (1983), + Cline et al. (1981).

The resultant acetyl-CoA is then utilised by acetyl-CoA carboxylase for the generation of malonyl-CoA for fatty acid synthesis. Considerable controversy has surrounded experiments on the nature of the plastid enzyme. This was because two distinct molecular structures for acetyl-CoA carboxylase have been identified in Nature (see Harwood and Russell, 1984). In bacteria such as E. coli acetyl-CoA carboxylase exists as a multienzyme complex from which biotin carboxylase, biotin carboxyl carrier protein and the transcarboxylase can be purified as separate proteins. By contrast, the mammalian acetyl-CoA carboxylase is a high molecular weight multifunctional protein. The two types of carboxylase also have quite different methods of regulation (Harwood and Russell, 1984). Early attempts to purify acetyl-CoA carboxylase from plants suggested that it was similar to that from E. coli or, possibly, of an intermediate form (see Stumpf, 1980). It is now clear that these conclusions were incorrect since high levels of proteinases in the tissues studied gave rise to fragments from the original carboxylase protein. Evidence has gradually accumulated that the plant acetyl-CoA carboxylase, like that from mammals, is a multifunctional protein (Harwood, 1988a).

In leaves, acetyl-CoA carboxylase has been shown to be localised in chloroplasts (e.g. Nikolau et al., 1981) where its activity, which is low in the dark, is increased markedly by illumination. Similar changes have also been noted for acetate activation and overall fatty acid synthesis. In view of the well-known activation of mammalian acetyl-CoA carboxylase by tricarboxylic acids, such as citrate, these compounds have been tested with various plant acetyl-CoA carboxylase preparations. However, no significant stimulation has been observed nor is there any evidence of a phosphorylation/dephosphorylation mechanism for control. Instead, in maize a combination of factors has been implicated for regulation. Thus, illumination brings about changes in stromal pH, Mg^{++}, ATP and ADP which, when reproduced in vitro, caused a 24-fold increase in acetyl-CoA carboxylase activity (Nikolau and Hawke, 1984). However, as pointed out by Harwood (1988a), differences have been observed in the effect of nucleotides, for example, on different plant carboxylases so that it is not, at present, possible to generalise about regulatory mechanisms.

Purification of acetyl-CoA carboxylases has been aided by the improvement in proteinase inhibitors as well as by the development of affinity chromatography on avidin-monomer-Sepharose 4B. The enzyme from a number of seed tissues is a multifunctional protein of 220-240 kDa (see Harwood, 1988a). However, the situation for the enzyme from maize leaves seems less clear. Although the native enzyme had a molecular mass of 500 kDa, SDS-PAGE revealed a subunit mass of 60 kDa (Nikolau and Hawke, 1984). Moreover, examination of biotin-containing enzymes by probing Western blots with [I^{125}]-Streptavidin showed that all plant species examined contained a positive band of around 62 kDa but no high molecular weight biotinyl proteins (Nikolau et al., 1984). In contrast, purifications of acetyl-CoA carboxylases with molecular masses of 220-240 kDa have been reported for rape leaves, maize leaves (Hellyer et al., 1986) and soya bean leaves (Charles and Cherry, 1986).

Fatty acid synthetase is also concentrated in chloroplasts (Weaire and Kekwick, 1975) but, unlike acetyl-CoA carboxylase, the enzyme is a Type II multienzyme complex. Work with three systems -

avocado, barley and spinach - showed that the enzymes catalysing the six partial reactions could also be purified independently (see Stumpf, 1987; Harwlod, 1988a). However, it had in fact been known since 1964 that the plant enzyme had such a structure because fatty acid synthesis required a heat-stable component with similar properties to acyl carrier protein (ACP) (Overath and Stumpf, 1964) and which was later isolated from several plant tissues (Simoni et al., 1967).

An interesting feature of results obtained with the purified component proteins of plant fatty acid synthesis is the presence of isoforms. In the case of acyl carrier protein, two isoforms have been purified from spinach leaves while three isoforms were found in barley leaves (Hoj and Svendsen, 1984). Because acyl carrier protein participates in a number of enzyme reactions, one possible reason for the presence of isoforms would be if individual enzymes used the different ACP isoforms preferentially (Ohlrogge, 1987). This possibility was tested directly with spinach ACP_1 and ACP_2 when substrates made with the two isoforms were used in in vitro assays for malonyl-CoA:ACP transacylase, oleoyl-ACP thioesterase and glycerol 3-phosphate acyltransferase. The results showed that the latter two enzymes did show a distinct preference for ACP isoforms (Guerra et al., 1986).

Isoforms of other component proteins of fatty acid synthetase have also been isolated from different plant tissues. The function of these separate isoforms is unknown at present. Of the individual enzymes acetyl-CoA:ACP transacylase has the lowest measured activity in vitro and has been suggested to be rate-limiting for the overall reaction (see Stumpf, 1984). The substrate specificity of this transacylase shows that it has only poor activity with substrates other than acetyl-CoA (Shimakata and Stumpf, 1983). It can be inhibited by typical -SH reagents such as p-hydroxymercuribenzoate, by arsenite and, possibly, by thiolactomycin (Nishida et al., 1984). Increasing the rate of acetyl-ACP formation, either by raising acetyl-CoA concentrations or by adding more transacylase, caused a shift in the amount of medium-chain fatty acid products. This has been suggested to be a possible mechanism for product chain-length regulation in vivo (see Pollard and Singh, 1987; Harwood, 1988a). Confirmation that this may, indeed, be the case has come recently from experiments with the herbicide, fluazifop (Walker et al., 1988). This compound acts by inhibiting acetyl-CoA carboxylase in sensitive plant species. It, therefore, raises the acetyl-CoA/ malonyl-CoA ratio in vivo, a process which causes an increase in medium-chain fatty acid products (Walker et al., 1988).

The various other proteins of the fatty acid synthetase have been purified, in most cases to homogeneity. Details will be found in Stumpf (1987) and Harwood (1988a). Of particular note are the β-ketoacyl-ACP synthetases. Two such condensing enzymes have been separated from spinach leaves. The second, β-ketoacyl-ACP synthetase 2, is sensitive to arsenite inhibition whereas β-ketoacyl-ACP synthetase 1 is relatively insensitive but is inhibited by cerulenin. β-Ketoacyl-ACP synthetase 2 is only active with myristoyl-ACP and palmitoyl-ACP substrates and, therefore, corresponds to the 'palmitate elongase' described previously which was sensitive to arsenite (Harwood and Stumpf, 1971). Although two forms of β-ketoacyl-ACP synthetase have only been separated from spinach leaves (Stumpf, 1987) and rape seed (A.R. Slabas, personal communication) the widespread nature of the inhibition of stearate synthesis by arsenite points to the presence of two condensing enzymes being general.   Indeed, parallels can be drawn with the E. coli fatty acid synthetase where a long chain-length specific condensing enzyme is coded for by the fab B gene (see Harwood and Russell, 1984).

Once stearoyl-ACP has been formed, it is rapidly desaturated by a soluble plastid Δ9-desaturase.   That stearoyl-ACP was the substrate for oleate formation in plants was first shown by experiments with Chlorella and spinach chloroplasts (Nagai and Bloch, 1968).   The desaturase has been purified about 200-fold from developing safflower seeds.   In its native state it exists as a dimer with a molecular mass of 68 kDa and was purified using a combination of ion-exchange and affinity chromatography (McKeon and Stumpf, 1982).   Several aspects of its properties are of interest, including its substrate specificity.   The desaturase shows a marked preference for stearate over palmitate and also for the ACP- rather than the CoA-ester (Stumpf, 1984).   These properties, undoubtedly, explain why stearate accumulates at such low levels in plants and why oleate, rather than palmitoleate, is the major monoenoic acid (Harwood, 1988a).

ORGANELLE COOPERATION IN OLEATE METABOLISM

As far as oleate, fatty acid synthesis is concentrated in the plastid.   Thereafter, two major routes can be followed.   The extent to which each is used depends on the plant species as well as the complex lipid concerned.   Basically, the insertion of further double bonds into a monoenoic fatty acid requires the activity of desaturases utilising complex lipid substrates.   This interesting fact was suggested first through labelling experiments by Nichols, James and coworkers (e.g. Nichols et al., 1967) and later extended by in vitro experiments utilising exogenous lipid substrates (Gurr et al., 1969).

The fatty acid products of de novo fatty acid synthesis (palmitate, oleate) can be transferred to glycerol 3-phosphate by the action of two plastid acyltransferases.   The first of these, which is specific for the sn-1 position and which prefers oleate is a soluble stromal enzyme.   It is present in isoforms in several species and the gene for one of the squash isoforms has been recently identified and sequenced (N. Murata - personal communication).   The second acyltransferase is located in the envelope membranes.   In contrast to the soluble enzyme, the

---

Table 3.   Positional distribution of labelled fatty acids in envelope phosphatidate.

	% Fatty acids incorporated					
	Total		sn-1		sn-2	
	16:0	18:1	16:0	18:1	16:0	18:1
Pea	54.4	43.6	14.2	85.8	96.6	3.4
Spinach	55.5	44.5	10.1	89.9	96.3	3.7

Taken from Joyard and Douce (1987)

lysophosphatidate:acyl-ACP acyltransferase prefers palmitoyl-ACP as substrate. Therefore, a typical chloroplast produces phosphatidate which will contain oleate at the sn-1 position and palmitate at the sn-2 position (Table 3). This is referred to as a 'prokaryotic' pattern and such a distribution will be preserved in the chloroplast lipids of plants where most fatty acid metabolism continues within the plastid. Typical plants of this type are the 16:3 plants. In such plants the phosphatidate formed can be rapidly dephosphorylated by phosphatidate phosphohydrolase (Price-Jones and Harwood, 1987). The resultant diacylglycerol is then a substrate for MGDG (and, possibly, SQDG) formation (Harwood, 1988b). Alternatively phosphatidate is used by a cytidylyltransferase to form CDP-diacylglycerol which is a precursor for the major chloroplast phospholipid, PG.

Instead of transferring fatty acyl groups to glycerol 3-phosphate, plastids also contain an active acyl-ACP thioesterase. It is present in the stromal phase of chloroplasts (Joyard and Stumpf, 1981) and has been purified from maturing seeds of safflower (McKeon and Stumpf, 1982). However, unesterified fatty acids do not accumulate because an active acyl-CoA synthetase is present in the plastid envelope (Joyard and Stumpf, 1981). By these means, therefore, fatty acids can be made available to the extra-plastidic compartment and esterification to molecules such as PC becomes possible. In adition, glycerol 3-phosphate acyltransferase and lysophosphatidate acyltransferase are present in microsomal (probably endoplasmic reticulum) membranes. In contrast to the plastid acyltransferases, these enzymes utilise acyl-CoA substrates and synthesise phosphatidate which is, predominantly, sn-1 saturated (palmitoyl-), sn-2 unsaturated (oleoyl-).

FURTHER DESATURATION OF MONOENOIC FATTY ACIDS

For plants which have exported their oleate from the plastid to the endoplasmic reticulum, PC acts as the major substrate for linoleate formation (see Harwood, 1988a). However, it must be emphasised that the premier role of PC does not, necessarily, exclude other potential complex lipid substrates. Evidence has accumulated in several tissues for the involvement of other lipids (Harwood, 1988a). Whether PC is a major substrate for α-linolenate synthesis in leaf tissues seems doubtful. Instead, considerable evidence from in vivo (Heinz and Harwood, 1977) and chloroplast labelling experiments (Roughan et al., 1979; Jones and Harwood, 1980) shows that MGDG is the main substrate for linoleate desaturation.

The use of two diferent lipid substrates, located in different parts of the 18:3-plant cell, for linoleate and linolenate formation raises obvious questions about the coordination of fatty acid and membrane synthesis. However, in 18:3-plants it has been proposed that the diacylglycerol backbone of PC is transferred to MGDG. In simplistic terms it has been assumed that a phospholipid transfer protein (Kader et al., 1982) transfers PC, containing linoleate, back to the plastid where the diacylglycerol is released for MGDG synthesis. Indeed, in vitro experiments have shown that ^{14}C-linoleoyl-PC can donate its linoleate to MGDG for further desaturation when chloroplasts are also incubated with a phospholipid transfer protein (Jones and Harwood, 1980; Ohnishi and Yamada, 1982). However, although the 16:3-/18:3-plant theories

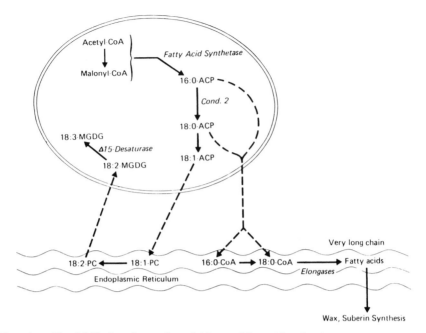

Fig. 1. Simplified scheme for fatty acid synthesis in an
'18:3-plant'.

require the maintenance of an intact diacylglycerol with specific
fatty acyl combinations, no evidence for an enzyme capable of
releasing diacylglycerol from PC has been detected in plastid
envelopes.   This severe problem has been discussed very well by
Joyard and Douce (1987) who also point out that the use of
transacylases may offer a solution.   In that regard it is worth
recalling that evidence was produced some years ago for a rapid
equilibration of oleate and linoleate between (chloroplastic) MGDG
and (non-chloroplastic) PC.   It was suggested that such an
equilibrium could have been due to transacylase activity (Harwood,
1979a; Wharfe and Harwood, 1978).

Compelling indirect evidence that MGDG is a major substrate for
linoleate desaturation comes from labelling experiments with leaf
pieces and chloroplasts (see Joyard and Douce, 1987; Harwood,
1988a). Direct demonstration that ^{14}C-linoleoyl-MGDG could act as a
substrate for Δ15-desaturation has been obtained with lettuce and
pea chloroplasts (Jones and Harwood, 1980) but the desaturase has
not been purified.   MGDG may also act as a substrate for oleate
desaturation for those plants which retain fatty acids within the
plastid (16:3-plants), and in such plants it has been proposed that
all chloroplast lipids are capable of acting as desaturase
substrates (see Joyard and Douce, 1987).

Thus, the formation of polyunsaturated fatty acids in leaves can
be linked to the formation of different complex lipids.   To a
greater or lesser extent, cooperation of organelles is involved so
that, in 18:3-plants de novo synthesised oleate is exported out of
the chloroplast, desaturated to linoleate, and then returned to the
plastid for further modification.   This generalised scheme is shown
in Figure 1.

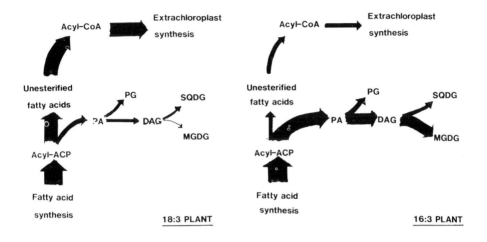

Fig. 2.   Comparison of the channelling of newly synthesised fatty
          acids in 16:3 and 18:3 plants (for further details see
          Joyard and Douce, 1987).

By contrast, in 16:3-plants the diacylglycerol backbone
generated by acylation of glycerol 3-phosphate in plastids is
retained within the organelle for incorporation into complex lipids
and for use in successive desaturations. Thus, for example,
hexadecatrienoate (present at the sn-2 position of MGDG in
16:3-plants) seems to be formed by three successive desaturations of
palmitate (e.g. Thompson et al., 1986). The channelling of carbon
through the 'prokaryotic' and 'eukaryotic' pathways is illustrated
in Figure 2, where the relevant enzymes are emphasised.

IMPORTANCE OF DIFFERENT PHOSPHOLIPID SYNTHETIC PATHWAYS

In plants, the zwitterionic phospholipids, PC and
phosphatidylethanolamine, are made predominantly by the CDP-base
pathway. Evidence for this statement and details of enzymes
concerned are given in various reviews (e.g. Harwood, 1979b; Mudd,
1980; Moore, 1982; Harwood, 1988b). The central enzyme of the
CDP-base pathway, the cytidylyltransferase, has been shown to be
rate-limiting (e.g. Price-Jones and Harwood, 1983). The enzyme's
activity appears to be controlled by protein turnover and by
allosteric regulation (Price-Jones and Harwood, 1986).

On the other hand, acidic phospholipids such as
phosphatidylinositol and PG are made from a CDP-diacylglycerol
precursor. The formation of PG has been shown to take place on

chloroplast envelopes (Sparace and Mudd, 1982; Andrews and Mudd, 1985). This is particularly interesting in view of the similar location of lysophosphatidate:acyl-ACP transacylase and phosphatidate phosphohydrolase (see Joyard and Douce, 1987). Thus, two enzymes compete for the phosphatidate generated and, in 18:3-plants a large portion of the phosphatidate formed is channelled towards PG synthesis. General characteristics of the pathway for PG formation seem to be similar to those described for other systems. The inhibition of phosphatidylglycerophosphate phosphatase by mercuric ions is typical (Andrews and Mudd, 1985).

Interestingly, little is known of the synthesis of the 'typical' mitochondrial constituent, diphosphatidylglycerol. This is because efforts to demonstrate the enzymes concerned in vitro have failed (see Moore, 1982).

Phosphatidylinositol has been known to be a significant constituent of plant tissues for many years. However, it is only recently that the polyphosphoinositides have been detected with certainty (see Harwood, 1988b). In view of the surge of interest in these lipids in mammalian tissues, similar agonist-stimulated second messenger functions (Berridge, 1984) have been sought in plant tissues. Several circumstantial observations have implied such a role (Harwood, 1988b) but it is only recently that a rapid turnover of the higher inositides has been seen under conditions which may have a physiological significance. Cultures of Cartharanthus roseus cease growing after a certain period and this arrest in G_1 phase is overcome by the addition of the synthetic auxin 2,4-D or of indole acetic acid. This recommenced growth and differentiation was accompanied by a rapid, transient breakdown of phosphatidylinositol 4-phosphate and phosphatidylinositol 4,5-bisphosphate to yield inositol 1,4-bisphosphate and inositol 1,4,5-trisphosphate (Ettlinger and Lehle, 1988). Recovery of the original levels of polyphosphoinositides then followed. Taken together with observations that inositol trisphosphate can cause calcium release in plants (e.g. Schumaker and Sze, 1987) these results show considerable parallels with the mammalian systems and offer exciting prospects for the future.

FORMATION OF GLYCOSYLGLYCERIDES - THE MAIN THYLAKOID LIPIDS

As already stated above, the major chloroplast thylakoid lipids are the glycosylglycerides - MGDG, DGDG and SQDG. Of these lipids only the pathway to MGDG is known with any certainty. In this case it is known that galactose is transferred from UDP-galactose to diacylglycerol acceptors in the chloroplast envelope. Early work (see Douce and Joyard, 1980) suggested that two galactosyl-transferases were present, one forming MGDG and the other involved in a second transfer to yield DGDG. However, efforts to purify the latter and demonstrate its physiological importance have, so far, proved fruitless. Nevertheless, there is considerable indirect evidence for the presence of two separate enzymes. For example, it is known that the fatty acyl compositions of MGDG and DGDG are rather different. In fact, in '16:3-plants' this difference is absolute for hexadecatrienoate which is confined to MGDG (Harwood, 1980a). In addition, a preference for unsaturated diacylglycerols for the first galactosyltransferase was noted from experiments in vitro (Mudd et al., 1969; Bowden and Williams, 1973). Other differences between the two galactosyltransferases include their differential solubility and the stereochemical specificity of the two galactose groups of DGDG which would sensibly require specific enzymes for their attachment (Harwood, 1988b).

Douce and coworkers showed first the importance of the chloroplast envelope for galactolipid formation (see Joyard and Douce, 1987). The '18:3-plant' pea has been shown to have its UDP-galactose:diacylglycerol galactosyltransferase located in the outer membrane (Cline and Keegstra, 1983). In contrast, the '16:3-plant' spinach has the enzyme confined to the inner envelope membrane (Block et al., 1983). This difference may reflect the fact that diacylglycerol in '16:3-plants' is generated by envelope phosphatidate phosphohydrolase whereas that in '18:3-plants' is thought to originate from the extra-chloroplastic compartment (e.g. Heinz and Roughan, 1983). Very recently some sucess has been achieved in solubilising and partly purifying the UDP-galactose diacylglycerol galactosyltransferase from spinach (Coves et al., 1988).

DGDG can also be produced by the activity of a galactolipid:galactolipid galactosyltransferase (GGGT; see Fig. 3). This enzyme was first found in spinach envelopes (von Besouw and Wintermans, 1978). In vitro the enzyme is capable of forming other (non-physiological) higher homologues of galactolipids. Although GGGT has high activity and can be easily studied, its role in DGDG formation is still controversial. For example, in spite of evidence that only GGGT activity could account for DGDG in spinach chloroplast envelopes (Heemskerk et al., 1983), problems with its purported role have been detailed (Joyard and Douce, 1987).

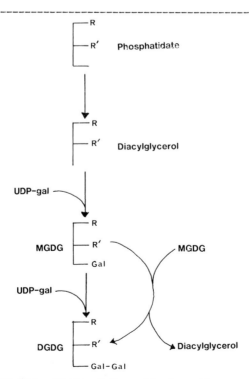

Fig. 3. Pathways for galactosylglyceride formation.

The third glycosylglyceride is the plant sulpholipid, SQDG. The biosynthetic route for sulpholipid formation has been a fertile ground for speculation for over 20 years. Benson (1963) first suggested that it could be formed by a modification of the glycolytic pathway. He tested a number of tissues, including higher plants, for the presence of appropriate ^{35}S-intermediates and identified a number of relevant compounds (see Harwood, 1980b). Sulpholactate, sulpholactaldehyde and sulphopropanediol were all isolated from Chlorella (Shibuya and Benson, 1961). Later experiments with Euglena showed that cysteic acid was an effective precursor of SQDG (Davies et al., 1966). In alfalfa seedlings ^{35}S-cysteic acid was shown to be a more efficient precursor of SQDG than ^{35}S-sulphate (Harwood, 1975) but, in contrast, in spinach seedlings evidence was produced that ^{35}S from cysteate was only incorporated into SQDG after breakdown to sulphate (Kleppinger-Sparace et al., 1985).

The effect of a number of reagents on incorporation of radioactivity from ^{35}S-sulphate into SQDG is shown in Table 4. It will be seen that molybdate caused a consistent decrease in labelling in all tissues, implying the use of PAPS or APS in the biosynthetic pathway. Sulphite was a good inhibitor in most systems, perhaps indicating the SQDG synthesis involved nucleophilic displacement of O-acetylserine to yield cysteic acid (Harwood, 1980b). In addition, sulpholactic acid has been found to reduce ^{35}S-incorporation from sulphate into SQDG in alfalfa, in agreement with results predicted for a sulphoglycolytic pathway (see Harwood, 1980b). Various other postulated reactions in the scheme have also been tested (Harwood, 1980b) although it seems clear that in spinach the pathway does not seem to operate (Mudd and Kleppinger-Sparace, 1987). The latter plant may, however, be an exception (Table 4). However, there seems no evidence that sulphoquinovose itself is an intermediate in SQDG formation in those plants tested (Harwood, 1980; Kleppinger-Sparace et al., 1985).

Table 4.   Effect of additions on ^{35}S-sulphate incorporation into SQDG.

Addition	Plant system				
	Euglena	Alfalfa seed	Spinach leaf	Wheat seed	Brussels sprout seed
Molybdate (1 mM)	62	52	–	62	59
Cysteate (various concs.)	44	62	93	6	33
Sulphite (0.1 mM)	–	62	8	25	–
Sulphate (0.1 mM)	–	105	–	118	–

Results expressed as % control incorporation. Taken from Harwood (1975, 1980b), Mudd and Kleppinger-Sparace (1987) and Y. Ellis and J.L. Harwood (unpublished data).

Some years ago we developed a silver nitrate-t.l.c. system which could be used to separate molecular species of intact SQDG. Use of this technique revealed that relatively saturated molecular species of SQDG were rapidly labelled, in contrast to the more prevalent polyunsaturated species (Heinz and Harwood, 1977; Nicholls and Harwood, 1979). It was postulated that these two groups of sulpholipid molecules might have distinct types of functions. Thus, the fast turning-over species could play a dynamic (e.g. enzymatic) role while the slowly metabolised trienoic species might have a structural function (see Harwood, 1980b). Interestingly, rapidly-labelled SQDG is associated with purified chloroplast CF_0-CF_1 ATPase and this sulpholipid is primarily a saturated species (Pick et al., 1985). Moreover, SQDG has also been found to prevent proton leakage in reconstitution experiments (Pick et al., 1987) i.e. it had a structural role. Thus, although the molecular species labelling experiments can also be interpreted by considering the use of SQDG as a substrate for fatty acid desaturation (Joyard and Douce, 1987), the above experiments provide evidence in favour of the 'two functions' hypothesis (Harwood, 1980b).

FUTURE PROSPECTS

It will be clear from the foregoing that a great deal more work is required before we have even a cursory picture of lipid metabolism in plant tissues. The lack of knowledge about even the pathways for formation of such major compounds as DGDG and SQDG is embarrassing. Moreover, we know very little about the regulation of pathways for lipid synthesis. Improvement in this situation can only come, in my opinion, from the careful application of sound subcellular fractionation and enzymological methods. Once purified enzymes are available, it should be possible to use molecular biology to help in the elucidation of regulatory steps and pathways. If overexpression of relevant genes can be obtained in appropriate model systems, then it may be possible to study enzymic mechanisms. Considering the novelty of many plant lipid pathways, then the latter possibility is exciting indeed.

REFERENCES

Andrews, J. and Mudd, J.B. (1985) Plant Physiol. **79**, 259-265
Benson, A.A. (1963) Adv. Lipid Res. **1**, 387-394
Berridge, M.J. (1984) Biochem. J. **220**, 345-360
van Besouw, A. and Wintermans, J.F.G. (1978) Biochim. Biophys. Acta **529**, 44-53
Block, M.A., Dorne, A-J., Joyard, J. and Douce, J. (1983) J. Biol. Chem. **258**, 13281-13286
Bowden, B.N. and Williams, P.M. (1973) Phytochemistry **11**, 1059-1064
Charles, D.J. and Cherry, J.H. (1986) Phytochemistry **25**, 1067-1071
Cline, K. and Keegstra, K. (1983) Plant Physiol. **71**, 366-372
Cline, K., Andrews, J., Mersey, B., Newcomb, E.H. and Keegstra, K. (1981) Proc. Natl. Acad. Sci. U.S.A. **78**, 3595-3599
Coves, J., Joyard, J. and Douce, R. (1988) Proc. Natl. Acad. Sci. U.S.A., in press
Davies, W.H., Mercer, E.I. and Goodwin, T.W. (1966) Biochem. J. **98**, 369-373
Dorne, A-J., Joyard, J., Block, M.A. and Douce, R. (1985) J. Cell Biol. **100**, 1690-1697
Ettlinger, C. and Lehle, L. (1988) Nature **331**, 176-178

Gounaris, K., Barber, J. and Harwood, J.L. (1986) Biochem. J. **237**, 313-326

Guerra, D.J., Ohlrogge, J.B. and Frentzen, M. (1986) Plant Physiol. **82**, 448-453

Gurr, M.I., Robinson, P. and James, A.T. (1969) Eur. J. Biochem. **9**, 70-78

Harwood, J.L. (1975) Biochim. Biophys. Acta **398**, 224-230

Harwood, J.L. (1979a) Phytochemistry **18**, 1811-1814

Harwood, J.L. (1979b) Prog. Lipid Res. **18**, 55-86

Harwood, J.L. (1980a) in Biochemistry of Plants (Stumpf, P.K. and Conn, E.E., eds.), vol. 4, pp. 1-55, Academic Press, New York

Harwood, J.L. (1980b) in Biochemistry of Plants (Stumpf, P.K. and Conn, E.E., eds.), pp. 301-320, Academic Press, New York

Harwood, J.L. (1988a) Ann. Rev. Plant Physiol. **39**, 101-138

Harwood, J.L. (1988b) CRC Critical Reviews, in press

Harwood, J.L. and Russell, N.J. (1984) Lipids of Plants and Microbes, Allen and Unwin, London

Harwood, J.L. and Stumpf, P.K. (1971) Arch. Biochem. Biophys. **142**, 281-291

Heemskerk, J.W.M., Bogemann, G. and Wintermans, J.F.G.M. (1983) Biochim. Biophys. Acta **754**, 181-189

Heinz, E. (1977) in Lipids and Lipid Polymers in Higher Plants (Lichtenthaler, H.K. and Tevini, M., eds.), pp. 102-120, Springer-Verlag, Berlin

Heinz, E. and Harwood, J.L. (1977) Hoppe-Seyler's Z. Physiol. Chem. **38**, 897-908

Heinz, E. and Roughan, P.G. (1983) Plant Physiol. **72**, 273-279

Hellyer, A., Bambridge, H.E. and Slabas, A.R. (1986) Biochem. Soc. Trans. **14**, 565-568

Hoj, P.B. and Svendsen, I.B. (1984) Carlsberg Res. Commun. **49**, 483-492

Jones, A.V.M. and Harwood, J.L. (1980) Biochem. J. **190**, 851-854

Joyard, J. and Douce, R. (1987) in Biochemistry of Plants (Stumpf, P.K. and Conn, E.E., eds.), vol. 9, pp. 215-274, Academic Press, New York

Joyard, J. and Stumpf, P.K. (1981) Plant Physiol. **67**, 250-256

Kader, J.C., Douady, D. and Mazliak, P. (1982) in Phospholipids (Hawthorne, J.N. and Ansell, G.B., eds.), pp. 279-311, Elsevier, Amsterdam

Kleppinger-Sparace, K.F., Mudd, J.B. and Bishop, D.G. (1985) Arch. Biochem. Biophys. **240**, 859-865

Liedvogel, B. (1987) in Metabolism, Structure and Functions of Plant Lipids (Stumpf, P.K., Mudd, J.B. and Nes, W.D., eds.), pp. 509-512, Plenum Press, New York

McKeon, T.M. and Stumpf, P.K. (1982) J. Biol. Chem. **257**, 12141-12147

Moore, T.S. (1982) Ann. Rev. Plant Physiol. **33**, 235-259

Mudd, J.B. (1980) in Biochemistry of Plants (Stumpf, P.K. and Conn, E.E., eds.), vol. 4, pp. 249-282, Academic Press, New York

Mudd, J.B. and Kleppinger-Sparace, K.F. (1987) in Biochemistry of Plants (Stumpf, P.K. and Conn, E.E., eds.), vol. 9, pp. 275-289, Academic Press, New York

Mudd, J.B., van Vliet, H.H.D.M. and van Deenen, L.L.M. (1969) J. Lipid Res. **10**, 623-630

Murata, N. (1983) Plant Cell Physiol. **24**, 81-86

Nagai, J. and Bloch, K. (1968) J. Biol. Chem. **243**, 4626-4633

Nichols, B.W., James, A.T. and Breuer, J. (1967) Biochem. J. **104**, 486-496

Nicholls, R.G. and Harwood, J.L. (1979) Phytochemistry **18**, 1151-1154

Nikolau, B.J. and Hawke, J.C. (1984) Arch. Biochem. Biophys. **228**, 86-96

Nikolau, B.J., Hawke, J.C. and Slack, C.R. (1981) Arch. Biochem. Biophys. **211**, 605–612

Nikolau, B.J., Wurtels, E.S. and Stumpf, P.K. (1984) Plant Physiol. **75**, 895–901

Nishida, I., Kawaguchi, A. and Yamada, M. (1984) Plant Cell Physiol. **25**, 265–268

Ohlrogge, J.B. (1987) in Biochemistry of Plants (Stumpf, P.K. and Conn, E.E., eds.), vol. 9, pp. 137–157, Academic Press, New York

Ohnishi, J-I. and Yamada, M. (1982) Plant Cell Physiol. **23**, 767–773

Overath, P. and Stumpf, P.K. (1964) J. Biol. Chem. **239**, 4103–4220

Pick, U., Gounaris, A., Admon, A. and Barber, J. (1985) Biochim. Biophys. Acta **808**, 415–420

Pick, U., Weiss, M., Gounaris, K. and Barber, J. (1987) Biochim. Biophys. Acta **891**, 28–39

Pollard, M.R. and Singh, S.S. (1987) in Metabolism, Structure and Function of Plant Lipids (Stumpf, P.K., Mudd, J.B. and Nes, W.D., eds.), pp. 455–463, Plenum Press, New York

Price-Jones, M.J. and Harwood, J.L. (1983) Biochem. J. **216**, 627–631

Price-Jones, M.J. and Harwood, J.L. (1986) Biochem. J. **240**, 837–842

Price-Jones, M.J. and Harwood, J.L. (1987) in Phosphatidate Phosphohydrolase (Brindley, D.N., ed.), vol. 2, pp. 2–37, CRC Press, Boca Raton

Roughan, P.G. (1987) in The Metabolism, Structure and Function of Plant Lipids (Stumpf, P.K., Mudd, J.B. and Nes, W.D., eds.), pp. 247–254, Plenum Press, New York

Roughan, P.G., Mudd, J.B., McManus, T.T. and Slack, C.R. (1979) Biochem. J. **184**, 571–574

Schumaker, K.S. and Sze, H. (1987) J. Biol. Chem. **262**, 3944–3946

Shibuya, I. and Benson, A.A. (1961) Nature **192**, 1186

Shimakata, T. and Stumpf, P.K. (1983) J. Biol. Chem. **258**, 3592–3598

Simoni, R.D., Criddle, R.S. and Stumpf, P.K. (1967) J. Biol. Chem. **242**, 573–581

Sparace, S.A. and Mudd, J.B. (1982) Plant Physiol. **70**, 1260–1264

Stitt, M. and ap Rees, T. (1979) Phytochemistry **18**, 1905–1911

Stumpf, P.K. (1980) in Biochemistry of Plants (Stumpf, P.K. and Conn, E.E., eds.), pp. 177–203, Academic Press, New York

Stumpf, P.K. (1984) in Fatty Acid Metabolism and its Regulation (Numa, S., ed.), pp. 155–199, Elsevier, Amsterdam

Stumpf, P.K. (1987) in Biochemistry of Plants (Stumpf, P.K. and Conn, E.E., eds.), vol. 9, pp. 121–136, Academic Press, New York

Thompson, G.A., Roughan, P.G., Browse, J.A., Slack, C.R. and Gardiner, S.E. (1986) Plant Physiol. **82**, 357–362

Walker, K.A., Ridley, S.M. and Harwood, J.L. (1988) Biochem. J. in press

Weaire, P.J. and Kekwick, R.G.O. (1975) Biochem. J. **146**, 425–437

Wharfe, J. and Harwood, J.L. (1978) Biochem. J. **174**, 163–169

Processing peptidases of higher plant chloroplasts

Peter D. ELDERFIELD, Janet E. MUSGROVE, Patricia M. KIRWIN and Colin ROBINSON

Department of Biological Sciences, University of Warwick, Coventry, CV4 7AL, U.K.

SYNOPSIS

The proteolytic processing of imported chloroplast proteins is discussed in terms of the locations, characteristics and specificities of the peptidases involved.

INTRODUCTION

Most chloroplast proteins are imported into the organelle after synthesis in the cytoplasm. Imported proteins are initially synthesised as precursors containing aminoterminal pre-sequences; the precursors are subsequently transported into the chloroplast by a post-translational, ATP-dependent mechanism. Receptors in the chloroplast envelope are probably involved in binding the precursors in the early stages of transport.

THE STROMAL PROCESSING PEPTIDASE

Imported stromal proteins are rapidly processed to the mature size by a stromal peptidase which has been extensively characterised after partial purification from pea chloroplasts (Robinson and Ellis, 1984). The enzyme has a native molecular weight of ca. 180 kDa, a pH optimum of 8.5 - 9 and is inhibited by metal-chelating compounds such as 1,10-phenanthroline and EDTA. Significantly, the peptidase displays a high degree of reaction specificity: the enzyme processes the precursor of Rubisco small subunit to the mature size but no further (Fig 1) and fails to cleave any non-chloroplast proteins tested to date. The precursor of ferredoxin-NADP-oxidoreductase, another stromal protein is also processed to the mature size but no further.

The processing activity appears to be conserved throughout higher plants, since the purified pea chloroplast enzyme efficiently processes precursors of other dicot, and also monocot, chloroplast proteins. The activity also appears to be present in other plastid types; we have found that soluble extracts of non-photosynthetic plastids (from Ricinus communis seed endosperm tissue) contain an

----

ABBREVIATIONS: Pre-PC and PCi precursor and processing intermediate of plastocyanin, respectively.
The financial support of the SERC is gratefully acknowledged.

Fig. 1.   Elution of stromal processing activity from a DEAE-Sephacel
          column.
The autoradiogram shows processing to the mature size of the
precursor of pea Rubisco small subunit.

----

apparently identical processing activity (Halpin, Musgrove, Lord and
Robinson, submitted).
     The basis for the high degree of reaction specificity of this
peptidase is poorly understood. Analysis of the cleavage sites
within a number of precursors reveals little primary sequence
homology which could serve as a recognition signal. The enzyme may
instead recognise a defined three-dimensional structure in the region
of the cleavage site.

IMPORT OF THYLAKOID LUMEN PROTEINS

     The biogenesis of these proteins is complex because three
membranes lie between the thylakoid lumen and the cytoplasm.
Analysis of plastocyanin (PC) has shown that the import of this
protein can be divided into two phases. Initially pre-PC is
transported into the stroma and processed into an intermediate form
by the stromal peptidase described above. This intermediate is
subsequently transferred into the thylakoid lumen and processed to
the mature size by a second, thylakoid peptidase (Smeekens et al.,
1985; Hageman et al., 1986).
     The peptidase has been extensively purified after extraction
from pea thylakoids using Triton X-100. The partially purified
enzyme processes the PC processing intermediate to the mature size
(but no further) and, at a much lower rate, processes pre-PC to the
mature size in a single step. Removal of the first section of the
pre-sequence by the stromal peptidase is thus not a pre-requisite for
recognition of the second processing site by the thylakoidal
peptidase (Kirwin et al., 1987). The thylakoidal peptidase, in
common with the stromal peptidase, is highly specific in that foreign
proteins tested to date are not cleaved.
     The precise location of the thylakoidal peptidase has been
examined in some detail. Experiments involving differential
centrifugation of Triton extracts have shown that the peptidase is an
integral membrane protein. The organisation of the peptidase was
further studied by sucrose density gradient centrifugation of the
partially purified enzyme. Fig 2A shows the polypeptide profiles of
the gradient fractions; Photosystem I particles migrate towards the

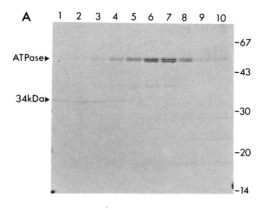

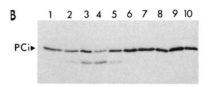

Fig. 2.   Sucrose density gradient centrifugation of the thylakoidal
          processing peptidase.
A: Coomassie - stained gel of gradient fractions (left = top). The α
and β ATPase subunits, and the 34 kDa extrinsic PS II protein, are
indicated. B: Assay of gradient fractions for processing of the
plastocyanin intermediate to the mature size.

----

bottom of the gradient, whereas the ATP synthase is found in the
middle section. Immunoblotting showed the cytochrome b6/f complex to
be in the upper section of the gradient with a peak in fractions 5-6
(not shown). Fig 2B shows an assay of the fractions for processing
activity using an artificial PC intermediate, PCi, described in
Hageman et al., 1986. Peak processing activity is found in fractions
4-5, showing that the peptidase is not associated with any of the
major thylakoidal protein complexes (Photosystem II particles are
removed at an earlier purification step). The result also suggests
that the peptidase is not itself in the form of a large complex.
     Additional experiments have been carried out to analyse the
distribution network. These studies have shown that the peptidase is
exclusively located in non-appressed lamellae of the membrane, i.e.
those in direct contact with the stroma.

CONCLUSIONS

     Chloroplast protein precursors undergo highly specific

processing reactions after import into the organelle. Imported
stromal proteins are rapidly processed to the mature size by a
soluble stromal peptidase, whereas thylakoid lumen proteins undergo
successive cleavages by a stromal and a thylakoidal peptidase. Both
peptidases exhibit a considerable degree of reaction specificity but
further work is required to determine the molecular basis for this
specificity.

REFERENCES

Hageman, J., Robinson, C., Smeekens, S. & Weisbeek, P. (1986) Nature
   324, 567-569.
Kirwin, P.M., Elderfield, P.D. & Robinson, C. (1987) J. Biol. Chem.
   262, 16386-16390.
Robinson, C. & Ellis, R.J. (1984) Eur. J. Biochem. 142, 337-342.
Smeekens, S., Bauerle, C., Hageman, J., Keegstra, K. & Weisbeek, P.
(1986) Cell 46, 365-375.

The synergistic effect of light and heat stress on the inactivation of photosystem II

Gadi SCHUSTER, Susanna SHOCHAT, Noam ADIR, Dena EVEN, Dvorah ISH-SHALOM, Bernhard GRIMM*, Klaus KLOPPSTECH* and Itzhak OHAD

Department of Biological Chemistry, The Hebrew University, Jerusalem, Israel and *Institute of Botany, Hannover, University, Herrenhauser Strasse 2, D-3000 Hannover, FRG.

SYNOPSIS

Photoinhibition of Photosystem II at high light intensities correlates with an increased turnover of the D1 protein induced by an alteration of the reaction center II. The D2 protein is affected but to a lesser degree. Photo- inhibition is accelerated by heat stress which affects the degradation process. Heat shock proteins protect Photosystem II against the synergistic effect of heat and high light intensity.

INTRODUCTION

It is now generally accepted that the process of photo-inhibition in vivo occurs when the chloroplasts are exposed to light fluencies exceeding those required to saturate photosynthesis (Powels, 1984). Loss of photosynthetic activity appears to be localised at the level of Photosystem II (PSII) reaction center (Bjorkman, 1987; Ohad et al., 1988), and seem to be due to an alteration of the D1 protein followed by its degradation and replacement with a newly synthesized molecule (Kyle et al., 1984). When the rate of D1 damage exceeds the rate of its replacement with a functional molecule, PSII is inactivated (Schuster and Ohad, 1988). We have recently demonstrated that under heat stress conditions (heat shock, HS) the process of PSII inactivation in the light is drastically accelerated while the D1 degradation is inhibited (Schuster et al., 1988). Under these conditions, D1 as well as other thylakoid proteins became aggregated (polymerised) and under extreme conditions the process is irreversible. The loss of PSII activity during heat shock is transiently protected if heat shock proteins whose synthesis is induced by this treatment are present in the cells. In Chlamydomonas reinhardtii (Kloppstech et al., 1985) as well as in higher plants (Vierling et al., 1986) nuclear coded heat shock proteins translated in the cytoplasm are transported into the chloroplast and integrated into the thylakoid membrane (Kloppstech et al., 1985; Schuster et al., 1988).

The financial support of the Israeli National Academy of Sciences (I.O.), the DFG and the Israeli-German Program of the Council for R&D (I.O. and K.K.) is gratefully acknowledged.

During the past few years, we have investigated the process of PSII inactivation under heat and light stress conditions and the results obtained are reviewed below.

DEGRADATION OF REACTION CENTER II PROTEINS DURING PHOTOINHIBITION

The D1 protein, now considered to participate together with the D2 protein in the formation of reaction center II (Trebst and Debka, 1985) turns over in the light (Mattoo et al., 1984) and it was suggested that the light induced turnover of this protein underlies the mechanism of photoinhibition (Kyle et al., 1984; Ohad et al., 1988). The question arose as to whether this is the only PSII protein which turns over during photoinhibition. To answer this question a systematic study of the fate of several PSII proteins including the LHCII and CP29 antennase polypeptides, the core complex 44 and 51 kDa proteins, the water oxidising complex proteins of 17, 23 and 29 kDa and the D1 and D2 reaction center proteins was investigated during photoinhibition of Chlamydomonas reinhardtii cells. Using a variety of techniques including dot blotting, Western blotting, immunogold labeling and quantitative electron microscopy as well as pulse-chase radioactive tracing, we could demonstrate that only the D1 and to some extent the D2 protein turn over during photoinhibition. The half-life of the D1 protein under our experimental conditions was 90 min and that of D2 about 120 min. No detectable turnover of the D2 protein could be demonstrated in cells exposed to non-photo-inhibitory light intensities (Schuster et al., 1988).

The initial process of the inactivation of the PSII reaction center was studied using the thermoluminescence technique which permits detection of the presence of the S2QB- and S2QA- pairs induced by one saturating light flash of dark adapted cells in the absence or presence of Diuron.

The method also enables to measure both the relative content of the pairs on a chlorophyll basis and their stability (Ohad et al., 1988).

Exposure of Chlamydomonas cells to photoinhibitory conditions completely abolished the S2QB- signal while at least 30-50% of the S2QA signal was still detectable (Ohad et al., 1988).

The stability of the S2QB- pair was altered in cells exposed even to light intensities which did not inactivate the PSII electron flow in continuous light, as demonstrated by a significant shift in the emission temperature from about $30°$ C to $15 - 17°$ C and a reduction in the half life of about 50% at room temperature (Ohad et al., 1988). Cells exposed to light intensities which do not inactivate PSII electron flow but alter the S2QB- properties respond by an accelerated synthesis of the D1 protein. This increase in the D1 synthesis is maintained for at least 40 min even if the cells are transferred back to low light intensities indicating that the alteration induced in the reaction center II is recognised as a signal for the turnover of the D1 protein.

Recovery of the S2QB- signal properties and stability occurred in cells exposed to normal growth light intensity for at least 2 to 3 hours. Cells exposed to light intensity 2-3 fold that required to saturate photosynthesis for 20-35 min. and in which the S2QB- signal was shifted to a lower emission temperature did not exibit any significant change in the water splitting activity as demonstrated by a normal operation of the S states detected by measurments of oxygen yield per flash following a train of saturating flashes. These data were interpreted as evidence for a light induced alteration of the reaction center II during its operation in the light expressed primarily at the level of the D1 (QB) component of the D1-D2 heterodimer (Marder et al., 1987).

Analysis of the D1 protein properties as revealed by use of cross-linking reagents demonstrated that this protein is not subject to cross linking by a variety of reagents including Glutaraldehyde unless the protein is denatured by treatment with detergents (Adir and Ohad, 1986; Adir and Ohad, 1988). This property of the D1 protein is found in thylakoids of all types of higher plants, as well as in green algae and cyanobacteria.

The detergents able to induce cross-linking of the D1 protein are specific and are characterised by presence of non-ionic hydrophylic groups and alkyl chains 7-9 carbon long (Adir and Ohad, 1988).

The resistence of the D1 protein to crosslinking reagents is not restricted to the mature protein and is exhibited by the non-processed precursor as well as by the protein in its transient palmytoilated state (Adir and Ohad, 1988).

These results are interpreted as an indication that the D1 protein assumes a stable and well defined spatial conformation from the onset of its insertion into the stroma thylakoids by the the polyribosomes bound to the membrane involved in its synthesis.

This property of the D1 protein could be relevant to the process of its easy removal and replacement into the reaction center II during its light induced turnover (Herrin et al., 1981).

SYNTHESIS AND LOCALISATION OF THE HEAT SCHOCK PROTEINS IN CHLAMYDOMONAS.

We have demonstrated before that heat shock induces in Chlamydomonas the synthesis of several nuclear coded proteins including proteins of 60-90 kDa, 40-55 kDa as well as several proteins of lower molecular weight (Kloppstech et al., 1985), as also reported for higher plants (Vierling et al., 1986).

Among the low apparent molecular mass proteins, one, HSP22, is transported into the chloroplast and is associated with the grana lamellae region of the thyalakoids (Kloppstech et al., 1985; Schuster et al., 1988).

The HSP22 mRNA transcription is induced already at $38°C$, its synthesis being maximal after 1-2 hr of heat shock.

The translation product of this RNA in vitro has the same apparent molecular mass as the protein synthesised in vivo.

A similar protein has been found in pea plants chloroplasts which however is synthesised as a precursor and is processed when transported into the chloroplast (Kloppstech et al., 1985). In the pea chloroplast the HSP22 became also integrated into the chloroplast membranes in vivo as well as in vitro when the chloroplasts used for in vitro transport are obtained from heat treated plants. Transport of the HS protein into chloroplasts obtained from control plants results into its processing to the mature form but the protein remained in the soluble stroma phase of the organelle (Kloppstech et al., 1985).

A cDNA clone was obtained using poly (A)+ RNA from heat shock treated Chlamydomonas cells. The cDNA clone was used for both sequencing and the construction of an expression vector. The fusion protein so obtained was used to prepare polyclonal antibodies which reacted specifically with the HSP22 of Chlamydomonas but not with

that of pea plants (Ish-Shalom et al., ms. in preparation).
      The expression of the HSP22 in vivo could be monitored by both
radioactive tracers and immunoblotting. The HSP22 did not accumulate
in cells heat treated in the dark or low light intensity (5-10
W/m2). The accumulation of the HSP22 proceeded as a function of the
heat shok time for at least 2 hr, if the heat treatment was carried
out in the light at about 30 W/m2. However the protein turned over in
the light and its degradation was comparable to or faster than that
of the D1 protein (Even, 1986).
      Analysis of the derived amino acid sequence of the
Chlamydomonas HSP22 (Grimm et al., in preparation) shows similarities
with small molecular weight HSPs of soy beans (Nagao et al., 1985).
The sequence does not indicate that it contains extensive hydrophobic
regions in agreement with previous findings demonstrating that HSP22
in Chlamydomonas is a membrane extrinsic protein (Schuster et al.,
1988).

LOW LIGHT INDUCED LOSS OF THE PSII ACTIVITY DURING HEAT SHOCK OF
CHLAMYDOMONAS CELLS: PROTECTION BY HEAT SHOCK PROTEINS

      Since in natural conditions in the field high light intensity
is usually accompanied by elevated temperatures, it was of interest
to examine the process of photoinhibition under conditions of heat
stress.
      The results of experiments devised to test the stability of
photosynthetic electron flow under heat shock conditions in
Chlamydomonas demonstrated that the activity is not drastically
affected if the heat shock is carried out in the dark (Schuster et
al., 1988). However if the cells are exposed to the light during the
treatment a rapid inactivation of oxygen evolution ocurs. Measurments
of partial reactions of the photosynthetic electron transfer chain
indicated that the damage is localised in Photosystem II as also
indicated by measurments of fluorescence kinetics (Schuster et al.,
1988).
      Examination of the thylakoid polypeptide pattern disclosed that
the D1 protein is altered and became aggregated (polymerised) if the
heat shock treatment is carried out at light intensities which cause
no damage at growth temperature [80-200 W/m^2 (Schuster et al.,
1988)]. The aggregation process is rather specific for the D1 protein
at low light (35-50 W/m^2) and became non specific, the whole array
of the thylakoid proteins forming a high molecular weight mass at
higher light intensities [150-250 W/m^2, (Schuster et al., 1988)].
      Prevention of the heat shock protein synthesis by inhibitors of
cytoplasmic protein translation such as cycloheximide, accelerated
the dammaging process considerably while preadaptation of the cells
by exposure to 40° C in low light protected against the loss of
photosynthetic activity as monitored by fluorescence kinetics
measurments. Preadaptation under conditions preventing the synthesis
and accumulation of the heat shock proteins did not result in a
protective effect (Schuster et al., 1988).
      The results thus far reviewed strongly support the hypothesis
that HSPs might be involved in the protection of the photosynthetic
electron transfer chain from the combined harmful effects of heat and
light stress.

CONCLUSIONS

The following mechanism is suggested as a working hyopothesis to explain the observed phenomena.

The charge separation induced by photon excitation of the Photosystem II reaction center induces as yet a non- identified alteration in the properties of the D1 (QB) protein. This alteration is recognised and triggers the specific degradation of the protein which according to recent data may contain a specific amino acid sequence susceptible for proteolytic activity found in fast turning over proteins (Greenberg et al., 1987). At high photon flux densities, the rate of D1 alteration might exceed the ability of the degradation system to remove the altered protein. The replacement of the inactive D1 might require its removal and thus the rate of de novo synthesis and/or accumulation of the D1 protein might be limited by the rate of its removal.

Under these conditions, i.e. high rates of D1 turnover, the D2 protein might remain exposed to the activity of the proteases which recognise non assembled PSII rection center-core proteins and degrade them as it appears to be the case in mutants in which the synthesis of only one specific polypeptide is impaired but results in the inability to accumulate other PSII components (Jensen et al., 1986).

The spatial conformation of the D1 protein which seems to be achieved and preserved through all the stages of its synthesis transport along the thylakoid membrane and its integration in the reaction center II might facilitate the traffic into and away from the reaction center-core complex which initially may assume a stable conformation once assembled even if the D1 protein is transiently removed during the process of its light dependent turnover. Only when this transient condition lasts for longer periods of time as the case may be during photoinhibition, the rest of the reaction center proteins might became partially disassembled and/or exposed to proteolytic activity as it seems to be the case for the D2 protein. The rate of light dependent damage to PSII can be estimated as one reaction center inactivated for about 10^7 photon events and under the experimental conditions of photoinhibition as carried out in our work the half life of PSII activity is about 45 min. Under the same conditions as mentioned above the half life of the D1 and D2 proteins estimated by immunological or radioactive tracing techniques, is 90 min. and 120 min. respectively.

Thus a situation can be expected, at relatively low light intensities at which the turnover of the D2 protein can not be detected, and no detectable loss of PSII activity occurs while the D1 protein is altered and exchanged continuously. At slightly higher photon flux densities, the rate of removal of the D1 protein might became a limiting factor and inactive reaction centers impaired in the D1 (QB) function characterised by an increasse in the intrinsic fluorescence can be detected (Ohad et al., 1988). At even higher light intensities and exposure time, complete inactivation of the reaction center II might occur, affecting the QA (D2) protein as well and characterised by a reduction in the maximal fluorescence as observed (Kyle et al., 1984; Ohad et al., 1984). The heat shock condition brings an additional damage to the system, namely that of eventually impairing the system involved in the removal of the altered D1 protein. In this case irreversible aggregation (polymerisation) of the D1 and eventually other thylakoid proteins occurs, which may be at least transiently protected by the presence of heat shock induced proteins. One candidate for this protective role is obviously the HSP22 localised in the grana region where most

of the PSII centers are localised as well. Induction of all the
events ascribed to the heat shock and light conditions can be due to
the formation of free radicals in the reaction center II when exposed
to the light. Free radicals can induce polymerisation of polypeptides
and lipids (Wolf et al., 1986), and can be the major culprit in the
process of light induced alteration of the D1 protein even in non
photoinhibitory conditions. Therefore, an additional function of D1
besides those already established is that of a "suicide" protein
which has evolved together with the development of oxygenic
photosynthesis. Its removal by a specific degradation process is thus
a necessary step in order to prevent the formation of aggregates
which might cause an irreversible inactivation of the photosynthetic
apparatus. Certainly the explanation of the turnover, photoinhibition
and heat shock effects, on the D1, and the reaction center II, are
speculative, only parts of the entire chain of events being based on
observed experimental facts. However a large part of these ideas
provide a solid basis for further experimental work and it is well
understood by these authors that only additional experimental proof
can transform hypotheses into accepted facts.

REFERENCES

Adir, N. and Ohad, I. (1986) Biochim. Biophys. Acta, 850, 264-274.
Adir, N. and Ohad, I. (1988) J. Biol. Chem. 263, 283-289.
Bjorkman, O. (1987) in Progress in Photosynthesis Research (Biggins,
  J., ed.) Vol.4, pp. 1-10, Martinus Nijhoff, Dordrecht.
Even, D. (1986) Thesis for the master degree, The Hebrew University,
  Jerusalem, Israel.
Greenberg, B. M., Gaba, V., Mattoo,A. K. and Edelman, M. (1987) The
  EMBO J. 6, 2865-2869.
Herrin, D. Michaels, A. and Hickey, E. (1981) Biochim. Biophys. Acta,
  655, 136-145.
Jensen, K. H., Herrin, D. L., Plumley, F. G. and Schmidt, G. W.
  (1986) J. Cell Biol. 103, 1315-1325.
Kloppstech, K., Meyer, G., Schuster, G. and Ohad, I. (1985) The EMBO
  J. 1, 1901-1909.
Kyle, D. Ohad, I. and Arntzen, C. J. (1984) Proc. Natl. Acad. Sci.
  USA 81, 4070-4074.
Marder, J. B., Chapman, D. J., Telfer, A., Nixon, P. J. and Barber,
  J. (1987) Plant Mol. Biol., Martinus Nijhoff Publishers, Dordrecht,
  9, 325-333.
Mattoo, A. K., Hoffman-Falk, H., Marder, B. and Edelman, M. (1984)
  Proc. Natl. Acad. Sci. USA 81, 1380-1384.
Nagao, R. T. Czarnecka, E. Gurley, W. B. Schoffl, F. and Key, J. L.
  (1985) Mol. Cell Biol. 5, 3417-3428.
Ohad I, Koike, H., Shochat, S. and Inoue, Y. (1988) Biochim. Biophys.
  Acta, in press.
Powels, S. B. (1984) Ann. Rev. Plant Phys. 35, 15-44.
Schuster, G., Even, D. Kloppstech, K. and Ohad, I. (1988) The EMBO J.
  7, 1-6.
Schuster, G. and Ohad, I. (1988) J. Cell Biol., submitted.
Trebst, A. and Debka, B. (1985) Springer Ser. Chem. Phys. 42, 210-
  224.
Vierling, E., Mishkind, M., Schmidt, G. W. and Key, J. L. (1986)
  Proc. Natl. Acad. Sci. USA, 83, 361-365.
Wolff, S. P., Garner, A. and Dean, R. T. (1986) Trends Biochim. Sci.
  11, 27-31.

The molecular genetics of thylakoid proteins

Tristan A. DYER

Institute of Plant Science Research, Cambridge Laboratory, Maris Lane, Cambridge CB2 2LQ, U.K.

SYNOPSIS

Progress that has been made in isolating and characterizing the genes for the thylakoid proteins involved in photosynthesis is briefly described.

INTRODUCTION

Thylakoid proteins provide the essential three dimensional organization of the components involved in the light reactions of photosynthesis. However, they are difficult to study. One reason for this results from the hydrophobic properties of most of the thylakoid proteins which causes problems in their fractionation and characterization. Also, because there are many of them it is often impossible to resolve them adequately using available fractionation techniques. Furthermore, their respective amounts apparently differ appreciably and it is consequently a problem to detect those present in relatively small quantities.

Both an indication of the complexity of the problem and a partial resolution of it have come from studies of the coding sequences of the thylakoid proteins. Intensive research over the past decade has shown that many of the genes for the thylakoid proteins are located in chloroplast genome while the rest are in the nucleus. In fact it seems that the primary function of the chloroplast genome is to encode the primary amino acid sequences of about half of the integral thylakoid proteins and of many of the components of the transcription and translation apparatus involved in the synthesis of these thylakoid proteins (see Dyer, 1985). Recently, with the use of expression vectors such as λgt11 (e.g. Tittgen et al., 1986), rapid progress has been made in the isolation of cDNA for nuclear-encoded thylakoid proteins. From the coding sequences the amino acid sequences of the protein products may be readily deduced and from this data many of the features of the mature protein can be predicted. Much of what we know about the properties of the thylakoid proteins has been derived in this way rather than from a study of the proteins themselves.

A brief account is given here of the status of current research on the molecular genetics of the thyalkoid proteins.

```
-----------------------------------------------------------------------
```
Table 1. Photosystem I protein components.
```
-----------------------------------------------------------------------
```
Component	Mr	Mc*	Gene and Location	
```
-----------------------------------------------------------------------
```

Core proteins

Subunit IA ⎤ Reaction	66-67 k	83.2 k	psaA	(cp)
IB ⎦ centre		82.5	psaB	(cp)
II	21			(n)
III	17			(n)
IV	13			(n)
V	11			(n)
VI	9			(n)
EA (electron acceptor)	8	8.9	psaC	(cp)
protein			(frxA)+	

Antenna proteins

| LHCI | 20,22,23 | 24 | Cab | (n) |
```
-----------------------------------------------------------------------
```
* molecular weight calculated from coding sequence
+ initial designation

The photosystem I complex

     The subunit composition of the photosystem I (PSI) complex (Table
1) has not been well defined due to variation in the number and size
of these components in different PSI preparations. All, however,
contain two large reaction centre polypeptides which bind the primary
electron acceptors as well as chlorophyll a molecules. The genes for
these proteins are near to each other and probably co-transcribed in
the chloroplasts of higher plants (Fish et al., 1985; Kirsch et al.,
1986; Lehmbeck et al., 1986) and of Euglena (Cushman et al., 1988a).
The cyanobacterium Synechococcus has a similar pair of genes (Cantrell
et al., 1987) which resemble the corresponding plant genes to a very
high degree, as is found for the D1 and D2 reaction centre proteins of
PSII (see below). In Chlamydomonas one of these genes is in three
widely dispersed fragments, the transcripts of which are probably
joined together by a trans-splicing mechanism (Kück et al., 1987).
The two proteins have substantial sequence homology ($\approx45\%$) and
probably function in pairs. The gene psaC for another chloroplast-
encoded component codes for the apoprotein of the PSI iron-sulphur
centres A and B (Hayashida et al., 1987; Oh-oka et al., 1987; Høj et
al., 1987).
     The other PSI core proteins, for which there is now some protein
sequence data, are apparently all nuclear-encoded (Dunn et al., 1988).
     There are probably at least three types of proteins in the
peripheral chlorophyll a/b-binding (high harvesting) complex of PSI
(see Williams & Ellis, 1986; Høyer-Hansen et al., 1988). These
proteins are quite distinct from those of PSII. Furthermore, whereas
the cDNAs for the LHCII apoproteins were relatively easy to isolate,
those for LHCI have been difficult to obtain. However, there are
claims now that this has been achieved (Pichersky et al., 1987;
Stayton et al., 1987) but these findings will need to be
substantiated.

Table 2. Photosystem II (PSII) protein components.

Component	Mr	Mc	Gene and Location	
Core proteins				
D1] Reaction	32 k	37.2 k	psbA	(cp)
D2] centre	32	39.4	psbD	(cp)
Cytochrome b_{559} (α & β)	10	9.4,4.4	psbE,F	(cp)
Antenna proteins				
Chlorophyll a-binding	47	56.2	psbB	(cp)
antenna proteins	44	51.8	psbC	(cp)
LHCII Type 1] Chl a/b-	25-28	25.2	Cab	(n)
LHCII Type 2] binding			Cab	(n)
CP29] protein	29			
Oxygen-evolving complex				
OE1	33	26.5	woxA	(n)
OE2	23	20.2	woxB	(n)
OE3	16	16.5	woxC	(n)
Others				
Phosphoprotein	10	7.8	psbH	(cp)
Intrinsic proteins	24			(n)
	22			(n)
	10			(n)

Phytosystem II

Photosystem II (PSII) appears to have by far the most protein
components of the thylakoid complexes (Table 2). The hydrophobic
components of the reaction centre D1 (see Erickson et al., 1985), D2
(Rasmussen et al., 1984; Alt et al., 1984) and the α and β components
of cytochrome b_{559} are all coded for by the chloroplast genome as are
the two chlorophyll a-binding antenna proteins. The three extrinsic
proteins (M_r 33, 23 and 16 k) which are components of the oxygen
evolving apparatus, are all nuclear encoded (Tyagi et al., 1987;
Jansen et al., 1987) and must traverse both the nuclear envelope and
the thylakoid membrane to reach their site of action in the thylakoid
lumen. These three proteins are, in turn, in close structural
association with three other proteins (M_r 24, 22 and 10 k) which are
also nuclear-encoded but intrinsic membrane components (Ljungberg et
al., 1986). A chloroplast-encoded 10 kDa phosphoprotein possibly
forms a link between the core and the chlorophyll a/b-binding antenna
proteins. In most plants the latter are of two main types (see
Pichersky et al., 1987) and these are coded for by a large number of
nuclear genes (about 16 in petunia for example; Dunsmuir et al.,
1983). Many of the genes for these proteins have been characterized.
There is also a small amount of another chlorophyll a/b-binding
antenna protein designated CP29 (see Green, 1988). In addition to all

these proteins there are also at least five small proteins of 5 kDa or
less which are coded for by nuclear and chloroplast genes.
      The genes for the chloroplast-encoded components are scattered in
the chloroplast genome. While the gene for the D1 protein (psbA) is
possibly co-transcribed with a tRNA gene (Stern & Gruissem, 1987),
psbB the gene for the M_r 47 k chlorophyll a-binding protein (Morris
& Herrmann, 1984) and psbH (Hird et al., 1986a; Westhoff et al.,
1986), the gene for the 10 kDa phosphoprotein are co-transcribed with
components of the cytochrome b_6/f complex (Westhoff & Herrmann, 1988;
Kohchi et al., 1988). The genes psbC and psbD for the M_r 43 k
antenna protein and for the D2 core protein respectively are
co-transcribed and overlap one another (Alt et al., 1984; Bookjans et
al., 1986) even in Synechocystis (Chisholm & Williams, 1988). The
genes for the α and β components of the cytochrome b_{559} (psbE, psbF)
are very close to one another and also co-transcribed (Herrmann et
al., 1984; Hird et al., 1986b; Cushman et al., 1988b; Carillo et al.,
1986). Small open reading frames are associated and co-transcribed
with each of these transcription complexes. Probably some of these
code for small proteins of the type described above as, more often
than not, it is genes of the same complex that are co-transcribed.

Cytochrome b_6/f complex

      There is still some uncertainty as to the number of protein
subunits in the chloroplast cytochrome b_6/f complex. Well
authenticated components include the apoproteins of cytochrome b_6 and
f, the Rieske protein and "subunit IV." Apart from the Rieske
protein, all are chloroplast encoded (Table 3). The recent work of
Lemaire et al. (1986) with Chlamydomonas mutants suggests that in
addition there is a component of M_r 19.5 k which is probably nuclear
encoded while L. Bogorad (unpublished results) has suggested that
there is a small chloroplast-encoded peptide of 37 amino acids in this
complex as well. The b_6 apoprotein and subunit IV genes (petB and
petD) both have introns (Tanaka et al., 1987; Rock et al., 1987) and
are co-transcribed along with some PSII genes (see above). However,
the translation of the PSII and cytochrome genes is regulated
differently, the cytochrome complex components being made in the dark
and light while the PSII proteins are only made on illumination
(Westhoff and Herrmann, 1988). The derived amino acid sequences of
the cytochrome b_6 and subunit IV genes show homologies to different

Table 3. Cytochrome b_6/f protein components.

Component	Mr	Mc	Gene and Location	
Cytochrome f apoprotein	31.3 (41) k	31.9 (35.3)* k	petA	(cp)
Cytochrome b_{563}	23	23.7	petB	(cp)
Rieske protein	18–20	18.8 (26.8)*		(n)
Subunit IV	15.2–17.5	15.2	petD	(cp)
Subunit V	19.5			(n)

* precursor sizes

Table 4. ATP synthase components.

Component			Mr	Mc	Gene and Location	
F_1	α	catalytic	58 k	55.4 k	atpA	(cp)
	β		57	53.9	atpB	(cp)
	γ	S-S regulation	38	–	atpC	(n)
	δ	Fo link	25	–	atpD	(n)
	ε	regulation	14	14.7	atpE	(cp)
Fo	a (IV)	proton pore	19	20 (27.1)*	atpI	(cp)
	I	F_1 anchor	18	19 (20.9)*	atpF	(cp)
	II		16	–	atpG	(n)
	III	proton-translocating	8	8	atpH	(cp)

* precursor sizes
F_1 stoichiometry $\alpha_3\ \beta_3\ \gamma_1\ \delta_1\ \varepsilon_1$
Fo stoichiometry $a_1\ I_1\ II_1\ III_{10}$

domains of the cytochrome b of mitochondria (Widger et al., 1984).
Highly conserved open reading frames which may be co-transcribed with
the cytochrome f gene (Willey, 1984b) have also been observed but it
is not known yet whether these are translated. The chloroplast pet
genes for a number of different plants have been sequenced (petA:
Willey et al., 1984a,b; Alt & Herrmann, 1984: petB,D: Heinemeyer et
al., 1984; Fukuzawa et al., 1987) as has the cDNA of the spinach
Rieske protein (Steppuhn et al., 1987).

ATP synthase

The thylakoid ATP synthase catalyses ATP synthesis/hydrolysis coupled
with transmembrane proton transport. It has a hydrophilic part (CF_1)
which contains the nucleotide-binding sites and a hydrophobic part
(CF_O) which is imbedded in the membrane providing a proton channel
and anchor for CF_1 on the membrane. CF_1 has five subunits (α, β, γ, δ
and ε) and while the α, β and ε components are chloroplast encoded,
the γ and δ components are nuclear encoded (Table 4). The CF_O part
seems to have four subunits: a, I, II and III with all except II being
coded for by chloroplast genes. Subunit a is synthesized as a
component of 247 amino acids (Cozens et al., 1986) but apparently
loses the 18 amino-terminal amino acids during maturation (Fromme et
al., 1987) while subunit I loses 17 amino terminal residues (Bird et
al., 1985). All the nuclear-encoded components have transit peptides
which are removed once they have entered the chloroplast. The
chloroplast genes of several different types of plant (e.g. Hennig &
Herrmann, 1986) and cDNAs of the γ, δ and subunit II nuclear-encoded
components of spinach have been cloned and sequenced (see Hermans et
al., 1988).
    The chloroplast ATP synthase genes (atpA, B, E, F, H & I) of
vascular plants are in two operons containing 4 and 2 genes
respectively. One group has the genes atpI, H, F & A in a single
operon (see Hennig & Herrmann, 1986; Hudson et al., 1987), while atpB
and E are in the other operon. This arrangement is very similar to
that found in the cyanobacterium Synechococcus and this similarity
extends to the sequences themselves too (Cozens & Walker, 1987). The

**Table 5. Genes for putative protein components of a chloroplast
NAD(P)H quinone oxidoreductase.**

Chloroplasts		Equivalent mitochondrial component	
Marchantia	Tobacco	Gene	Mc
ndh1	ndhA	ND1	24 k
2	B	2	25
3	C	3	6
4	D	4	36-39
4L	E	4L	3.5
5	F	5	51
6	G*	6	18

* ORF 138 + ORF 99B - see Ohyama et al., 1988.

atpF gene in vascular plants, however, has an intron (Bird et al.,
1985) and in most plants aptB and atpE overlap one another and are
probably also co-transcribed (Zurawski et al., 1982; Shinozaki et al.,
1983; Howe et al., 1985; Zurawski et al., 1986; Kobayashi et al.,
1987). In Chamydomonas there have been substantial rearrangements of
the ATP synthase genes (Woessner et al., 1987).
      It would seem that all the genes for the chloroplast ATP synthase
protein components have now been identified.

ndh genes

One of the great surprises to come out of the complete sequencing of
the chloroplast genome (Shinozaki et al., 1986; Ohyama et al., 1986)
has been the discovery that it contains open reading frames (ORFs)
coresponding to the seven ndh genes found also in animal mitochondria
and which code in the latter for components of NADH ubiquinone
oxidoreductase (NADH dehydrogenase - see Table 5). The chloroplast
genes are transcribed (Matsubayashi et al., 1987) and, one must
therefore assume, have protein products. This, in turn, implies that
chloroplasts have a NAD(P)H quinone reductase which may be involved in
respiratory processes. Several recent papers suggest that there is
such an activity (chlororespiration) in chloroplasts (Bennoun, 1982;
Peltier et al., 1987; Peltier & Sarrey, 1988). However, the complex
in mitochondria contains about 25 components (Chomyn et al., 1986) and
it is difficult to see how a complex of that size would have gone
undetected unless present only in very small amounts. Preliminary
evidence suggests that a gene previously designated as coding for a
PSII component (psbG - Steinmetz et al., 1986) may in fact be an
additional ndh gene (P.J. Nixon, K. Gounaris, S.A. Coomber, C.N.
Hunter, J. Barber, unpublished results). This is consistent with it
being co-transcribed and overlapping ndhC.

Genes of other thylakoid proteins

Thylakoids have attached to them via an anchor protein a proportion of
the ferredoxin-NADP+ oxidoreductase (FNR) present in chloroplasts.
The FNR is involved in cyclic electron flow and may be associated with
either the b_6/f or PSI complex (Clark et al., 1984). It is nuclear
encoded and its cDNA has recently been characterized (Newman & Gray,

1988). The copper protein plastocyanin, which is an electron carrier between the cytochrome b_6/f complex and PSI, is encoded by a single nuclear gene (Rother et al., 1986) and its cDNA has also been characterized (Smeekens et al., 1985; Nielsen & Gaussing, 1987).

The complete sequencing of the chloroplast genomes of tobacco and Marchantia has revealed the presence of numerous open reading frames for putative membrane proteins. One such sequence frxB probably codes for a 4Fe-4S) ferredoxin-like protein. As many of the other putative coding sequences have yet to be designated, we may yet have to revise substantially our concepts of the structure of thylakoid protein complexes as new components are discovered and characterized.

Nuclear genes for thylakoid proteins are not so accessible and many may still be unknown. Thylakoid proteins have been detected such as cytochrome b-560 (Rolfe et al., 1987) and protein kinases (Coughlan & Hind, 1986) for which coding sequences (probably nuclear) have yet to be determined. Also certain nuclear gene may act in a regulatory capacity, directly affecting the translation of even chloroplast-encoded components (Mayfield et al., 1987; Kuchka et al., 1988). Thus in the future we may expect to discover that there is a complex interplay between the nuclear and organelle genomes during the biosynthesis of the photosynthetic apparatus.

REFERENCES

Alt, J. & Herrmann, R. G. (1984) Curr. Genet. **8**, 551-557

Alt, J., Morris, J., Westhoff, P. & Herrmann, R. G. (1984) Curr. Genet. **8**, 597-606

Bennoun, P. (1982) Proc. Natl. Acad. Sci. USA **79**, 4352-4356

Bird, C. R., Koller, B., Auffret, A. D., Huttly, A. K., Howe, C. J., Dyer, T. A. & Gray, J. C. (1985) EMBO J. **4**, 1381-1388

Bookjans, G., Stummann, B. M., Rasmussen, O. F. & Henningsen, K. W. (1986) Plant Mol. Biol. **6**, 359-366

Cantrell, A. & Bryant, D. A. (1987) Plant Mol. Biol. **9**, 453-468

Carillo, N., Seyer, P., Tyagi, A. & Herrmann, R. G. (1986) Curr. Genet. **10**, 619-624

Chisholm, D. & Williams, J. G. K. (1988) Plant Mol. Biol. **10**, 293-301

Chomyn, A., Cleeter, M. W. J., Ragan, C. I., Riley, M., Doolittle, R. F. & Attardi, G. (1986) Science **234**, 614-618

Clark, R. D., Hawkesford, M. J., Coughlan, S. J., Bennett, J. & Hind, G. (1984) FEBS Lett. **174**, 137-142

Coughlan, S. J. & Hind, G. (1986) J. Biol. Chem. **268**, 14062-14068

Cozens, A. L. & Walker, J. E. (1987) J. Mol. Biol. **194**, 359-383

Cozens, A. L., Walker, J. E., Philips, A. L., Huttly, A. K. & Gray, J. C. (1986) EMBO J. **5**, 217-222

Cushman, J. C., Hallick, R. B. & Price, C. A. (1988a) Curr. Genet. **13**, 159-171

Cushman, J. C., Christopher, D. A., Little, M. C., Hallick, R. B. & Price, C. A. (1988b) Curr. Genet. **13**, 173-180

Dunn, P. P. J., Packman, I. C., Pappin, D. & Gray, J. C. (1988) FEBS Lett. **228**, 157-161

Dunsmuir, P., Smith, S. M. & Bedbrook, J. (1983) J. Mol. Appl. Genet. **2**, 285-300

Dyer, T. A. (1985) Oxford Surveys of Plant Mol. & Cell. Biol. 2, 147-177

Erickson, J. M., Delepelaire, P. & Rochaix, J.-D. (1985) in Molecular Biology of the Photosynthetic Apparatus (Steinback, K., Bonitz, S., Arntzen, C. J. & Bogorad, L., eds.), pp. 53-66, Cold Spring Harbor

Fish, L. E., Kück, U. & Bogorad, L. (1985) J. Biol. Chem. **260,**
     1413-1421
Fromme, P., Gräber, P. & Salnikow, J. (1987) FEBS Lett. **218,** 27-30
Fukuzawa, H., Yoshida, T., Kohchi, T., Okumura, T., Sawano, Y. &
     Ohyama, K. (1987) FEBS Lett. **220,** 61-66
Green, B. R. (1988) Photosyn. Res. **15,** 3-32
Hayashida, N., Matsubayashi, T., Shinozaki, M., Inoue, K. & Hiyama,
     T. (1987) Curr. Genet. **12,** 247-250
Heinemeyer, W., Alt, J. & Herrmann, R. G. (1984) Curr. Genet. **8,**
     543-549
Hennig, J. & Herrmann, R. G. (1986) Mol. Gen. Genet. **203,** 117-128
Hermans, J., Rother, Ch., Bichler, J., Steppuhn, J. & Herrmann, R. G.
     (1988) Plant Mol. Biol. **10,** 323-330
Herrmann, R. G., Alt, J., Schiller, B., Widger, W. R. & Cramer, W. A.
     (1984) FEBS Lett. **176,** 239-244
Hird, S. M., Dyer, T. A. & Gray, J. C. (1986a) FEBS Lett. **209,**
     181-186
Hird, S. M., Willey, D. L., Dyer, T. A. & Gray, J. C. (1986b) Mol.
     Gen. Genet. **203,** 95-100
Høj, B. B., Svendsen, I., Scheller, H. V. & Moller, B. L. (1987) J.
     Biol. Chem. **262,** 12676-12684
Howe, C. J., Fearnley, I. M., Walker, J. E., Dyer, T. A. & Gray, J.
     C. (1985) Plant Mol. Biol. **4,** 333-345
Høyer-Hansen, G., Bassi, R., Honberg, L. S. & Simpson, D. J. (1988)
     Planta **173,** 12-21
Hudson, G. S., Mason, J. G., Holton, T. A., Koller, B., Cox, G. B.,
     Whitfeld, P. R. & Bottomley, W. (1987) J. Mol. Biol. **196,**
     283-298
Jansen, T., Rother, C., Steppuhn, J., Reinke, H., Beyreuther, K.,
     Jansson, C., Andersson, B. & Herrmann, R. G. (1987) FEBS Lett.
     216, 234-240
Kirsch, W., Seyer, P. & Herrmann, R. G. (1986) Curr. Genet. **10,**
     843-855
Kobayashi, K., Nakamura, K. & Asahi, T. (1987) Nucl. Acids Res. **15,**
     7177
Kohchi, T., Yoshida, T., Kamano, T. & Ohyama, K. (1988) EMBO J. **7,**
     885-891
Kuchka, M. R., Mayfield, S. P. & Rochaix, J.-D. (1988) EMBO J. **7,**
     319-324
Kück, U., Choquet, Y., Schneider, M., Dron, M. & Bennoun, P. (1987)
     EMBO J. **6,** 2185-2195
Lehmbeck, J., Rasmussen, O.F., Bookjans, G. B., Jepsen, B. R.,
     Stummann, B. M. & Henningsen, K. W. (1986) Plant Mol. Biol. **7,**
     3-10
Lemaire, C., Girard-Bascou, J., Wollman, F.-A. & Bennoun, P. (1986)
     Biochim. Biophys. Acta **851,** 229-238
Ljungberg, U., Åkerlund, H.-E. & Andersson, B. (1986) Eur. J.
     Biochem. **158,** 477-482
Matsubayashi, T., Wakasugi, T., Shinozaki, K., Yamaguchi-Shinozaki,
     K., Zaita, N., Hidaka, T., Meng, B. Y., Ohto, C., Tanaka, M.,
     Kato, A., Miruyama, T. & Sugiura, M. (1987) Mol. Gen. Genet.
     210, 385-393
Mayfield, S. P., Rahire, M., Frank, G., Zuber, H. & Rochaix, J.-D.
     (1987) Proc. Natl. Acad. Sci. USA **84,** 749-753
Morris, J. & Herrmann, R. G. (1984) Nucl. Acids Res. **12,** 2837-2850
Nielsen, P. S. & Gausing, K. (1987) FEBS Lett. **225,** 159-162
Ohyama, K., Fukuzawa, H., Kohchi, T., Shirai, H., Sano, T., Sano, S.,
     Umesono, K., Shiki, Y., Takeuchi, M., Chang, Z., Aota, S.,
     Inokuchi, H. & Ozeki, H. (1986) Nature **322,** 572-574
Ohyama, K., Kohchi, T., Sano, T. & Yamada, Y. (1988) TIBS **13,** 19-22

Oh-oka, H., Takahashi, Y., Wada, K., Matsubara, H., Ohyama, K. &
    Ozeki, H. (1987) FEBS Lett. **218,** 52-54
Peltier, G. & Sarrey, F. (1988) FEBS Lett. **228,** 259-262
Peltier, G., Ravenel, J. & Vermeglio, A. (1987) Biochim. Biophys.
    Acta **893,** 83-90
Pichersky, E., Hoffman, N. E., Bernatzky, R., Piechulla, B., Tanksley,
    S. D. & Cashmore, A. R. (1987) Plant Mol. Biol. **9,** 205-216
Rasmussen, O. F., Bookjans, G., Stummann, B. M. & Henningsen, K. W.
    (1984) Plant Mol. Biol. **3,** 191-199
Rock, C. D., Barkan, A. & Taylor, W. C. (1987) Curr. Genet. **12,** 69-77
Rolfe, S. A., Sanguansermsri, M. & Bendall, D. S. (1987) Biochim.
    Biophys. Acta **894,** 434-442
Rother, C., Jansen, T., Tyagi, A., Tittgen, J. & Herrmann, R. G.
    (1986) Curr. Genet. **11,** 171-176
Shinozaki, K., Deno, H., Kato, A. & Sugiura, M. (1983) Gene **24,**
    147-155
Shinozaki, K., Ohme, M., Tanaka, M., Wakasugi, T., Hayashida, N.,
    Matsubayashi, T., Zaita, N., Chunwongse, J., Obokata, J.,
    Yamaguchi-Shinozaki, K., Ohto, C., Torazawa, K., Meng, B. Y.,
    Sugita, M., Deno, H., Kamogashira, T., Yamada, K., Kusuda, J.,
    Takaiwa, F., Kato, A., Tohdoh, N., Shimada, H. & Sugiura, M.
    EMBO J. **5,** 2043-2049
Smeekens, S., de Groot, M., van Binsbergen, J. & Weisbeek, P. (1985)
    Nature **317,** 456-458
Stayton, M. M., Brosio, P. & Dunsmuir, P. (1987) Plant Mol. Biol.
    10, 127-137
Steinmetz, A. A., Castroviejo, M., Sayre, R. T. & Bogorad, L. (1986)
    J. Biol. Chem. **261,** 2485-2488
Steppuhn, J., Rother, C., Hermans, J., Jansen, T., Salnikow, J.,
    Hauska, G. & Herrmann, R. G. (1987) Mol. Gen. Genet. **210,**
    171-177
Stern, D. B. & Gruissem, W. (1987) Cell **81,** 1145-1157
Tanaka, M., Obokata, J., Chunwongse, J., Shinozaki, K. & Sugiura, M.
    (1987) Mol. Gen. Genet. **209,** 427-431
Tittgen, J., Hermans, J., Steppuhn, J. Jansen, T., Jansson, C.,
    Andersson, B., Nechushtai, R., Nelson, N. & Herrmann, R. G.
    (1986) Mol. Gen. Genet. **204,** 258-265
Tyagi, A., Hermans, J., Steppuhn, J., Jansson, Ch., Vater, F. &
    Herrmann, R. G. (1987) Mol. Gen. Genet. **207,** 288-293
Westhoff, P. & Herrmann, R. G. (1988) Eur. J. Biochem. **171,** 551-564
Westhoff, P., Farchaus, J. W. & Herrmann, R. G. (1986) Curr. Genet.
    11, 165-169
Widger, W. R., Cramer, W. A., Herrmann, R. G. & Trebst, A. (1984)
    Proc. Natl. Acad. Sci. USA **81,** 674-678
Willey, D. L., Auffret, A. D. & Gray, J. C. (1984a) Cell **36,** 555-562
Willey, D. L., Howe, C. J., Auffret, A. D., Bowman, C. M., Dyer, T.
    A. & Gray, J. C. (1984b) Mol. Gen. Genet. **194,** 416-422
Williams, R. S. & Ellis, R. J. (1986) FEBS Lett. **203,** 295-300
Woessner, J. P., Gillham, N. W. & Boynton, J. E. (1987) Plant Mol.
    Biol. **8,** 151-158
Zurawski, G., Bottomley, W. & Whitfeld, P. R. (1982) Proc. Natl.
    Acad. Sci. USA **79,** 6260-6264
Zurawski, G., Bottomley, W. & Whitfeld, P. R. (1986) Nucl. Acids
    Res. **14,** 3974

Properties of the photosystem II quinone binding region

Michael C.W. EVANS, Jonathan H.A. NUGENT, Julia A.M. HUBBARD, Christalla DEMETRIOU, Christopher J. LOCKETT and Andrew R. CORRIE

Department of Biology, University College London, Darwin Building, Gower Street, London WC1E 6BT, U.K.

SYNOPSIS

We have investigated the quinone-binding region of PSII using exchange reactions to replace key components. The changes were monitored using the esr signals from the Q_A iron-semiquinone and Fe^{3+} non-haem iron components. We have also investigated by optical techniques,the redox dependence of the us and ms reduction of P680 following excitation with a laser flash.

INTRODUCTION

In the PSII reaction centre complex , an electron is transferred across the membrane from the reaction centre chlorophyll,P680,to a pheophytin and then to two quinones, Q_A and Q_B. Q_A acts as a single-electron carrier operating between quinone and semiquinone forms whilst Q_B operates as a two-electron carrier. The quinol form of Q_B is used to transfer the electrons out of the reaction centre to the cytochrome b/f complex.

The reaction centre of PSII in higher plants is located on a membrane-protein complex containing two 32kDa chloroplast-encoded polypeptides termed D_1 and D_2 (Nanba & Satoh,1987). These polypeptides have considerable amino acid sequence homology to the L and M polypeptides of the reaction centre of purple photosynthetic bacteria (Michel & Deisenhofer,1988; Trebst, 1986,1987). The Q_A and Q_B binding sites are either side of the non-haem iron atom which is proposed to ligate to both D_1 and D_2 in an identical manner to the non-haem iron in the purple bacterial reaction centre.

ABBREVIATIONS: Chl, Chlorophyll; PSII, Photosystem II; mT, millitesla; Mes, 4-morpholineethanesulfonic acid; Hepes, 4-(2-hydroxyethyl) 1-piperazineethanesulfonic acid; DCMU, 3-(3,4-dichlorophenyl)-1,1-dimethylurea.

We would like to thank the U.K. Science and Engineering Research Council for financial support.The work on the D_1D_2 preparation was done in collaboration with Professor Jim Barber and Dr Alison Telfer at Imperial College.

The non-haem iron may be involved in two characteristics of PSII; **A**. A requirement for bicarbonate to achieve maximum rates of electron transport. In the absence of bicarbonate the oxidation of Q_A^- is slower.The protonation of Q_B^- and exchange of quinone/quinol with the membrane pool may also be slowed (Govindjee & van Rensen,1978; Stemler & Govindjee,1973; van Rensen et al., 1988; Vermaas & van Rensen,1981). **B**. Oxidation of the non-haem iron Fe^{2+} to Fe^{3+}, which is then able to act as an additional electron acceptor (Nugent & Evans,1980; Petrouleas & Diner,1986; Wraight,1985).

Electron spin resonance (esr) provides a useful probe of the quinone-binding region of PSII as signals are observed from; the pheophytin intermediate ,I^- (Klimov et al.,1980); Q_A^-Fe (Nugent et al.,1981,1982; Rutherford & Zimmerman,1984) and from the oxidised non-haem iron,Fe^{3+}(Nugent & Evans,1980; Petrouleas & Diner,1986; Zimmerman & Rutherford,1986).Two forms of iron-semiquinone ,Q_A^-Fe, spectra are observed,a "g=1.8" form (Nugent et al.,1981) seen at low pH and in the presence of formate (Vermaas & Rutherford,1984) and a "g=1.9" form seen at higher pH (Rutherford & Zimmerman,1984). The g=6 signal from the Fe^{3+} has a pH dependency which indicates that the pH conditions required for iron oxidation are similar to those required for the "g=1.9" form of iron-semiquinone signal (Diner & Petrouleas, 1987; Petrouleas & Diner,1987).

Although a simple model of the electron flow in the acceptor complex would require only the pheophytin and two quinones, Q_A and Q_B, experiments on the redox properties of the complex suggests that more components are involved. Titrations of fluorescence yield by a number of different techniques over many years have consistently suggested that quenchers (ie electron acceptors) are present with potentials around 0mV [Qh] and −300mV [Ql] . We have shown , that these steps can be related to two waves in the reduction of the iron-quinone using esr to detect the g=1.8 form of the iron-quinone, the split pheophytin radical and the spin polarised reaction centre triplet (Evans & Ford,1986; Evans et al.,1985). However, these experiments also suggested that an additional component was present with a potential of −420mV. Triplet formation was observed only after reduction of this component. The redox properties of the electron acceptors can also be investigated by observing the potential dependence of P680 oxidation and rereduction following flash excitation. We have now investigated the redox dependence of the μs and ms rereduction of P680 following excitation with a short (800ps) laser flash.In this paper we also describe esr experiments which support a simple model for the quinone-binding region linking many of the unusual properties of PSII

MATERIALS AND METHODS

Photosystem II was prepared by the method of Ford & Evans (1983) from market spinach (<u>Spinacea oleracea</u>) or green-house grown pea (<u>Pisum sativum</u> var Feltham First).

It was resuspended and stored at 77K in 20mM MES,5mM MgCl$_2$,15mM NaCl and 20% (v/v) glycerol pH 6.3 (Buffer A). Preparations with oxygen evolution rates of 400-1000 umoles O$_2$ / mg Chl/ hr were used.

Exchangeable protons were replaced by deuterium as given in Nugent (1987). These samples are referred to as 'deuterated' in the text. Four types of sample from the same initial PSII preparation were made 1. HPSII in 50mM Hepes,5mM MgCl$_2$,15mM NaCl and 20% [v/v] glycerol pH 7.5 (Buffer B) 2. DPSII in ^{2}H$_2$O Buffer B 3. HPSII in Buffer A and 4. DPSII in ^{2}H$_2$O Buffer A were made. Each preparation was illuminated for 10s then centrifuged and resuspended again in their final buffer. The standard formula of pD = pH + 0.4 was used to adjust ^{2}H$_2$O buffers.

Duplicate samples were prepared as follows: The HPSII or DPSII preparations were placed in esr tubes and dark adapted for 4 hours. Samples were then frozen to 77K in the dark. Dark adapted samples were illuminated at 77K using a 1000W light source with samples irradiated in liquid nitrogen in a silvered dewar for 5 min.

Esr spectrometry was performed at cryogenic temperatures using a Jeol X-band spectrometer with 100kHz field modulation and an Oxford instruments liquid helium cryostat. 0.3 ml samples in 3mm diameter calibrated quartz tubes were used. Chlorophyll concentrations of samples and esr conditions are described in the text.Note that the spectra shown are broad and the stated g-value of a spectral feature indicates the point at which it either reaches a peak or crosses the baseline. Spectra were stored and data manipulations performed using DEC PDP 11 and Tektronix microcomputers.

Optical measurements of P680 at 820nm and 680nm were made using a laser flash spectrophotometer as described in Mansfield et al,.1987 except the measuring device was a large area photodiode. Redox mediators [5-20μM] were used in redox titrations as in Evans,1987.

RESULTS

Iron-semiquinone esr signals

We examined the effects of deuterium for hydrogen exchange on the esr characteristics of Q$_A$$^-$, as it had previously been shown that pH affects the iron-semiquinone Q$_A$$^-$ signal (Rutherford & Zimmerman,1984).

The PSII iron-semiquinone spectra at two pH values are shown in Fig.1. Comparing the Q$_A$$^-$ spectra at pH 7.5 (fig 1A and B) showed that both the deuterated and protonated samples had the broad "g=1.9" form of the iron-semiquinone, Q$_A$$^-$, following illumination at 77K. Close examination of several spectra suggested that a shift to a slightly higher g-value occurred in deuterated samples.An inflexion near g=1.6 was also present which we have recently attributed to bound Q$_B$ interacting with Q$_A$ (Nugent et al, unpublished results). The spectra of the 4h dark adapted samples were examined at higher gain, (fig.2A and B). This showed dark stable iron-semiquinone to be present in each sample although there was considerably less dark stable iron-semiquinone in the

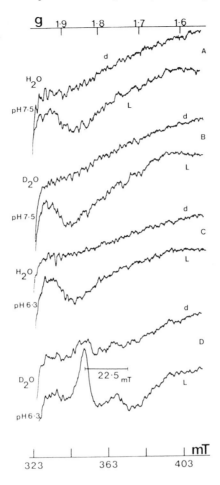

Figure 1. Esr spectra of the Q_A^-Fe iron-semiquinone electron acceptor in H_2O and D_2O buffers. [A] HPSII pH 7.5 [B] DPSII pH 7.5 [C] HPSII pH 6.3. [D] DPSII pH 6.3 . The upper spectrum of each pair is from a 4h dark adapted sample (d) and the lower spectrum is from the sample following 5 min illumination at 77K (1). Chlorophyll concentration 5mg/ml. E.s.r. conditions: modulation 1mT; microwave power 10mW; temperature 6K.

---

deuterated sample. There was no g=1.6 inflexion in the spectra of either dark adapted sample.

Comparing the pH 6.3 spectra (fig 1C and D) showed that the protonated sample retained the "g=1.9" form but the deuterated sample had the "g=1.8" form. The 22.5 mT separation indicated between the major features (g=1.84 peak and g=1.73 trough) in the deuterated sample is 50% less than that reported (Rutherford et al.,1984; Vermaas & Rutherford,1984) for "g=1.8" protonated samples at low pH [39-45 mT]. However it is only slightly narrower than that reported in samples where bicarbonate was displaced

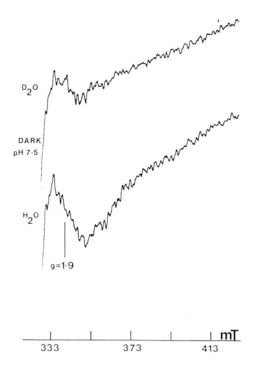

Figure 2. Esr spectra of [A] 4h dark adapted DPSII.pH 7.5 and [B] 4h dark adapted HPSII ,pH 7.5. Spectra shown are the average of 4 recorded spectra .two from each duplicate sample. Conditions as Fig.1.

by formate (Vermaas & Rutherford,1984) or in samples treated with herbicides which displace Q_B (Rutherford et al,1984). In the deuterated sample there was a dark stable signal which was broader and a slightly lower g-value than the signal observed following illumination at 77K.   This may be a Q_B^-Fe spectrum as Q_A^- would not be expected to be dark stable under these conditions.
        We suggest that the increased width of the Q_A^- signal observed by Rutherford et al (Rutherford et al.,1984)   at low pH and with herbicides such as DCMU, resulted   at   least partly from the mixture of the "g=1.9" and "g=1.8" spectral forms in their samples.

Non-haem Fe^{3+} iron g=6 esr signals

        The oxidation of the non-haem iron in PSII has been associated with a shift in the pK_a of a group from >8 to about 5 (Wraight,1985).This indicated a deprotonation linked to iron oxidation.   Recently a link between the pH conditions needed to observe the "g=1.9" form of Q_A^- and the g=6 esr signal of oxidised non-haem iron has been made (Diner & Petrouleas,1987; Petrouleas & Diner). Fig 3 shows the g=6 region in deuterated PSII samples at two pH values.No significant differences were observed between

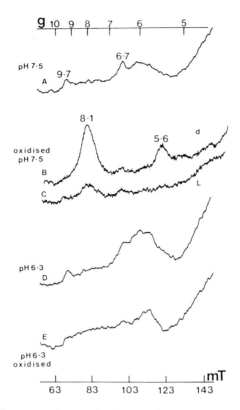

Figure 3. Esr spectra of the g=6 non-haem iron region of PSII.
[A] DPSII pH 7.5, 1h dark adapted. [B] DPSII pH 7.5 plus 5mM potassium ferricyanide,50 min dark adapted   [C] sample as b after 5 min illumination at 77K [D] DPSII pH 6.3, 1h dark adapted   [E] DPSII pH 6.3 plus 5mM ferricyanide, 50 min dark adapted. Chl concentration as Fig.1.   The g values of the major peaks discussed in the text are labelled.   Esr conditions: modulation width 1mT; temperature 4K; Microwave power A,D-E   5mW ,B-C 400uW .

----

protonated and deuterated samples. The peak near g=9.7 from 'rhombic' iron was present in all samples. The peak at g=6.7 was equivalent to that identified by Diner & Petrouleas,1987 as indicating centres with a bound quinone.
     The figure shows that the non-haem iron could be oxidised at pH 7.5 (fig.3B) but not at pH 6.3 (fig. 3E). The ability of the oxidised non-haem iron to act as an electron acceptor is shown by the removal of the signals near g=8 and g=5.6 by 77K illumination (Fig.3 B-C). We have confirmed the work of Petrouleas & Diner,1987 that at low pH , the appearance of the "g=1.8" form of Q_A^- (i.e.loss of bound bicarbonate) correlates wth the loss of the ability to generate the g=6 Fe^{3+} signal (compare similar samples in figs.1 and 3). However we observed that the ability to oxidise the non-haem iron developed

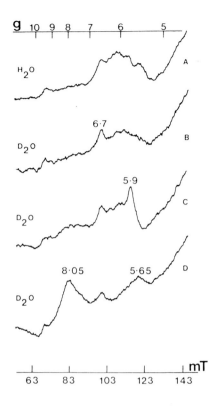

Figure 4. Esr spectra showing the effect of formate on the formation of Fe^{3+}
[A]HPSII pH 7.5 plus 100mM sodium formate,1h dark adapted. [B] DPSII pH 7.5 plus 100mM formate,1h dark adapted. [C] DPSII plus 100mM formate,45 min dark adapted then 5mM ferricyanide added and dark adapted for 45 min at 4°C. [D] DPSII plus 5mM ferricyanide,45 min dark adapted then 100mM formate added and dark adapted for 45 min. Conditions as fig.3 except Microwave power 5mW .

at a pH above that required to observe the "g=1.9" form of the iron-semiquinone in a duplicate non-oxidised sample.We did not investigate the iron-semiquinone region in oxidised samples due to the effects of ferricyanide on the baseline. Failure to oxidise the non-haem iron near the pK_a of the bound bicarbonate may result from loss of bicarbonate caused by a shift of the pK_a in the oxidised sample and/or by the prevention of deprotonation.

To investigate the effect of bicarbonate binding we have used samples at pH 7.5 and attempted to oxidise the non-haem iron before and after formate treatment. The formate treatment at this pH caused a change from the "1.9 form" to the "1.8 form" of Q_A^- in non-oxidised samples (not shown).Formate treatment at pH 7.5 does not change the esr spectrum in the g=6 region in untreated samples (fig. 3A and 4B). However the oxidation by ferricyanide of the iron was prevented.A small sharp peak

was induced at g=5.9 (fig.3C) but this was unaffected   by
77K   illumination. When the non-haem iron was oxidised
before addition of formate, the g=6 signal was   still
observed (fig.3D). This indicates that the Fe^{3+} state
protects   the   bicarbonate from   removal   by   formate.The
relationships between bicarbonate and the ability to
oxidise the non-haem iron therefore suggest a close
structural relationship in the quinone binding region.

Optical measurements on P680 reduction by   electron
acceptors

     We find two μs decay phases for P680 rereduction
following a laser flash. One decays with a $t_{1/2}$ = 400μs.
this component is present at oxidised potentials and is
lost between 0mV and -100mV apparently corresponding to
Qh. It is replaced at lower redox potentials by a faster
decay component with $t_{1/2}$ = 1-200μs. this is lost below
-400mV. This faster component corresponds to the lowest
potential component in the esr redox titration. There is
no μs phase directly corresponding to Ql. However there
are in our experiments decay components with a lifetime
of 1-2ms. These may reflect electron donation to centres
in which the electron has been transferred out of the
reaction centre. This slow phase of decay has two waves
in the titration ,the first between 0mV and -100mV and
the second near -300mV. These correspond to the expected
potentials of Qh and Ql. The extent of the 200us decay is
increased as the -300mV phase of the slow decay is
titrated. These experiments support the proposal that the
first electron acceptor following the pheophytin has a
potential of -420mV and that it may transfer electrons to
a component with a potential around -300mV.

DISCUSSION

     We observed that Q_B binding caused characteristic
changes in the Q_A^- esr signal and confirmed that the
characteristics of Q_A^- depend on bicarbonate binding.The
non-haem iron was oxidised only under conditions where
bicarbonate was bound.
     These results support a hypothesis which gives a
central role to bicarbonate in providing the conditions
for electron transfer in higher plant PSII. It is
suggested that bicarbonate binds at or close to the non-
haem iron, stabilising a proton at the Q_B binding site
and influencing the redox properties of Q_B.   The
structural changes caused by bicarbonate binding also
influence the characteristics of Q_A and stabilise the
Fe^{3+} non-haem iron.
     The proposed role of the potentially bidentate
bicarbonate ligand is to provide negative charge to
increase the pK_a of a specific protonated group needed
for electron transfer . A protonated group carrying a
positive charge may be required at the site of Q_B binding
so that the anionic species is stabilised. The positive
charge would be a protonated amino acid residue such as
histidine and the bicarbonate would bind to a second

positively charged group ,another amino acid residue or the non-haem iron. Loss of the positive charge at the Q_B site by deprotonation in the absence of bicarbonate or the presence of the monodentate formate ligand would decrease Q_B binding and impair electron flow. Exchange of the proton for a deuteron may be responsible for the changes seen in the exchange experiments.The bicarbonate may also be involved in the protonation reactions following reduction of Q_B.

Herbicides, especially the phenolic herbicides which carry a negative charge, compete for bicarbonate and Q_B binding sites (van Rensen,1984). Therefore some herbicides will interfere with electron flow by influencing Q_B through effects on bicarbonate binding.

The bicarbonate also plays a role in stabilisation of the Fe^{3+} non-haem iron. The presence of the negatively charged bicarbonate lowers the redox potential of the iron allowing the ferricyanide oxidation in PSII (Diner & Petrouleas,1987,Petrouleas & Diner, 1987). Bicarbonate affects the shape of the Q_A iron-semiquinone signal through its binding at or near the iron.When bicarbonate is absent the Q_A^-Fe signal is similar to that seen in purple bacteria. Allosteric effects caused by the structural changes resulting from bicarbonate binding and charge pairing may be involved in causing the change in esr characteristics.

The effects of bicarbonate binding on the redox potential of the iron suggest that both the Q_A and Q_B redox potentials could also be altered by bicarbonate binding. The maintainence of a positively charged Q_B binding site would raise the midpoint redox potentials of the Q/Q^{--} and Q^{--}/Q_2^- couples and lower the pK_a values providing the conditions for electron transport between Q_A and Q_B. It is also possible to suggest that the Q_L and Q_H forms of Q_A result from centres having combinations of free or bound bicarbonate and Q_B. Further redox titrations of Q_A under conditions where pH, bound bicarbonate and bound Q_B are controlled are now needed.

Interpretation of the optical results is difficult as they suggest the presence of two competent electron transfer paths, one functioning around 0mV, and one around -400mV. An approach to understanding this may be to use the more highly purified PSII preparations which are becoming available.

We have made a preliminary investigation of the redox properties of the D1/D2 reaction centre preparation (Telfer et al,1988). We confirm that the esr spin polarised reaction centre triplet could be observed in this preparation. The triplet is observed over a very wide redox range, +400mV to -500mV. There was no indication that any of the acceptors functioning between these potentials was present in the preparation. Below - 550mV the loss of triplet formation is thought to reflect chemical reduction of the pheophytin acceptor. Above +400mV loss of the triplet formation was accompanied by the appearance of a light induced g=2 radical which may reflect $P680^+$. This result suggests that at these potentials an electron acceptor which can function at low temperature is oxidised. The most likely candidate for this is the non-haem iron associated with the acceptor complex and which is retained in this preparation. We could observe esr signals from non-haem iron but not any light induced redox change. The signals are small and difficult to detect, further work is required to clarify this.

REFERENCES

Diner,B.A. & Petrouleas, V. (1987) Biochim. Biophys. Acta 893, 138-148

Evans,M.C.W. (1987) Biochim. Biophys. Acta 894, 524-533

Evans, M.C.W. & Ford, R.C. (1986) FEBS Lett. 195, 290-294

Evans, M.C.W. ,Atkinson,Y.E. & Ford,R.C. (1985) Biochim. Biophys. Acta 806,2472-254

Ford,R.C. & Evans,M.C.W. (1983) FEBS Lett. 160,159-163

Govindjee & Van Rensen,J.J.S. (1978) Biochim. Biophys. Acta 505,183-213

Klimov,V.V. ,Dolan,E., Shaw,E.R. & Ke.B. (1980) Proc. Natl. Acad. Sci. USA 77, 7227-7231

Mansfield,R.W., Hubbard,J.A.M.,Nugent,J.H.A. & Evans, M.C.W. (1987) FEBS Lett. 220, 74-78

Michel,H. & Deisenhofer,J.(1988) Biochemistry 27,1-7

Nanba,O. & Satoh,K. (1987) Proc. Natl. Acad. Sci. USA 84,109-112

Nugent,J.H.A. (1987) Biochim. Biophys. Acta 893,184-189

Nugent,J.H.A. & Evans, M.C.W. (1980) FEBS Lett. 112,1-4

Nugent,J.H.A., Diner,B.A. & Evans,M.C.W. (1981) FEBS Lett. 124,241-244

Nugent,J.H.A., Evans,M.C.W. & Diner,B.A. (1982) Biochim. Biophys. Acta 682, 106-114

Petrouleas,V. & Diner,B.A. (1986) Biochim. Biophys. Acta 849, 265-275

Petrouleas,V. & Diner,B.A. (1987) Biochim. Biophys. Acta 893, 126-137

Rutherford , A.W. & Zimmermann,J.L. (1984) Biochim. Biophys. Acta 767, 168-175

Rutherford,A.W. , Zimmermann,J.L. & Mathis,P. (1984) FEBS Lett. 165, 156-162

Stemler,A. & Govindjee (1973) Pl. Physiol. 52, 119-123

Telfer,A.. Barber,J. & Evans,M.C.W. (1988) FEBS Lett. in press

Trebst,A. (1986) Z. Naturforsch. 41c, 240-245

Trebst,A. (1987) Z. Naturforsch. 42c, 742-750

Van Rensen J.J.S. (1984) Z. Naturforsch. 39c, 374-377

Van Rensen,J.J.S. ,Tonk,W.J.M. & de Bruijn,S.M. (1988) FEBS Lett. 226,347-351

Vermaas,W.F.J. & Rutherford,A.W. (1984) FEBS Lett. 175,243-248

Vermaas,W.F.J. & Van Rensen, J.J.S. (1981) Biochim. Biophys. Acta 636,168-174

Wraight,C.A. (1985) Biochim. Biophys. Acta 809,320-330

Zimmerman, J-L. & Rutherford,A.W. (1986) Biochim. Biophys. Acta 851, 416-423

A veteran's look at the chloroplast H+-ATPase and photosystem I reaction center

Nathan NELSON

*Roche Institute of Molecular Biology, Roche Research Center, Nutley, New Jersey 07110, U.S.A.

SYNOPSIS

The study of protein complexes of photosynthetic membranes had reached maturity. The use of biochemistry, immunology and molecular biology advanced the photosynthetic research and elucidated some steps in the evolution of the chloroplast protein complexes. Gene duplication appears to be a crucial step in the evolution of energy transducing systems. More subunits were added to the protein complexes during the advancement in evolution. Some of them were acquired from unrelated systems and they have no homologs in the protein complex from primitive creatures.

Evolution of the chloroplast H+-ATPase

Proton-ATPases can be divided into three main classes: 1) Plasma membrane-type, which is present in the plasma membranes of plant and fungal cells and in the gastric mucose, 2) Eubacterial-type, which is present in eubacteria, mitochondria and chloroplasts, 3) Vacuolar-type, which is present in the vacuolar system of eukaryotic cells. While the first class of H+-ATPases is operating via phosphoenzyme intermediate, the other two classes operate without phosphoenzyme intermediate (Mellman et al., 1986; Bowman and Bowman, 1986; Nelson, 1987). Several evidences suggest that the plasma membrane H+-ATPase is part of the family of E_1-E_2 ion pumps, and it is likely that all of the enzymes belonging to this family have evolved from a common ancestral ATPase (Goffeau and Slayman, 1981; Serrano et al., 1986). Since the vacuolar-type and the eubacterial-type H+-ATPases operate without phosphoenzyme intermediate they are likely to share some common features in the mode of their catalytic activity. For example, it was shown that the phenomenon denoted as single site catalysis for the eubacterial enzymes is also operating in the H+-ATPases from yeast vacuoles (Uchida et al., 1988) and chromaffin granules (Moriyama and Nelson, 1987).

The function of the membrane sector of the eubacterial-type enzymes is to conduct protons across the membrane. One of the polypeptides that compose this sector binds DCCD, and is soluble in organic solvents (Nelson, 1981; Futai and Kanazawa, 1983; Hoppe and Sebald, 1984; Walker et al., 1984; Schneider and Altendorf, 1987; Sutton and Apps, 1981). A polypeptide with similar properties is present in all of the vacuolar-type H+-ATPases (Sutton and Apps, 1981; Nelson, 1987). These features may indicate that the vacuolar and eubacterial H+-ATPases evolved from a common ancestral enzyme (Nelson, 1988). Recently, the cDNA encoding the proteolipid subunit of the chromaffin granules H+-ATPase was cloned and sequenced (Mandel et al., 1988). The gene encodes a highly hydrophobic protein with a molecular weight of 15,849 which is double the size of the eubacterial proteolipids. Moreover, hydropathy plots revealed four transmembrane segments, which is twice the number of the transmembrane segments of the eubacterial proteolipids. While the proteolipids from *E. coli*, yeast mitochondria and spinach chloroplasts were matched to the N-terminal half of chromaffin granules proteolipid, the proteolipids of bovine and *Neurospora* mitochondria matched better to the C-terminal half. It was concluded that the proteolipids of the eubacterial and vacuolar H+-ATPases evolved from a common ancestral gene that existed prior to the segregation of the various organelles (Mandel et al., 1988).

Table 1.   Evolution of the eubacterial and vacuolar H+-ATPases

---

Eubacterial H+-ATPase			Ancestral H+-ATPase		Vacuolar H+-ATPase
β	50 kd (+ε 15 kd?)	←	A 70 kd	→	70 kd
α	55 kd	←	B 60 kd	→	60 kd
γ	35 kd	←	C 40 kd	→	40 kd
a	30 kd	←	D 30 kd	→	30 kd
c	8 kd	←	E 8 kd	→	16 kd

A similar evolutionary pathway was likely to take place in respect to some of the other subunits of the eubacterial and vacuolar H+-ATPases. Table 1 suggests the origin and the evolution of few of the subunits of the two classes of proton-ATPases. As suggested previously (Nelson, 1988), we anticipated that a vacuolar-type H+-ATPase would be found in archaebacteria functioning in ATP formation. It is logical to assume that the Mitchellian ATP formation at the expense of protonmotive force (Mitchell, 1968) evolved in primordial organisms living in an acidic environment. The archaebacterim *Sulfolobus acidocaldarius* may represent a rudiment of such an organism. This bacteria leaves in an acidic environment of about 3.5 pH units and it maintains a ΔpH of over 3 units (Lubben et al., 1987). Under these conditions ATP can be formed even if a stoichiometry of H+-ATPase is 2 (Bennett and Spanswick, 1984). This kind of H+ to ATP ratio was measured for vacuolar H+-ATPases (Johnson et al., 1982). Therefore, it may be that the primordial ATP synthases were closely related to the current vacuolar H+-ATPases, functioning in both proton secretion and ATP formation. The evolution of the eubacterial-type enzymes was concomitant with increasing the H+ to ATP ratio to 3 and obtaining higher coupling between proton conduction and ATP formation. This enabled the organisms to harness smaller protonmotive forces and led to the development of eukaryotic cells, bearing power stations in the form of mitochondria and chloroplasts. At the same time, the vacuolar-type H+-ATPases could no longer compete on the job of ATP producers, and started functioning in limited acidification of

internal organelles in the eukaryotic cell. For perfecting this
function, the coupling between the ATPase and proton pumping
activities of the enzymes became looser, and a proton slip was
introduced for obtaining better control over the internal pH of the
various organelles (Moriyama and Nelson, 1988). We propose that the
main event leading to the loose coupling was the duplication of the
gene coding for the proteolipid (Mandel et al., 1988). If this is
the case, we anticipate that the proteolipids of the H⁺-ATPase from
Sulfolobus, that function in ATP formation, would be in a size of
about 8 kd, the rest of the subunits maintaining closer relation to
the vacuolar H⁺-ATPases.

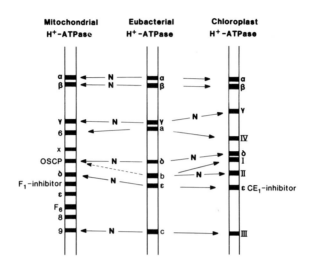

Fig. 1. Evolution of ATP-synthases
The genes that were transferred to the nucleus are marked by N. The
arrows show the origin of the various subunits. The origin of
several of the mitochondrial H⁺-ATPase subunits is not known.

        The eubacterial H⁺-ATPases evolved to their current form by
deleting unnecessary parts of certain subunits such as the α and β
and by adding new polypeptides to the complex. The basic subunit
structure of these enzymes is eight subunits, five for the catalytic
sector and thee for the membrane sector (Futai and Kanazawa, 1983).
As shown in Fig. 1, two main events took place when the chloroplast
H⁺-ATPase evolved. Subunit b of the membrane sector underwent gene
duplication (Cozens et al., 1986), and the genes coding for subunits
γ, δ and II were transferred to the nucleus (Nalin and Nelson,
1987). In addition, two specific properties were introduced to the
chloroplast H⁺-ATPase: regulation of the catalytic activity by light
and rendering the enzyme into a latent ATPase (Mills et al., 1980;
Mills and Mitchell, 1982; Nelson et al., 1972). The light regulation evolved by the introduction of two -SH groups into the γ
subunit which regulates the activity of the enzyme by oxidation-
reduction reactions (Nalin and McCarty, 1984; Ketcham et al., 1984).

In order to inhibit the ATPase activity of the enzyme at night time,
the chloroplast H^+-ATPase became a latent enzyme by converting the ϵ
subunit into CF_1-inhibitor (Nelson et al., 1972).

The rise, fall and resurrection of CF_1-inhibitor

    Isolated CF_1 is a latent ATPase that can be activated by a
number of treatments such as heat, proteases, ditiothreitol,
detergents and alcohols (McCarty and Nalin, 1986). Chloroplast
membranes are devoid of ATPase activity that can be elicited by some
of the above mentioned treatments. It was observed that addition of
the purified ϵ subunit to heat-activated CF_1 inhibited the ATPase
activity of the enzyme (Nelson et al., 1972). Moreover, antibody
against γ subunit prevented the inhibition of the ATPase activity by
isolated ϵ subunit (Nelson et al., 1973). Removal of the γ subunit
by trypsin treatment also prevented the inhibition by the
CF_1-inhibitor (Deters et al., 1975). Therefore, it was concluded
that the γ subunit is required for the activity of ϵ subunit as an
ATPase inhibitor (Nelson, 1976). It was proposed that this subunit
is responsible for the property that CF_1 is a latent ATPase, and the
inactivation of the ATPase reaction in the chloroplasts, was also
attributed to the same subunit. Activation of the enzyme, for both
ATPase and ATP-synthase activities, occurs via movements of ϵ
subunit but not by complete dissociation as was proposed for
F_1-inhibitor (Pedersen et al., 1981). This notion was challenged
and the release of tightly bound nucleotides was suggested as the
main regulatory factor in converting the latent enzyme into an
active ATP-synthase (Shavit, 1980; Graber et al., 1977; Reimer and
Selman, 1978). Recent studies by McCarty and his colleagues clearly
place the CF_1-inhibitor as the major *in situ* regulator of CF_1
(Richter et al., 1984; Richter and McCarty, 1987). They observed
that removal of the ϵ subunit from CF_1 renders it into an active
ATPase. The depleted enzyme can bind to the membrane but its
coupling activity became dependent on the addition of purified ϵ
subunit. Antibody against the ϵ subunit activated both the isolated
enzyme and the membrane bound enzyme. While incubation of thylakoid
membranes with the antibody in the dark had no effect on photophosphorylation, incubation during illumination strongly inhibited
the reaction. Therefore, the anticipated conformational changes in
the position of the ϵ subunit during activation of the enzyme
received a strong support.

Fifteen years with photosystem I reaction center - the unknown

    The function of photosystem I in the photosynthetic membranes
is to catalyze the photoreduction of NADP (Thornberg, 1986).
Plastocyanin serves as an electron donor to the system and ferredoxin is the electron acceptor. Therefore, photosystem I reaction
center must be defined as the membrane complex that catalyzes the
photoreduction of ferredoxin with plastocyanin as an electron donor.
However, there is no consensus on the more precise definition of
photosystem I reaction center. While we preferred to define the
system as the minimal structure that catalyzes the photoreduction of
ferredoxin (Bengis and Nelson, 1975, 1977), Mullet et al. (1980)
added to the system a few light harvesting chlorophyll protein
complexes, that specialized in light harvesting for photosystem I.

Fig. 2 shows schematically the differences between the two preparations. Later on, the two definitions were classified as a core
complex referring to the minimal structure, and a holo-PSI complex
for the wider interpretation including four more chlorophyll-
proteins denoted as LHCI (Thornberg, 1986). Ironically, when we
appeared with the first photosystem I reaction center (Nelson and
Bengis, 1974), the objections were that it contained too many
proteins and pigments. Today we know that photosystem I reaction
center contains eight different polypeptides denoted as subunits Ia,
Ib, II, II, IV, V, VI and VII in the order of decreasing molecular
weights of 83 kDa to about 8 kDa (Nelson, 1987). While the biophysical studies of photosystem I are advanced in comparison with other
membrane protein complexes (Malkin, 1982; Okamura et al., 1982),the
biochemistry of the reaction center is lagging in comparison with
complexes such as cytochrome oxidase and cytochrome b-c_1 complex
(Hauska, 1986). This is mainly due to lack of a ready transformation system for chloroplast containing cells.

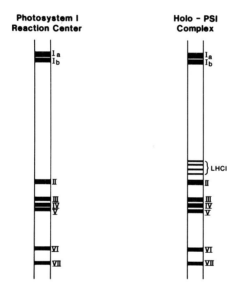

**Fig. 2. Schematic representation of the subunit structure of the
two different preparations of photosystem I reaction center**

The precise function of each individual subunit in the reaction
center is still largely unknown. The subcomplex of subunits Ia and
Ib denoted as P_{700} reaction center (Bengis and Nelson, 1975), or CPI
(Thornber, 1975), was isolated and studied quite extensively. This
complex contains the P_{700} pigment, about 40 chlorophyll a molecules,
a few β-carotene molecules, two phylloquinones and the bound
ferredoxin "X" (Goldbeck and Cornelius, 1986; Schoeder and Lokau,
1986; Nelson, 1987). This preparation is fully capable of catalyzing light dependent charge separation at room temperature. There-

fore, it was defined as P_{700} reaction center, that is the minimal
unit that can function in the primary photochemical processes of
photosystem I (Bengis and Nelson, 1975).

Subunit VII, which is a chloroplast gene product, was identified as the apoprotein of the bound ferredoxins A and/or B (Hoj et
al., 1987). The precise role of the rest of the subunits is largely
obscure. A photosystem I reaction center was isolated from cyanobacteria and green algae (Nechushtai and Nelson, 1981; Nechushtai et
al., 1983; Takahashi et al., 1982; Lundell et al., 1985). All of
these preparations are not capable of NADP-photoreduction, and to
date there is no preparation of resolved photosystem I from algae or
cyanobacteria that catalyzes the photoreduction of NADP. One
explanation to this phenomenon is that one or more of the subunits
of the reaction center from those sources is dissociated from the
complex during its preparation. The lack of active photosystem I
reaction center from cyanobacteria is one of the main obstacles in
the advanced studies of this complex. The genes coding for subunits
Ia and Ib from chloroplasts, chlamydomonas and a cyanobacterium were
cloned and sequenced (Fish et al., 1985; Kirsh et al., 1986; Kuck et
al., 1987; Bryant et al., 1987). Subunits Ia and Ib have been
highly conserved throughout evolution and there is a high homology
among the gene products from various sources. Subunits Ia and Ib
are also homologous to each other, suggesting that they may have
risen through gene duplication (Fig. 3). This event of gene
duplication may have been the decisive step towards the formation of
photosystem I, and it was proposed that these two polypeptides share
the pigments that participate in the primary photochemical activity
of the system (Bengis and Nelson, 1977). Subunit VI is marked by
the absence of methionine and cysteine in its amino acid composition. A subunit with similar molecular weight that also lacks
methionine and cysteine was identified in the preparation of
photosystem I reaction center from chlamydomonas (Nechushtai and
Nelson, 1981).

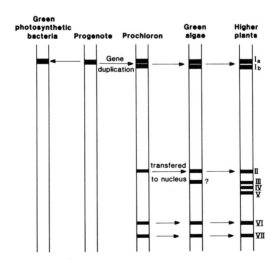

Fig. 3.   Evolution of photosystem I reaction center

Subunit II is the enigma of the reaction center. There is strong immunological cross-reactivity among subunit II from cyanobacteria, prochloron, through green algae up to higher plants (Nechushtai et al., 1983; Schuster et al., 1985). This indicates amino acid sequence homology and a conserved function throughout evolution. The conservation was maintained despite the transfer of the gene coding for this subunit into the nucleus of eukaryotic cells. Biochemical studies provided some evidences that subunit II may contain one of the bound ferredoxins A or B (Nechushtai and Nelson, 1981; Bonnerjea et al., 1985). This suggestion is supported by recent cross-linking experiments showing interaction between ferredoxin and subunit II (Zanetti and Merati, 1987). Sequences of the genes coding for this subunit in both prokaryotes and eukaryotes will provide a better insight into the function and evolution of this polypeptide.

Photosystem I reaction center from various sources have recently been crystallized (Ford et al., 1987; Witt et al., 1987; Reilly and Nelson, unpublished). Combination of molecular biology with advanced structural studies are likely to capture the imagination of the scientists in the field.

REFERENCES

Bengis, C. & Nelson, N. (1975) J. Biol. Chem. 250, 2783-2788.
Bengis, C. & Nelson, N. (1977) J. Biol. Chem. 252, 4564-4569.
Bennett, A.B., Spanswick, R.M. (1984) Plant Physiol. 74, 545-548.
Bonnerjea, J., Ortiz, W. & Malkin, R. (1985) Archives Biochem. Biophys. 240, 15-20.
Bowman, B.J. & Bowman, E.J. (1986) J. Membr. Biol. 94, 83-97.
Bryant, D.A., DeLorimer, R., Guglielmi, G., Stirewalt, V.L., Cantrell, A. & Stevens, S.E. (1987) in Progress in Photosynthesis Research (Biggins, J., ed.), vol. IV, pp. 749-755, Martinus Nijhoff Publishers, Dordrecht.
Cozens, A.L., Walker, J.E., Phillips, A.L., Huttly, A.K. & Gray, J.C. (1986) EMBO J. 5, 217-222.
Deters, D.W., Racker, E.,. Nelson, N. & Nelson, H. (1975) J. Biol. Chem. 250, 1041-1047.
Fish, L.K., Kuck, U. & Bogorad, L. (1985) J. Biol. Chem. 260, 1413-1421.
Ford, R.C., Picot, D. & Garavito, R.M. (1987) EMBO J. 6, 1581-1586.
Futai, M. & Kanazawa, H. (1983) Microbiol. Rev. 47, 285-312.
Goffeau, A. & Slayman, C.W. (1981) Biochim. Biophys. Acta 639, 197-223.
Goldbeck, J.H. & Cornelius, J.M. (1986) Biochim. Biophys. Acta 849, 16-24.
Graber, P., Schlodder, E. & Witt, H.T. (1977) Biochim. Biophys. Acta 461, 426-440.
Hauska, G. (1986) Encycl. Plant Physiol. New Series 19, 496-507.
Hoj, P.B., Svendsen, I., Scheller, H.V. & Moller, B.L. (1987) J. Biol. Chem. 262, 12676-12684.
Hoppe, J. & Sebald, W. (1984) Biochim. Biophys. Acta 768, 1-27.
Johnson, R.G., Beers, M.F., Scarpa, A. (1982) J. Biol. Chem. 257, 10701-10707.
Ketcham, S.R., Davenport, J.W., Warncke, K., & McCarty, R.E. (1984) J. Biol. Chem. 259, 7286-7293.
Kirsch, W., Seyer, P. & Herrmann, R.G. (1986) Curr. Genet. 10, 843-855.

Kuck, U., Choquet, Y., Schneider, M., Dron, M. & Bennoun, P. (1987) EMBO J. 6, 2185-2195.

Lubben, M., Lunsdorf, H. & Schafer, G. (1987) Eur. J. Biochem. 167, 211-219.

Lundell, D.J., Glazer, A.N., Melis, A. & Malkin, R. (1985) J. Biol. Chem. 260, 646-654.

Malkin, R. (1982) Annu. Rev. Plant Physiol. 33, 455-479.

Mandel, M., Moriyama, Y., Hulmes, J.D., Pan, Y.-C. E., Nelson, H. & Nelson, N. (1988) Submitted.

McCarty, R.E. & Nalin, C.M. (1986) Encycl. Plant Physiol. New Ser. 19, 576-583.

Mellman, I., Fuchs, R. & Helenius, A. (1986) Annu Rev. Biochem. 55, 663-700.

Mills, J.D., Mitchell, P. & Schurmann, P. (1980) FEBS Lett. 112, 173-177.

Mills, J.D. & Mitchell, P. (1982) FEBS Lett. 144, 67-67.

Mitchell, P. (1968) Chemiosmotic Coupling and Energy Transdcution, (Bodmin, Cornwall, England).

Moriyama, Y. & Nelson, N. (1987) J. Biol. Chem. 262, 14723-14729.

Moriyama, Y. & Nelson, N. (1988) in The Ion Pumps - Structure, Function and Regulation (Stein, W., ed.), in press.

Mullet, J.E., Burke, J.J. & Arntzen, C.J. (1980) Plant Physiol. 65, 814-822.

Nalin, C.M. & McCarty, R.E. (1984) J. Biol. Chem. 259, 7275-7280.

Nalin, C.M. & Nelson, N. (1987) Curr. Top. Bioenerg. 15, 273-294.

Nechushtai, R. & Nelson, N. (1981) J. Biol. Chem. 256, 11624-11628.

Nechushtai, R., Muster, P., Binder, A., Liveanu, V. & Nelson, N. (1983) Proc. Natl. Acad. Sci. USA 80, 1179-1183.

Nelson, N. (1976) Biochim. Biophys. Acta 456, 314-338.

Nelson, N. (1981) Curr. Top. Bioenerg. 11, 1-33.

Nelson, N. (1987) BioEssays 7, 251-254.

Nelson, N. (1987) in New Comprehensive Biochemistry (Amesz, J., ed.), vol. 15, pp. 213-231, Elsevier, Amsterdam.

Nelson, N. (1988) Plant Physiol. 86, 1-3.

Nelson, N., Nelson, H. & Racker, E. (1972) J. Biol. Chem. 247, 7657-7662.

Nelson, N., Deters, D.W., Nelson, H. & Racker, E. (1973) J. Biol. Chem. 248, 2049-2055.

Nelson, N. & Bengis, C. (1974) in Proceeding III International Congress on Photosynthetic Research, Rehovot, Israel (Abron, M., ed.), vol. I, pp. 609-620, Elsevier, Amsterdam.

Okamura, M.Y., Feher, G. & Nelson, N. (1982) in Photosynthesis (Govindjee, ed.), vol. I, pp. 195-272, Academic Press, New York.

Pedersen, P.L., Schwerzmann, K. & Citron, N. (1981) Curr. Top. Bioenerg. 11, 149-199.

Reimer, S. & Selman, B.R. (1978) J. Biol. Chem. 253, 7249-7255.

Richter, M.L. & McCarty, R.E. (1987) J. Biol. Chem. 262, 15037-15040.

Richter, M.L., Patrie, W.J. & McCarty, R.E. (1984) J. Biol. Chem. 259, 7371-7373.

Schneider, E. & Altendorf, K. (1987) Microbiol. Rev. 51, 477-497.

Schoeder, H.-H. & Lockau, W. (1986) FEBS Lett. 199, 23-27.

Schuster, G., Nechushtai, R., Nelson, N. & Ohad, I. (1985) FEBS Lett. 191,29-33.

Serrano, R., Kielland-Brandt, M.C. & Fink, G.R. (1986) Nature 319, 689-693.

Shavit, N. (1980) Annu. Rev. Biochem. 49, 111-138.

Sutton, R. & Apps, D.K. (1981) FEBS Lett. 130, 103-106.

Takahashi, Y., Koike, H. & Katoh, S. (1982) Archives Biochem. Biophys. 219, 209-218.

Thornber, J.P. (1975) Annu. Rev. Plant Physiol. 26, 127-158.

Thornber, J.P. (1986) Encl. Plant Physiol. New Series 19, 98-142.

Uchida, E., Ohsumi, Y. & Anraku, Y. (1968) J. Biol. Chem. 263, 45-51.

Walker, J.E., Saraste, M. & Nicholas, J.F. (1984) Biochim. Biophys. Acta 768, 164-200.

Witt, I., Witt, H.T., Gerken, S., Saenger, W., Dekker, J.P. & Rogner, M. (1987) FEBS Lett. 221, 260-264.

Zanetti, G. & Merati, G. (1987) Eur. J. Biochem. 169, 143-146.

Lipid-protein interactions and membrane function

Kleoniki GOUNARIS, David J. CHAPMAN and James BARBER

AFRC Photosynthesis Research Group, Department of Pure and Applied Biology, Imperial College, London SW7 2BB, U.K.

INTRODUCTION

The polar lipid composition of thylakoid membranes has been highly conserved through the evolution of oxygenic photosynthetic organisms. Four major lipid classes comprise the total polar lipid content of the thylakoids of higher plant chloroplasts, eukaryotic algae and cyanobacteria. These are the two galactolipids, monogalactosyldiacylglycerol (MGDG) and digalactosyldiacylglycerol (DGDG), the glycolipid sulphoquinovosyldiacylglycerol, (SQDG), also known as the plant sulpholipid, and the phospholipid phosphatidylglycerol (PG). The two galactolipids are electroneutral and account for approximately 75% of the total polar lipid content of thylakoid membranes. The remaining 25% is made up by the negatively charged lipids PG and SQDG. It has always been accepted that the prime role of acyl membrane lipids is the provision of a stable matrix essential for enzymatic function and the "solvation" of protein complexes. The fact that thylakoid membrane lipids have been conserved during the course of evolution of oxygenic photosynthesis may also suggest that their specific properties are required for the organisation and function of the thylakoids. In this communication we briefly describe work that has been carried out in order to provide insight concerning the hydrophobic requirements of the protein complexes of the thylakoid membranes of higher plants.

POLAR LIPID DISTRIBUTION

The highly asymmetric distribution of the protein components of the higher plant thylakoids had raised the question whether the polar lipids are also asymmetrically distributed. Experiments have been carried out in several laboratories examining both the transmembrane and lateral distribution of the polar lipids. Regarding the former, three experimental approaches have essentially been employed, namely enzymatic hydrolysis of the lipids, chemical modification of the polar head groups and identification by means of lipid specific antibodies. The data obtained from these various experiments has been summarised in [1]. There seems to be poor agreement in the results

- -

ABBREVIATIONS: MGDG, monogalactosyldiacylglycerol; DGDG, digalacto-syldiacylglyerol; SQDG, sulphoquinovosyldiacylglycerol; PG, phosphatidyl-glycerol; PC, phosphatidylcholine; PS, phosphatidylserine; PS2, photosystem 2; DPQ, decylplastoquinone; PQ-9, plastoquinone-9; PQH_2, plastoquinol-9.

The financial support of the AFRC and SERC is gratefully acknowledged.

derived from the different approaches regarding the precise proportions of lipids in the two leaflets of the bilayer. In no case, however, has absolute asymmetry of either the two galactolipids or phosphatidylglycerol been observed. It is worth pointing out that the transmembrane distribution of the plant sulpholipid is unknown. Enzymatic hydrolysis studies [2] as well as localization by means of antibodies [3] have suggested that large amounts of the total sulpholipid is inaccessible to external probing. It has been suggested [4] that the plant sulpholipid may interact closely with integral membrane complexes and is not therefore, exposed to the membrane surface.

Experiments designed to examine the lateral distribution of polar lipids in the thylakoids have been carried out on either mechanically fragmented membranes or on fragments derived from detergent treatments [5-10]. There appears to be an agreement on the enrichment of MGDG compared to DGDG in the appressed regions of the thylakoids. The MGDG to DGDG ratio has always been found to be higher in the granal membranes in all species examined and it seems to be independent of the method employed in obtaining membrane fragments. The lateral distribution of the acidic lipids is not as yet clarified.

In conclusion, up to date evidence shows that the thylakoid membrane lipids are not following an asymmetric distribution similar to that of the protein components. This however by no means exclude the possibility that small amounts of specific lipids classes with perhaps specific fatty acyl compositions are required for optimal activity of enzymatic complexes.

THE ROLE OF LIPIDS IN MEMBRANE FUNCTION

As previously noted, lipids provide a hydrophobic environment within which enzymatic processes occur. It is well established that the physical properties of the lipids are critical to the functioning of the intrinsic membrane protein complexes, but this does not necessitate the presence of a specific chemical structure of the lipid molecule as long as appropriate characteristics such as fluidity is maintained. In relation to protein conformation the ratios of the hydrophilic to lipophilic portions of the lipid molecules has been shown to be of importance.

A number of approaches have been employed in order to examine the role of lipids in the function of the thylakoid membranes. Destabilisation of lipid-protein interactions in the thylakoids leads to the phase-separation of MGDG and the formation of non-bilayer structures [11,12]. It has thus been postulated [13] that this lipid may be involved in the incorporation and the structural stabilisation of the protein complexes in the membrane. Functional roles of lipids have been suggested through experiments in which inhibition of protein function is monitored after treatment with lipases [e.g. 14-16]) or lipid specific antibodies [17,18]. The association of certain lipids with protein complexes after isolation from the membrane could also provide insight into possible interactions in the in vivo system [e.g. 19-22].

Additional information on the involvement of lipids in particular enzymatic processes and the specificity of the lipid requirement has been gained through reconstitution studies of isolated protein complexes. It has been recently shown [23] that, as judged by 77K fluorescence emission spectra and P-700 oxidation kinetics, MGDG is efficient in restoring energy transfer from the light-harvesting protein complex to the reaction centre of photosystem 1. In fact, it was previously shown [20] that MGDG was the only thylakoid lipid that could partially restore the 77K fluorescence emission spectrum once thylakoids had been solubilised.

Extensive studies have been carried out on the hydrophobic requirements and lipid specificity of the thylakoid membrane CF_0-CF_1 ATP synthase protein complex. It was initially noted [24] that the activity of the chloroplast coupling factor 1 (CF_1) of the ATPase enzyme could be modified by amphiphilic molecules such as detergents and fatty acids. It was observed that the non-ionic detergent octylglucoside in particular, could activate Mg-ATPase of the CF_1, at concentrations exceeding the critical micellar concentration, in a cooperative manner. It was suggested that the detergent could induce this activation by maintaining the native conformation of a hydrophobic site on the enzyme which became exposed once the CF_1 was no longer membrane bound.

Experiments on the purified CF_0-CF_1 ATPase complex have revealed a great deal concerning the lipid requirements of this protein complex [25-27]. It was demonstrated that thylakoid lipids have a specific effect on the catalytic activity of the CF_0-CF_1. This effect was expressed in the kinetics of ATP hydrolysis and of ATP-Pi exchange reactions. For both reactions it was shown that the enzyme exhibited a higher activity for ATP when reconstituted in thylakoid lipids. Table 2 summarises the effect of thylakoid lipids on the activation of CF_0-CF_1 complex. It is obvious that both ATP hydrolysis and ATP-Pi exchange require the presence of MGDG in the reconstituted systems. The poorest activators were found to be the acidic lipids SQDG and PG [27].

Table 1. Activation of CF_0-CF_1 by thylakoid lipids

Lipid used in the reconstitution	Reaction rate (nmol/mg protein/min)	
	ATP hydrolysis	ATP-Pi exchange
None	42	0
Total thylakoid lipid extract	480	45
MGDG	465	10
DGDG	160	0
SQDG:PG (1:1)	25	0

The very low rates of the ATP-Pi exchange observed in reconstituted systems using thylakoid lipids were found to be due to the high permeability of these lipid vesicles to protons. Figure 1 shows that the optimal ratio of the two galactolipids for this reaction is about 70:30 for MGDG:DGDG. A further stimulation of the reaction rate could be obtained by the addition of the acidic phospholipid, phosphatidylserine. The effect was due to this lipid reducing the permeability of the proteoliposomes to protons. Interestingly, it was found that PG was ineffective in producing this type of stimulation. The other naturally occurring acidic lipid of the thylakoids, SQDG, was most effective as shown in Fig. 2. It is also worth noting that SQDG was found to be a tightly-bound non-exchangeable lipid in CF_0-CF_1 preparations [26] suggesting that this lipid may fulfill both a structural and a functional role in its association with this enzyme. Excess of acidic lipids strongly inhibits the CF_0-CF_1 reconstituted systems.

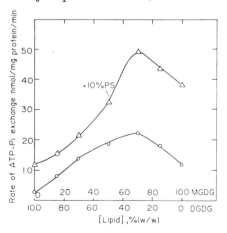

Fig. 1.   Optimal lipid concentration for reconstitution of the ATP-Pi exchange of isolated CF_0-CF_1 complex.

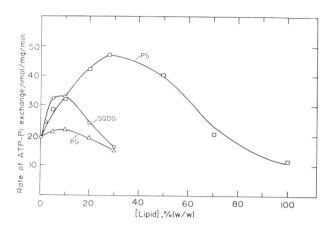

Fig. 2. The effect of acidic lipids on reconstitution of ATP-Pi exchange.
         CF_0-CF_1 was reconstituted with mixtures of MGDG:DGDG (2:1)
         and PS,PG or SQDG were added at the indicated concentrations.

Reconstitution experiments have also been carried out in attempts to
identify whether lipids play a role in the functioning of photosystem 2 (PS2)
in thylakoids of higher plants. It was observed that PS2 enriched membrane
fragments derived by multiple detergent (Triton X-100) treatment were
partially delipidated [6] but retained the capacity to evolve oxygen. A
stimulation of the rates of oxygen evolution could be achieved by the addition
of thylakoid lipids but complete inhibition of the oxygen evolution was
observed when a mixture of the two acidic lipids, PG and SQDG, was employed
[6]. It was found, as shown in Fig. 3. that a most profound effect was
obtained by SQDG alone [28]. It was also observed [29] that thylakoid lipids,
in cholate derived PS2 enriched membrane fragments, can induce a stimulation
of electron transport within PS2 when artificial electron donors are used.
More recently [30] it has been suggested that thylakoid lipids are required in
order to achieve stable charge separation within the reaction centre itself.

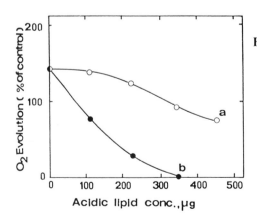

Fig. 3. The inhibition of O_2
         evolution by acidic
         lipids,
         (a) PG; (b) SQDG.
         Values were obtained
         by substituting acidic
         lipids for phosphati-
         dylcholine as to
         maintain a constant
         lipid to chlorophyll
         ratio.

Another component of the PS2 complex that has been shown to be affected by lipids is the cytochrome b559. It was shown [31,32] that the high potential of this cytochrome which is lost during detergent treatment of thylakoid membranes can be at least partially restored by reconstitution in thylakoid lipids. It is now possible to isolate a minimal unit of PS2 which is capable of primary charge separation [33,34] as well as support net electron transport with artificial electron donors and acceptors [35]. This isolated PS2 reaction centre is composed of two chlorophyll binding proteins, namely D1 and D2, and the two apoproteins of cytochrome b559, but binds neither the primary nor the secondary quinone acceptors. It was shown [35] that on addition of the relatively hydrophilic quinone, decylplastoquinone (DPQ), to the PS2 reaction centre a light-induced signal was observed which corresponded to the reduction of approximately 30% of the cytochrome b559 present in the complex. The observed signal was sensitive to herbicides, such as DCMU, suggesting that decylplastoquinone probably associated with the quinone binding sites in the complex. Under these conditions, the naturally occurring quinone, plastoquinone-9, was ineffective in inducing the reduction of the cytochrome.

In very recent experiments it was however possible to demonstrate that alterations of the assay medium by amphiphiles, such as detergents or lipids, induce changes in the isolated PS2 reaction centre which are briefly described below. It was observed that inclusion of detergents in the assay medium results in a significant increase in the amount of cytochrome b559 that can be reduced by decylplastoquinone. In addition, under these conditions plastoquinone-9 could mediate the light-dependent reduction of the cytochrome as shown in Fig. 4.

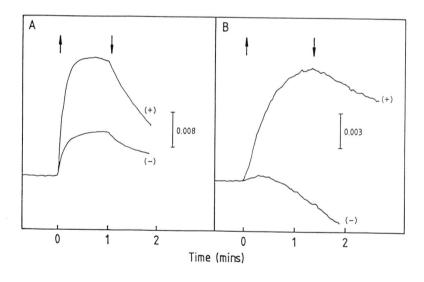

Fig. 4.   Light-induced absorbance changes at 430 nm. Samples contained PS2 reaction centre and $MnCl_2$ as the electron donor. In A, decylplastoquinone was added and in B, plastoquinone-9. (-) in the absence of detergent, (+) in the presence of detergent. Light on (↑) and light off (↓) as indicated.

It was found that only certain types of non-ionic detergents were able to induce this effect. In addition, ionic detergents were inhibitory dependent on their concentration in the dispersion. These data were taken to indicate that specific moieties of the amphiphiles may be required in order to mediate the observed reaction.

Under the conditions of its isolation [35] the PS2 reaction centre complex contains the cytochrome b559 in its low potential dithionite reducible form. It was observed that at alkaline pH values (pH > 8) the cytochrome could be reduced by ascorbate. It was however seen that inclusion of amphiphiles into the assay medium rendered most of the cytochrome ascorbate reducible over a wide pH range (Fig.5). One possible explanation of the data is that in the isolated PS2 reaction centre complex the cytochrome is in an environment that is inaccessible to the reductant, and that the amphiphiles cause changes in the complex such that access of the redox reagent is allowed. An alternative explanation is that in the presence of the amphiphiles the mid-point potential of the cytochrome is raised so that it can now be reduced by ascorbate. It was after the same amphiphile additions that the light-dependent reduction of the cytochrome by PQ-9 was observed. Interestingly, it was shown that in such an assay system it was the reduced form of the plastoquinone (PQH$_2$) that reduced the cytochrome b559 in a dark reaction (Fig.6).

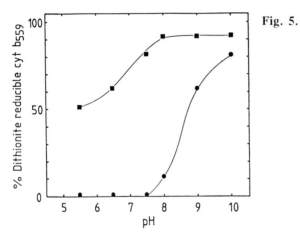

Fig. 5.   The reduction of cytochrome b559 by acorbate as a function of pH. The amount of cytochrome reduced is expressed as a percentage of the total dithionite reducible.
(●) in the absence of detergent.
(■) in the presence of detergent.

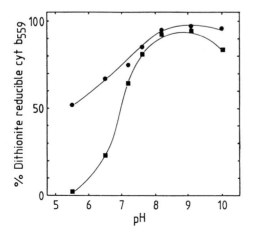

Fig 6.   The reduction of cytochrome b559 at differenct pH values by ascorbate (●) or PQH$_2$ (■) in the presence of detergenet. The amount of cytochrome reduced is expressed as a percentage of the total dithionite reducible.

If instead of detergents, lipids were used for reconstitution of the PS2 reaction centre the cytochrome can be reduced by PQH_2 in the dark. The lipid concentration dependency is shown in Fig. 7.

Preliminary experiments into which the concentration of the zwitterionic lipid phosphatidylchloline (PC) was titrated against the negatively charged lipid phosphatidylserine (PS) indicated that the reaction monitored is sensitive to the amounts of charged lipid present in the reconstituted systems (Fig. 8). The reaction was found to be inhibited when the concentrations of charged lipid exceeded approximately 20% of the total lipid and was abolished at concentrations exceeding 60%. Similar data were obtained when the light-dependent reduction of the cytochrome mediated by PQ-9 was monitored.

Whether these observations are directly related to the hydrophobic requirements of photosystem 2 remains to be established.

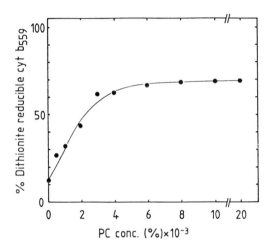

Fig. 7. The effect of PC concentrations on the reduction of cytochrome b559 by PQH_2. PS2 reaction centres were reconstituted in PC liposomes containing PQH_2 at various lipid to chlorophyll concentrations.

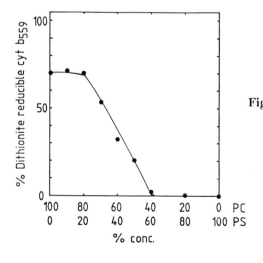

Fig. 8.   The inhibition of the PQH_2 reducible cytochrome b559 by the acidic lipid PS.

CONCLUSION

It has become apparent that membrane lipids can be directly involved in the function of membrane protein complexes and thus specifically influence the function of the membrane as a whole. Experiments involving the reconstitution of protein complexes and the monitoring of the resultant enzymatic activities have already provided insight into the lipid requirements of the proteins. Studies on thylakoid membrane complexes have emphasised the requirement for a critical balance between specific lipid classes. Such in vitro systems, due to the nature of the components, are necessarily complex. Careful characterisation however of the assay systems and examination of the amphiphiles which are effective in vitro will help us to identify the physico-chemical requirements for the functioning of the membrane and the type of interactions involved.

REFERENCES

1.   Gounaris, K., Barber, J. & Harwood, J.L. (1986) Biochem.J. 237, 313-326
2.   Rawyler, A., Unitt, M.B., Giroud, C., Davies, H., Mayer, J.-P., Harwood, J.L. & Siegenthaler, P.-A. (1987) Photosynth. Res. 11, 3-13
3.   Radunz, A. (1980) Z. Naturforsch. 35c, 1024-1031
4.   Barber, J. & Gounaris, K. (1986) Photosynth. Res. 9, 239-249
5.   Gounaris, K., Sundby, C., Andersson, B. & Barber, J. (1983) FEBS Lett. 156, 170-174
6.   Gounaris, K., Whitford, D. & Barber, J. (1983) FEBS Lett. 163, 230-234
7.   Henry, L.E.A., Mikkelsen, J.D. & Moller, B.L. (1983) Carlsberg Res. Commun. 48, 131-148
8.   Chapman, D.J., De-Felice, J. & Barber, J. (1984) in Developments in Plant Biology, Vol.9, Structure, Function and Metabolism of Plant Lipids (Siegenthaler, P.-A. & Eichenberg, W. eds.), pp.457-464, Elsevier, Amsterdam
9.   Murphy, D.J. & Woodrow, I.E. (1983) Biochim. Biophys. Acta 725, 104-112
10.   Mackender, R.O. & Gillanders, B.W. (1984) in Developments in Plant Biology, Vol.9, Structure, Function and Metabolism of Plant Lipids (Siegenthaler, P.-A. & Eichienberg, W. eds.), pp.401-404, Elsevier, Amsterdam
11.   Gounaris, K., Brain, A.P.R., Quinn, P.J. & Williams, W.P. (1983) FEBS Lett. 155, 47-52
12.   Gounaris, K., Brain, A.P.R., Quinn, P.J. & Williams, W.P. (1984) Biochim. Biophys. Acta 766, 198-208
13.   Williams, W.P., Gounaris, K. & Quinn, P. (1984) in Advances in Photosynthesis Research (Sybesma, C. ed.), Vol.3, pp.123-130, Martinus Nijhoff/Dr W. Junk, The Hague
14.   Rawyler, A. & Siegenthaler, P.-A. (1981) Biochim. Biophys. Acta 635, 348-358
15.   Krupa, Z. (1983) Photosynth. Res. 4, 229-239
16.   Jordan, B.R., Chow, W.S. & Baker, A.J. (1983) Biochim. Biohys. Acta 725, 77-86
17.   Radunz, A. (1981) Bev. Deutsch Bot. Ges. 94, 477-489
18.   Radunz, A., Bader, K.P. & Schmid, G.H. (1984) Z. Pflanzenphysiol. 114, 227-231
19.   Rawyler, A., Henry, L.E.A. & Siegenthaler, P.-A. (1980) Carlsberg. Res. Commun. 45, 443-451
20.   Siefermann-Harms, D., Ross, J.W., Kaneshiro, K.H. & Yamamoto, H.Y. (1982) FEBS Lett. 149, 191-196

21. Remy, R., Tremolieres, A., Duval, J.C., Ambard-Breteville, F. & Dubacq, J.P. (1982) FEBS Lett. 137, 271-275
22. Gounaris, K. & Barber, J. (1985) FEBS Let. 188, 68-72
23. Siefermann-Harms, D., Ninnemann, H. & Yamamoto, H.Y. (1987) Biochim. Biophys. Acta 892, 303-313
24. Pick, U. & Bassilian, S. (1982) Bichemistry 24, 6144-6152
25. Pick, U., Gounaris, K., Admon, A. & Barber, J. (1984) Biochim. Biophys. Acta 765, 12-20
26. Pick, U., Gounaris, K., Weiss, M. & Barber, J. (1985) Biochim. Biophys. Acta 808, 415-420
27. Pick, U., Weiss, M., Gounaris, K. & Barber, J. (1987) Biochim. Biophys. Acta 891, 28-39
28. Gounaris, K., Pick, U. & Barber, J. (1984) in Developments in Plant Biology, Vol.9, Structure, Function and Metabolism of Plant Lipids (Siegenthaler, P.-A. & Eichenberger, W. eds.), pp.451-455, Elsevier, Amsterdam
29. Akabori, K., Imaoka, A. & Toyoshima, Y. (1984) FEBS Lett. 173, 36-40
30. Eckert, H.-J., Toyoshima, Y., Akabori, K. & Dismukes, G.C. (1988) Photosynth. Res. 14, 31-41
31. Matsuda, H. & Butler, W..L. (1983) Biochim. Biophys. Acta 725, 320-324
32. Whitford, D., Gounaris, K. & Barber, J. (1984) in Advances in Photosynthesis Research (Sybesma, C. ed.), Vol.1, pp 497-500, Martinus Nijhoff/Dr W. Junk, The Hague
33. Nanba, O. & Satoh, K. (1987) Proc. Natl. Acad. Sci. USA 84, 101-112
34. Barber, J., Chapman, D.J. & Telfer, A. (1987) FEBS Lett. 220, 67-73
35. Chapman, D.J., Gounaris, K. & Barber, J. (1988) Biochim. Biophys. Acta 933, 423-431

Plasma membrane H$^+$-ATPase

Robert T. LEONARD

Department of Botany and Plant Sciences, University of California, Riverside, CA 92521.

SYNOPSIS

Progress in understanding the structure and function of the plasma membrane H$^+$-ATPase is summarized. Emphasis is given to information from research measuring the activity of the enzyme as K$^+$-stimulated phosphate released from ATP, transfer of radioactive phosphate from ATP to ATPase polypeptides, and ATP-dependent electrogenic transport of H$^+$ into inside-out vesicles derived from the plasma membrane. Attention is also given to the question of whether or not the H$^+$-ATPase also directly pumps K$^+$.

INTRODUCTION

The plasma membrane of plant cells contains an ATP-driven proton pump which plays a central role in energy conservation for ion and metabolite transport (Leonard, 1984; Sze, 1985; Serrano, 1988). The enzymatic name for the pump is ATP phosphohydrolase or, simply, ATPase. Since the ATPase generates a gradient in free energy of H$^+$ (charge and pH gradient) across the plasma membrane, it is usually referred to as an H$^+$-ATPase. The H$^+$ free energy gradient is available to energize a variety of secondary transport systems for mineral nutrients and also for the recovery of metabolites, such as sugars and amino acids, which have been lost to the cell wall space (Reinhold & Kaplan, 1984; Lucas & Madore, 1988). In addition, regulation of the H$^+$-ATPase may also be important for certain developmental responses of plants, because changes in ion and/or electrical gradients across the plasma membrane appear to be involved in certain types of signal transduction at the cell surface (Schaller & Sussman, 1988).

The H$^+$-ATPase of the plasma membrane is structurally and mechanistically different from H$^+$-ATPases of the tonoplast membrane and of the inner membrane of the mitochondrion (Table 1). The tonoplast-type H$^+$-ATPase is clearly structurally, mechanistically, and evolutionarily related to the F$_1$F$_0$-type ATPase of mitochondria, chloroplast, and certain bacteria (Nelson, 1988; Zimniak et al., 1988). Although the information available on the structure and function of the plasma membrane H$^+$-ATPase is incomplete, it appears that the plant enzyme is closely related to cation pumping ATPases of animal and fungal cells (Sze, 1985; Serrano, 1988). These latter types of ATPases are characterized by the formation and subsequent breakdown of a phosphorylated intermediate during the course of ATP

Table 1. Some properties of proton pumping ATPases of plant cells

Property	Plasma Membrane	Tonoplast-Type	Mitochondria
Substrate Km (MgATP)	MgATP 0.3–0.5 mM	MgATP (GTP) 0.1–0.3 mM	MgATP (GTP) 0.8 mM
Ion stimulation	Cations K=NH$_4$>Rb>>Na>Cs>Li	Anions Cl=Br>I>HCO$_3$>SO$_4$	Anions HCO$_3$>>Cl=Br>I=SO$_4$
pH Optimum	6.5	7.0–8.0	8.0–9.0
Inhibitors	Vanadate (10 µM) --- DCCD*	Nitrate (5–10 mM) --- DCCD	Nitrate (5–10 mM) Oligomycin (azide) DCCD
Structure catalytic subunit other subunits	Integral (>200 kDa) 100 kDa	Peripheral (400–500 kDa) 70 kDa 60 kDa 16 kDa ?	Peripheral (400–500 kDa) β 55 kDa α 58 kDa γ 36 kDa δ 22–26 kDa ε 14 kDa
Reaction Mechanism	Phosphorylated intermediate E$_1$-E$_2$-type	No phosphorylated intermediate F$_1$F$_0$-type	No phosphorylated intermediate F$_1$F$_0$-type
Function	Electrogenic H$^+$/K$^+$ exchange	Electrogenic H$^+$ transport	ATP synthesis and electrogenic H$^+$ transport

For references, see Leonard (1983, 1984), Sze (1985), Nelson (1988), and Zimniak et al. (1988).
* DCCD, N,N'-dicyclohexylcarbodiimide.

hydrolysis. They are inhibited by vanadate ($H_2VO_4^-$) because, as an analog of phosphate ($H_2PO_4^-$), it interferes with the activity of enzymes, such as certain ATPases and phosphatases, which undergo phosphorylation and dephosphorylation during the catalytic cycle.

The activity of the plasma membrane H^+-ATPase can be measured in different ways. Phosphate released from ATP can be determined using simple colorimetric techniques. The transfer of radioactive phosphate of gamma labeled $AT^{32}P$ to trichloroacetic acid-precipitable protein can be quantitated to study the reaction mechanism and to identify the catalytic subunit. And, the enzyme can be assayed by measuring ATP-driven formation of a charge and/or pH gradient in sealed, inside-out plasma membrane vesicles. Regardless of how the activity of the ATPase is measured, K^+ has been found to have specific effects on its activity. This has led to the suggestion that the H^+-ATPase is also directly involved in K^+ transport (Leonard, 1983, 1985).

In this paper, the results of research using these assay techniques is summarized in relation to the understanding of the structure and function of the plasma membrane H^+-ATPase. Particular attention is given to the question of whether or not the enzyme is directly involved in K^+ transport through the plasma membrane of plant cells.

PHOSPHATE RELEASE FROM ATP AS A MEASURE OF H^+-ATPase ACTIVITY

The activity of the plasma membrane ATPase, as measured by release of phosphate from ATP, is similar from tissue to tissue and from species to species. The activity ranges between 0.5 and 1.0 μmol phosphate released per mg protein per minute at pH 6.5 and 38 C. The addition of a monovalent cation to the assay is necessary for maximum ATPase activity with an order of preference of $K^+=NH_4^+>Rb^+>>Na^+>Cs^+>Li^+$, which is similar to that observed for the transport of these cations into roots (Leonard, 1983; O'Neil & Spanswick, 1984b). This property of the enzyme has been interpreted as supporting a direct role for the ATPase in K^+ transport. However, the increment of ATPase activity produced by addition of K^+ relative to the activity without added K^+ can be small and, in fact, difficult to demonstrate for plasma membrane preparations from certain plant species (Leonard, 1983). Even though the effect of K^+ on the ATPase becomes more apparent with detergent extraction and purification (e.g., O'Neil & Spanswick, 1984b; Singh et al., 1987), most researchers remain quite skeptical about whether or not K^+-stimulated ATPase activity is representative of the direct action of the H^+-ATPase in K^+ transport.

The ATPase shows simple Michaelis–Menten kinetics with respect to the substrate, MgATP. However, the application of enzyme kinetic analysis to K^+-stimulated ATPase activity resulted in complex kinetic data which are characteristic of a multisubunit enzyme showing negative cooperativity or half-of-the-sites reactivity (Leonard, 1983). Since the kinetic data of K^+ transport into roots are similarly complex, this property of the ATPase is also consistent with, but not proof for, the direct involvement of the enzyme in K^+ transport. This feature of the enzyme can also be documented after solubilization with detergent and reconstitution into liposomes (Singh et al., 1987). It has not been determined if this characteristic is retained by the purified enzyme.

MEASUREMENT AND IDENTIFICATION OF H$^+$-ATPase BY PHOSPHOENZYME
FORMATION

   The plasma membrane H$^+$-ATPase of plants, like cation-transport-
ing ATPase of animals and fungi, forms a covalent phosphorylated
intermediate during the course of ATP hydrolysis and the associated
ion pumping activity. The phosphorylation/dephosphorylation cycle
is believed to be responsible for the change in conformation necessary for alternating the access to the ion-binding site(s) between
the cytoplasm and the outside of the cell. The fact that the
enzyme forms a phosphorylated intermediate has allowed researchers
to obtain information about the reaction mechanism (Briskin &
Leonard, 1982; Briskin, 1986, 1988) and to identify the catalytic
subunit by polyacrylamide gel electrophoresis (Briskin & Leonard,
1982; Briskin & Poole, 1983; Gallagher & Leonard, 1987).
   Early attempts to demonstrate that the ATPase forms a phosphorylated intermediate were confounded by the presence of protein
kinase activity in the plasma membrane fraction (Briskin & Leonard,
1982). Treating the plasma membrane fraction from corn roots with
a relatively low concentration of detergent [0.1% (w/v) deoxy-
cholate] removed or inactivated most of the protein kinase activity.
The detergent-extracted plasma membrane preparation was used to
demonstrate the transfer of ^{32}P from gamma-labeled AT^{32}P to trichloroacetic acid insoluble protein. At ice temperature, phosphoenzyme formation and subsequent breakdown reached steady state in
seconds. The covalent bond formed between phosphate and protein
had characteristics expected for an acyl-phosphate bond which was
acid stable, but heat or alkaline labile with marked sensitivity
to hydroxylamine.
   Polyacrylamide gel electrophoresis at acidic pH and low temperature revealed one major size class of radioactive polypeptides
with a molecular mass of about 100 kDa (Briskin & Leonard, 1982).
Phosphorylation of the 100 kDa polypeptides was dependent on the
presence of Mg^{2+} and was substrate specific for ATP. It had the
expected property of rapid turnover as evidenced by chasing with
unlabeled ATP. The rate of phosphoenzyme formation was inhibited
by vanadate or other chemicals known to inhibit plasma membrane
ATPase activity. Monovalent cations stimulated phosphoenzyme breakdown with the same order of preference as stimulation of ATPase
activity.
   It was later shown that the acyl-phosphate bond was formed
with the beta carboxyl group on an aspartic acid residue on the
catalytic subunit of the ATPase (Walderhaug et al., 1985). The
sequence of amino acids at the active site of phosphorylation was
determined to be -Cys-(Ser/Thr)-Asp(P)-Lys-, which is identical to
the active site sequence for this type of ATPase of animal and
fungal cells. There is a striking similarity in the autoradiograph
of phosphopeptides obtained from Pronase digested phosphorylated
ATPase preparations for plant plasma membrane H$^+$-ATPase and the H$^+$,
K$^+$-ATPase of gastic mucosa (Walderhaug et al., 1985). The latter
enzyme directly catalyzes an electroneutral exchange of H$^+$ and K$^+$
across the plasma membrane of mucosal cells.
   Recent results of Briskin (1986, 1988) with the red beet plasma
membrane ATPase have confirmed these observations and substantially
extended the information available on the reaction mechanism. He
found that phosphoenzyme formation was accelerated at acidic pH and
that a histidyl residue appeared to be involved in this process.
There appeared to be two phosphoenzyme forms (E_1, E_2) which differed
in sensitivity to ADP and K$^+$. The E_1 form of the phosphoenzyme was

discharged by ADP with production of ATP. The E_2 form was unaffected by ADP, but its breakdown to release phosphate was stimulated by K^+. Conversion of E_1 to E_2 was also shown to be stimulated by K^+. These results clearly place the plant plasma membrane ATPase into the E_1, E_2 type of cation pump found in animal and fungal cells. They also indicate that the stimulation of H^+-ATPase activity by K^+ results from a stimulation of the steps in the reaction sequence (and the associated changes in conformation) that are expected for an ATPase which is directly involved in K^+ transport.

The fact that the H^+-ATPase forms a phosphorylated intermediate has allowed the identification, by polyacrylamide gel electrophoresis, of the phosphorylated (catalytic) subunit. Sodium (Briskin & Leonard, 1982) or lithium (Briskin & Poole, 1983) dodecyl sulfate polyacrylamide gel electrophoresis at low temperature and pH showed radioactive polypeptides with the expected characteristics at about 100 kDa. This observation was confirmed (Gallagher & Leonard, 1987a) using a phenol-urea-acetic acid gel system (Gallagher & Leonard, 1987b) that has superior resolution at the acidic pH and low temperature required for stability of the relatively labile acyl-phosphate bond.

Antibodies to 100 kDa polypeptides of plant and fungal H^+-ATPase (Gallagher, 1985; Surowy & Sussman, 1986; Nagao et al., 1987; Oleski & Bennett, 1987; DuPont et al., 1988) are available for use in screening expression libraries for cDNA sequences coresponding to mRNA coding for the catalytic subunit of the ATPase. This will soon allow detailed comparison of the sequence homology between plant, animal, and fungal ATPases. The anticipated availability of cDNA probes will also be important for studies on the developmental regulation of the plasma membrane ATPase.

Radiation inactivation of vanadate-sensitive ATPase activity of the plasma membrane fraction from red beet has been used to estimate the minimum molecular size of the membrane associated enzyme complex. The target molecular size was calculated to be at least 228 kDa, indicating that the functional ATPase complex in the membrane is at least a dimer (Briskin et al., 1985). The solubilized ATPase appears to be at least a trimer of 100 kDa subunits (Anthon & Spanswick, 1986; Schaller & Sussman, 1988).

PURIFICATION OF PLASMA MEMBRANE H^+-ATPase

There have been several published attempts to solubilize and purify the plant plasma membrane H^+-ATPase. A purified preparation is needed to determine the subunit composition of the functional ATPase complex and to settle the question of whether or not the enzyme is an electrogenic H^+/K^+ exchange pump. While the procedures used have been very similar, the level of purity reported has, in certain cases, been disturbingly different. Substantial purification of the solubilized ATPase has been reported for membranes isolated from roots of oat (Serrano, 1984), corn (Nagao et al., 1987), tomato (Anthon & Spanswick, 1986), and mung bean (Kasamo, 1986), as well as from storage tissue of red beet (Singh et al., 1987).

The procedure begins with a microsomal or plasma membrane-rich fraction that has been isolated from about 200 g of tissue. Weakly associated membrane proteins are extracted with Triton X-100 or sodium deoxycholate at a relatively low concentration (about 1 mg detergent/mg protein) in the presence of 0.3 to 0.5 M KCl or KBr. In one case, the Triton X-100 extracted pellet was again extracted

with a relatively low concentration of octylglucoside and deoxycholate (Anthon & Spanswick, 1986). For detergent extraction
and subsequent purification, the membrane preparation is usually
suspended in 20 to 30% glycerol. ATPase is solubilized from the
extracted membranes by treatment with lysolecithin (5 mg detergent/
mg protein) or Zwittergent 3-14 (1 mg detergent/mg protein). The
solubilized preparation of presumably integral membrane proteins is
centrifuged in a linear glycerol gradient to further purify the
ATPase. In one case, it was partially purified by removing the
detergent in the presence of a sonicated phospholipid preparation
to selectively reconstitute the ATPase into vesicles (Singh et al.,
1987). Fractions with the highest ATPase activity are combined and
usually diluted and centrifuged to concentrate the purified enzyme.
        The purification procedure results in a preparation which is
greatly enriched in 100 kDa polypeptides and has ATPase activity of
about 5 to 6 μmol/mg protein/min as compared to 0.5 to 1.0 μmol/
mg/min for the beginning membrane preparation. There has been one
remarkable exception where the final preparation consisted of only
100 kDa polypeptides and an activity of 17.6 μmol/mg/min (Anthon
& Spanswick, 1986). However, in view of the data presented by
Schaller & Sussman (1988), it is difficult to understand how the
final purification step of centrifugation in a linear glycerol
gradient can result in this level of purification. Schaller &
Sussman (1988) showed that the various integral membrane proteins
migrate in the glycerol gradient at rates that are related to the
size of the lipid-protein aggregate formed by detergent solubilization rather than to the molecular weight of the denatured polypeptide. They showed SDS gels for all fractions from the glycerol
gradient. Those fractions enriched in 100 kDa polypeptides contained several other polypeptides as has been reported by others
(Serrano, 1984; Kasamo, 1986) and as we have observed in our unpublished attempts to reproduce the Anthon & Spanswick procedure
with both corn and tomato microsomal membranes. In view of these
results, we have concluded that it is, at best, very difficult to
use glycerol gradient centrifugation of detergent-washed and solubilized membrane preparations to completely purify the ATPase. Perhaps, there are mysteries in the technique that are not routinely
reproducible, given published protocols. Nagao et al. (1987)
reported complete purification of the H$^+$-ATPase from corn roots
while using the same procedure as Serrano (1984) and Schaller &
Sussman (1988) whose final preparations from oat roots are clearly
only partially purified with respect to ATPase polypeptides.
        Polyacrylamide gel electrophoresis of plasma membrane fractions
from several plant species indicates that the catalytic subunit of
the ATPase accounts for only about 3 to 6% of the Coomassie blue
staining polypeptides (Gallagher & Leonard, 1987a). Comparative
figures for plasma membrane preparations from animal and fungal
cells are 30 to 40% (or higher), and 15%, respectively. A plasma
membrane preparation from a plant tissue or species that has unusually high ATPase activity or unusually high relative abundance of
ATPase polypeptides has not been reported. After removing weakly
associated polypeptides with low concentration of detergent, 100
kDa ATPase polypeptides account for about 15% of Coomassie blue
staining polypeptides (Gallagher & Leonard, 1987a), and this value
was increased to at least 70% following glycerol gradient centrifugation of the detergent-solubilized ATPase (Serrano, 1984). The
fact that ATPase polypeptides are initially present in the plasma
membrane in relatively low abundance has added to the difficulty of
purifying the enzyme. To completely purify the H$^+$-ATPase, it is

necessary to remove about 95% of the plasma membrane protein. If
accomplished, about a 20-fold increase in ATPase activity, relative
to the plasma membrane preparation, would be expected.

While the results of Anthon & Spanswick (1986) and of Nagao et
al. (1987) are consistent with the conclusion that plant plasma
membrane H^+-ATPase consists of only 100 kDa polypeptides, further
research on this question seems justified. As discussed below, a
partially purified H^+-ATPase preparation has been reconstituted
into liposomes to demonstrate ATP-dependent H^+ transport activity
(Serrano, 1984), as well as ATP-dependent K^+ transport in the
absence of an H^+-free energy gradient (Giannini et al., 1987b).
There are no reports of attempts to use solubilized and completely
purified ATPase to demonstrate electrogenic H^+/K^+ exchange in
reconstituted vesicles. Therefore, the definitive experiment on
whether or not the H^+-ATPase also pumps K^+ has not been conducted.

H^+-ATPase MEDIATED ION TRANSPORT IN ISOLATED PLASMA MEMBRANE VESICLES

Since the plasma membrane ATPase is an electrogenic H^+ pump, a
third way to assay for the activity of the enzyme is to measure ATP-
driven formation of a charge and/or pH gradient across the membrane
of vesicles derived from the plasma membrane. The vesicles need to
be relatively sealed to sustain the H^+-free energy gradient, and
they should be inside-out so that ATP added to the suspension of
vesicles has access to what was originally the cytoplasmic surface
of the ATPase. It has proven difficult to obtain preparations
of plasma membrane vesicles which could be used to demonstrate
vanadate-sensitive, ATP-driven H^+ transport. This has recently
been accomplished for both native and reconstituted plasma membrane
vesicles.

The formation of an ATPase-generated charge and pH gradient
across a vesicle membrane can be detected in several ways (Sze,
1985). In relatively sealed vesicles, the development of the H^+-
free energy gradient will limit the activity of the ATPase. Ionophores, such as nigericin or gramicidin, which discharge the gradient by increasing H^+ conductance (in exchange for a cation), should
stimulate ATPase activity. Hence, ionophore-stimulated ATPase
activity is taken as evidence for formation of ionic gradients
across the vesicle membrane. Radioactive molecules which are
relatively lipid permeant and distribute across the vesicle membrane
in response to a charge or pH gradient can be used with the Nernst
equation to quantitate the activity of the H^+-ATPase. However, by
far the most popular method for measuring the ion pumping activity
of an ATPase involves the use of fluorescent molecules which
distribute across the vesicle membrane in response to a charge or
pH gradient. [See Pope & Leigh (1988) for potential problems with
this method.] The optical signal from these fluorescent probes is
quenched in response to accumulation within the interior of the
vesicle. ATP-driven quenching of a fluorescent signal which is
sensitive to vanadate and reversed by addition of an ionophore
is taken as representative of the activity of an H^+-ATPase.

The first demonstration of ATP-driven H^+ transport was with a
solubilized and reconstituted preparation of plasma membrane ATPase
from oat roots (Serrano, 1984) and red beet storage tissue (O'Neil
& Spanswick, 1984a; Singh et al., 1987). The quenching of acridine
orange fluorescence was used to demonstrate ATP-dependent H^+ transport which had the ion and inhibitor sensitivities expected for

the plasma membrane H$^+$-ATPase. K$^+$ and other monovalent cations
stimulated H$^+$ transport activity with an order of preference that
was similar to their effects on enzyme activity (O'Neil & Spanswick,
1984a,b). However, it was not determined if there was ATP-driven
K$^+$ transport in the reconstituted vesicles or if K$^+$ stimulation of
H$^+$ transport was related to a direct involvement of the ATPase in K$^+$
transport.

Recently, there has been a significant improvement in the
ability to obtain native, inside-out plasma membrane vesicles which
are relatively sealed and can be used to demonstrate ATP-dependent
transport of H$^+$ (DeMichaelis & Spanswick, 1986; Giannini et al.,
1987a; DuPont et al., 1988), K$^+$ (Giannini et al., 1987b), and even
NO$_3$$^-$ (Ruiz-Cristin & Briskin, 1988). Giannini et al. (1987a) have
provided a most intriguing technical breakthrough in this research
area. They found that addition of 0.25 M KCl to the medium used to
homogenize red beet tissue resulted in H$^+$ transport activity in
microsomal vesicles that had the characteristics expected for the
tonoplast-type ATPase. However, if 0.5 M KI was substituted for
the KCl, then vesicles were obtained which had the H$^+$ transport
activity expected for the plasma membrane ATPase. Hence, it was
possible to selectively produce plasma membrane vesicles that were
active in ATP-dépendent H$^+$ transport. These vesicles have since
been shown to be useful for studying ATP-dependent K$^+$ (Giannini et
al., 1987b) and NO$_3$$^-$ (Ruiz-Cristin & Briskin, 1988) transport.

The availability of plasma membrane vesicles that are active
in K$^+$ transport allows experiments to be conducted to determine if
the transport of K$^+$ is directly or indirectly energized by ATP. If
K$^+$ transport depends on the ATPase-generated H$^+$-free energy gradient,
then elimination of the free energy gradient with a proton conduct-
ing ionophore should completely inhibit ATP-driven K$^+$ transport.
This experiment has been conducted by Briskin and colleagues using
both native and reconstituted plasma membrane vesicles from red
beet (Giannini et al., 1987b). They found that there is ATP-
dependent K$^+$ transport even when the H$^+$-free energy gradient has
been eliminated, suggesting that K$^+$ transport in the vesicles is
directly driven by an ATPase. This finding, particularly in view
of the other results which indicate that K$^+$ has specific effects
on H$^+$-ATPase activity, strongly suggests that the H$^+$-ATPase is also
a K$^+$ pump. However, it is still conceivable that there is an additional ATPase in the plasma membrane preparation which is directly
involved in K$^+$ transport. To finally resolve this question, it
will be necessary to confirm the observation of ATP-driven K$^+$ transport in the absence of an H$^+$-free energy gradient for a purified and
reconstituted preparation consisting only of polypeptides of the
plasma membrane H$^+$-ATPase.

CONCLUSIONS

The plasma membrane H$^+$-ATPase is a large (molecular mass in
excess of 228 kDa) lipid-requiring intrinsic membrane protein complex. The functional ATPase complex consists of similar (if not
identical) 100 kDa catalytic subunits which form at least a dimer
in the membrane. There does not appear to be other kinds of
subunits in the functional ATPase complex, but further research
is needed to confirm this point. During ATP hydrolysis, the
terminal phosphate of ATP is transferred to phosphorylate the
catalytic subunit of an aspartic acid residue at the ATP-binding
site. K$^+$ stimulates ATPase activity by increasing the rate of

dephosphorylation of the catalytic subunit and the conversion of the phosphoenzyme from an ADP-sensitive E_1 form to the K^+-sensitive E_2 form. The phosphorylation and dephosphorylation events in the catalytic cycle of the enzyme appear to be associated with conformation changes leading to electrogenic proton transport. In view of specific effects of K^+ on H^+-ATPase activity and the recent data with native and reconstituted vesicles showing ATP-dependent transport of K^+ in the absence of an H^+-free energy gradient, there is reason to conclude that the pump functions in electrogenic exchange of H^+ and K^+. However, reconstitution experiments with the completely purified H^+-ATPase are needed to test this conclusion.

REFERENCES

Anthon, G.E. & Spanswick, R.M. (1986) Plant Physiol. 81, 1080-1085
Briskin, D.P. (1986) Archiv. Biochem. Biophys. 248, 106-115
Briskin, D.P. (1988) Plant Physiol. 87, (in press)
Briskin, D.P. & Leonard, R.T. (1982) Proc. Natl. Acad. Sci. USA 79, 6922-6926
Briskin, D.P. & Poole, R.J. (1983) Plant Physiol. 71, 507-512
Briskin, D.P., Thornley, W.R. & Roti-Roti, J.L. (1985) Plant Physiol. 78, 642-644
DuPont, F.M., Tanaka, C.K. & Hurkman, W.J. (1988) Plant Physiol. 86, 717-724
DeMichaelis, M.I. & Spanswick, R.M. (1986) Plant Physiol. 81, 542-547
Gallagher, S.R. (1985) Ph.D. Dissertation, pp. 180-194, University of California, Riverside
Gallagher, S.R. & Leonard, R.T. (1987a) Plant Physiol. 83, 265-271
Gallagher, S.R. & Leoanrd, R.T. (1987b) Anal. Biochem. 162, 350-357
Giannini, J.L., Gildensoph, L.H. & Briskin, D.P. (1987a) Archiv. Biochem. Biophys. 254, 621-630
Giannini, J.L., Gildensoph, L.H., Ruiz-Cristin, J. & Briskin, D.P. (1987b) Plant PHysiol. Suppl. 83, s55
Kasamo, K. (1986) Plant Physiol. 80, 818-824
Leonard, R.T. (1983) in Metals and Micronutrients: Uptake and Utilization by Plants (Robb, D.A. & Pierpoint, W.S., eds.), Phytochemical Society of Europe Symposium Series No. 21, pp. 71-86, Academic Press, London
Leonard, R.T. (1984) in Advances in Plant Nutrition (Tinker, P.B. & Lauchli, A., eds.), vol. 1, pp. 209-240, Praeger, New York
Leonard, R.T. (1985) in Potassium in Agriculture (Munson, R.D., ed.), pp. 327-335, American Society of Agronomy, Crop Science of American, Soil Science Society of America, Madison, WI
Lucas, W.J. & Madore, M.A. (1988) in The Biochemistry of Plants (Preiss, J., ed.), pp. 35-84, Academic Press, New York
Nagao, T., Sasakawa, H. & Sugiyama, T. (1987) Plant Cell Physiol. 28, 1181-1186
Nelson, N. (1988) Plant Physiol. 86, 0001-0003
Oleski, N. & Bennett, A.B. (1987) Plant Physiol. 83, 569-572
O'Neil, S.D. & Spanswick, R.M. (1984a) J. Membrane Biol. 79, 231-243
O'Neil, S.D. & Spanswick, R.M. (1984b) J. Membrane Biol. 79, 245-256
Pope, A.J. & Leigh, R.A. (1988) Plant Physiol. 86, 1315-1322.
Reinhold, L. & Kaplan, A. (1984) Annu. Rev. Plant Physiol. 35, 45-83

Ruiz-Cristin, J. & Briskin, D.P. (1988) Plant Physiol. Suppl. 86, s79

Schaller, G.E. & Susman, M.R. (1988) Planta 173, 509-518

Serrano, R. (1984) Biochem. Biophys. Res. Comm. 121, 735-740

Serrano, R. (1988) Biochim. Biophys. Acta 947, 1-28

Singh, S.P., Sudershan Kesav, B.V. & Briskin, D.P. (1987) Physiol. Plant. 69, 617-626

Surowy, T.K. & Sussman, M.R. (1986) Biochim. Biophys. Acta 848, 24-34

Sze, H. (1985) Annu. Rev. Plant Physiol. 36, 175-208

Walderhaug, M.O., Post, R.L., Saccomani, G., Leonard, R.T. & Briskin, D.P. ;(1985) J. Biol. Chem. 260, 3852-3859

Zimniak, L., Dittrich, P., Gogarten, J.P., Kibak, H. & Taiz, L. (1988) J. Biol. Chem. (in press)

Intracellular cannibalism in higher plant cells

Roland DOUCE*, Richard BLIGNY*, Albert-Jean DORNE* and Claude ROBY[+]

CEN-G and University Joseph Fourier, DRF/PCV* and RMBM[+], 85X, F38041, Grenoble-cédex, FRANCE

SYNOPSIS

Plant cells, owing to the presence of intracellular pools of carbohydrate and to their ability to control an autophagic process, can survive adequately for a long period of time without receiving any external supply of organic carbon.

INTRODUCTION

A number of morphological observations suggest that in higher plant cells various cell organelles including mitochondria may be engulfed by the tonoplast membrane (for a review see Matile and Wiemken, 1976). However the biochemical counterpart of this autophagic activity is lacking.

In this article we intend to characterize the biochemical changes occuring during a prolonged period of sucrose starvation in higher plant cells. Our data suggest that plant cells owing to their ability to control an autophagic process can survive adequately for a long period of time without receiving any external supply of organic carbon.

EXPERIMENTAL

Sycamore (*Acer pseudoplatanus* L.) cells were grown in a nutrient medium as described previously (Bligny, 1977) and maintained in exponential growth by frequent subcultures. Cells harvested from the culture medium were rinsed three times by successive resuspensions in fresh culture medium devoid of sucrose and incubated at zero time in flasks containing sucrose-free culture medium. At each time cells were harvested for perchloric extract (Roby et al., 1987) and for the measurement of starch (Journet et al., 1986) sucrose (Journet et al., 1986) fatty acid deriving from polar lipids (Journet et al., 1986) dry weight and fresh weight. ^{31}P-NMR spectra of neutralized PCA extracts were measured on a Bruker NMR spectrometer (AH400) equipped with a 10-mm multinuclear probe tuned at 162 MHz. The deuterium resonance of D_2O was used as a lock signal and the spectra were recorded under conditions of broad-band proton decoupling. Spectra of standard solutions of known phosphate compounds at pH 7.8 were compared with that of PCA extracts of sycamore cells (Roby et al., 1987). The definitive attributions were made after running a series of spectra obtained

by addition of the authentic compounds to the PCA extracts. The
theory of NMR methods and relevant experimental techniques are
fully discussed in monograph by Roberts and Jardetzky (1981).

Cytochrome oxidase, polar lipids and cardiolipin measurements
were carried out according to Journet et al (1986).

Sycamore cell respiration was measured at 25°C in their culture
medium (Journet et al., 1981).

Mitochondria from sycamore protoplasts were isolated and
purified as described by Nishimura et al (1982) using discontinuous
Percoll gradients. The mitochondria subsequently concentrated by
differential centrifugation were better than 95 % as judged by
their impermeability to cytochrome c (Douce et al., 1972).

RESULTS

Effect of sucrose starvation on cell weight and on polar lipid
fatty acids of sycamore cells

Figure 1 summarizes the effect of sucrose starvation
of sycamore cells on intracellular fatty acid content and cell
weight (wet and dry weights). In the absence of sucrose growth
stopped completely because the cell wet weight per ml of culture
medium appeared to be constant for at least 70 h (fig. 1). When
sucrose was omitted from the nutrient medium, the dry weight
decreased to 50 % of the control value within 30 h. The marked
decline in the cell dry weight was attributable to a progressive
disappearance of sucrose from the vacuolar reservoir (80 ± 10 mM)
and starch from plastids (65 ± 10 mM expressed as glucose) (fig. 1
and 2). During this time, the intracellular fatty acid content
remained constant (fig. 1). However after 40 h of sucrose starvation,
when almost all the intracellular carbohydrate pool (starch +
sucrose) has disappeared, the cell fatty acid content declined
progressively (fig. 1). Analysis of cell phospholipids has indicated
that a long period of sucrose starvation also induced a progressive
disappearance of all the cell polar lipids (Journet et al., 1986)
[we have shown the major phospholipids of sycamore cells to be
phosphatidylcholine (40-45 %) and phosphatidylethanolamine
(25-27 %)]. For example, phosphatidylcholine decreased to 30 %
of the control value (1.3 µmol phosphatidylcholine g^{-1} wet weight)
within 70 h of sucrose starvation. Finally, the total protein
(including the enzymes of the glycolytic pathway in cytosol) and
galactolipid of the cell also decreased sharply during the
starvation period (Journet et al., 1986).

Effect of sucrose starvation on the rate of O_2 consumption by sycamore cells

The rate of respiration of cells deprived of sucrose appeared
to be constant for at least 24 h (Fig. 3). Thereafter, the rate
of O_2 consumption decreased with time. After 50 h of starvation,
their capacity to utilize O_2 was reduced to less than 50 % of that
of normal growing cells. Fig. 3 also indicates that the uncoupled
rate of O_2 consumption obtainable with 2 µM carbonyl cyanide p-tri-
fluoromethoxyphenylhydrazone (FCCP) decreased after 24 h in the
same ratio as the rate of respiration without uncoupler. A careful
comparison between Fig. 2 and 3 indicates that the rate of
O_2 consumption began to decline when the intracellular sucrose
has been consumed. Under these conditions starch content was reduced
to less than 30 % of that of normal cells.

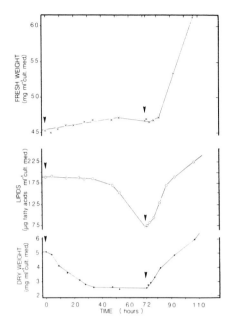

Fig. 1. Effect of sucrose starvation and replenishment on cell weight (wet and dry weight) and on intracellular fatty acids content of sycamore cells

Cells harvested from the culture medium were rinsed three times by successive resuspensions in fresh culture medium devoid of sucrose and incubated at zero time (45 mg wet weight ml^{-1}) in flasks containing sucrose-free culture medium. At intervals, cells were harvested on a fibreglass filter (15 s ; pressure of suction, 0.2 bar) to ascertain cell fresh weight. Cell dry weight was measured after lyophilization of the preceding samples. Total cell fatty acid content was measured as described under Materials and Methods. At arrow 50 mM sucrose was added to the culture medium. The complete oxidation of one glucose molecule requires 6 O_2 and sycamore cells exhibited an O_2 uptake rate of 42 nmol O_2 h^{-1} mg^{-1} wet weight (Rébeillé et al., 1985). From the knowledge of both these parameters one can calculate that 1.26 µg glucose h^{-1} mg^{-1} wet weight was required in order to sustain the respiration rate of sycamore cells. This value matches the measured rate of dry weight diminution (1.5 µg h^{-1} mg^{-1} wet weight).

The fact that the rate of O_2 consumption during sucrose starvation was always lower than the uncoupled rate (Fig. 3) strongly suggests that the rate at which sycamore cells respire is limited by the availability of ADP for either oxidative phosphorylation (Jacobus et al., 1982) or glycolysis in plastids and cytosolic phase of sycamore cells (ap Rees, 1985). In support of this suggestion we have observed that the absolute concentration of cytoplasmic ADP was not detectable by ^{31}P nmr (Rébeillé et al., 1985 ; Roby et al., 1987). The decrease in the uncoupled rate of O_2 consumption during the course of sucrose starvation is not easily interpretable. However, such an observation could be attributable to a progressive diminution of the number of mitochondria/cell. We therefore measured the levels of two specific mitochondrial markers (cardiolipin and

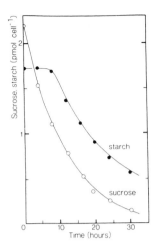

**Fig. 2. Effect of sucrose starvation of sycamore cells on intracellu-
lar sucrose and starch levels**
Cells harvested from the culture medium were rinsed three times
by successive resuspensions in fresh culture medium devoid of sucrose
and incubated at zero time (45 mg wet weight ml^{-1}) into flasks
containing sucrose-free culture medium. At each time cells were
harvested and sucrose and starch contents were measured as described
under Materials and Methods. The starch content is expressed as
micromoles of glucose per 10^6 cells. For 10^6 cells there are: wet
weight, 28 mg ; dry weight, 3.3 mg ; protein, 0.55 mg.
----

cytochrome aa_3) inside the cells during sucrose deprivation. Indeed
we have already demonstrated that in higher plant cells cardiolipin
and cytochrome aa_3 are exclusively localized in the mitochondrial
inner membranes (Bligny and Douce, 1980).

Quantitative determination of cytochrome aa_3 and cardiolipin in
sycamore cells and their mitochondria

        The values of the cytochrome aa_3 and cardiolipin contents
in sycamore cells and in purified sycamore cell mitochondria are
given in Table I. It appeared that the cardiolipin and cytochrome
aa_3 content of sycamore mitochondria was invariant to sucrose depletion. In marked contrast, expressed in terms of nanomole/10^6 cells,
the content of cytochrome aa_3 and cardiolipin was dramatically
lower in the sucrose-deficient cells after 50 h of sucrose starvation: less than half of the normal value. It is noteworthy that
the lag phase observed for cardiolipin or cytochrome aa_3 evolution
(Fig. 4) was comparable with that observed for O_2 uptake evolution
(Fig. 3). Furthermore, a comparison between Fig. 3 and 4 indicates
that the respiration rates decreased progressively in the same
ratio as the decrease in intracellular cardiolipin or cytochrome
aa_3. Identical results were obtained by using fumarase activity
as a mitochondrial matrix marker. In this case, on a cell number
basis the total activities found in normal and 50-h sucrose-starved
cells were 26 and 12 nmol $min^{-1}/10^6$ cells, respectively. Again the
fumarase specific activity of sycamore mitochondria was invariant

Table 1. Cytochrome oxidase and cardiolipin content of sycamore cells (normal and sucrose-starved) and sycamore cell mitochondria (normal and sucrose starved)

Cytochrome $\underline{aa_3}$ contents were measured from low-temperature difference spectra according to Journet et al. (1986). Cardiolipin contents were measured according to Journet et al. (1986). Normal cells : growing cells were harvested from the culture medium. Sucrose-starved cells : normal cells were rinsed three times by successive resuspension in fresh culture medium devoid of sucrose and incubated during 50 h into flasks containing sucrose-free culture medium. Normal mitochondria : mitochondria from normal cells ; sucrose-starved mitochondria : mitochondria from sucrose-starved cells. The S.D. is given for 95 %, the number of experiments is shown in parentheses.

Mitochondrial marker	Cells		Mitochondria	
	Normal	sucrose-starved	Normal	Sucrose-starved
	pmol 10^6 cells		nmol mg^{-1} protein	
Cytochrome $\underline{aa_3}$	13 ± 2(6)	5.8 ± 1.5(4)	0.37 ± 0.02(6)	0.35 ± 0.03(4)
Cardiolipin	1050 ± 150(6)	520 ± 80(4)	29 ± 2(6)	27 ± 2(4)

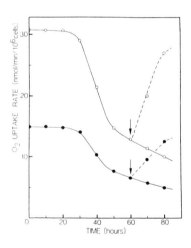

Fig. 3. Time course of sycamore cell respiration rate when cells are grown in a medium devoid of sucrose

Cells harvested from the culture medium were rinsed three times by successive resuspensions in fresh culture medium devoid of sucrose and incubated at zero time (20 mg wet weight ml^{-1}) into flasks containing sucrose-free culture medium. At each time, cells were harvested and the O_2 consumption rate was measured as described under "Materials and Methods" ; ●, normal respiration ; O, uncoupled respiration ; the dotted line corresponds to the enhancement of cell respiration that is observed when 50 mM sucrose is added to the culture medium (arrows).

to sucrose depletion (not shown). All these results demonstrate
that, after a long period of sucrose starvation, the progressive
decrease in the uncoupled rate of O_2 consumption by sycamore cells
was attributable to a progressive diminution of the number of
mitochondria/cell.

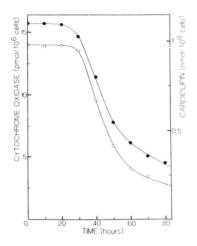

**Fig. 4. Effect of sucrose starvation of sycamore cells on intracellu-
lar cardiolipin (●) and cytochrome aa₃ (O) contents**
Cells harvested from the culture medium were rinsed three times
by successive resuspensions in fresh culture medium devoid of sucrose
and incubated at zero time (20 mg wet weight/ml) into flasks containing sucrose-free culture medium. At each time cells were harvested
and cardiolipin and cytochrome aa₃ contents were measured as described (Journet et al., 1986).
----

^{31}P NMR of neutralized Perchloric acid extracts: effect of sucrose
deprivation

Figure 5A,B shows a typical undecoupled ^{31}P NMR spectrum
of a perchloric acid extract of sycamore cells (9 g wet weight)
at pH 7.8 in the presence of a large excess of CDTA. The resonances
which were observed were sharp. From literature data (Satre et
al., 1986 ; Navon et al., 1979) and by adding known compounds to
the extract, the peaks obtained were identified as signals from
sugar phosphates including glucose 6-P, glucose 1-P, and fructose
6-P (unfortunately fructose 6-P is obscured by the resonance of
P-ethanolamine) ; Pi ; γ-, α-, and β-ATP ; γ-, α-, and β-UTP ;
β-, and α-ADP ; β- and α-UDPG ; β- and α-UDP-gal. As expected ATP
is the most abundant triphosphate, but other nucleotides are present
in significant amounts. Since it was possible to detect separately
the triplets of the β-phosphates of ATP, GTP, and (CTP + UTP),
a typical sycamore cell had the following nucleotide composition:
65 % ATP and 35 % UTP (+CTP). GTP was barely detectable. All the
phosphorylated compounds present at concentrations lower than 100 µM
(to estimate intracellular concentrations, we assumed that 6 g of
wet cells contains 1 ml of cytoplasm) were not distinguishable
from the background noise. This means that there are many phosphate
metabolites that are known to be present but are not seen in the
NMR spectra of the extract. Among these are fructose 1,6 bis-P,

pyrophosphate, and NADP (or NADPH). Interestingly, the peak at
3.3 ppm close to that of NADPH (2'-P) and NMP was assigned to
P-choline. The NAD (+NADP) peaks (originating from both reduced
and oxidized pyridine nucleotides) which yields a complex multiplet
centered at -11.16, -11.28, -11.45, -11.57 ppm on the high-field
side of the α-NTP peak were barely detectable.

The high NTP/NDP ratio as seen under *in vivo* conditions
(Roby et al., 1987) is also observed for cell extracts (Fig. 5).
A low level of NDP relative to NTP indicated that little if any
hydrolysis of NTP had occurred during the extraction procedure.
Furthermore, the spectrum of a cell extract in the presence of
CDTA showed the β-phosphorus signal of ATP at higher field (-21.15
ppm) than did the spectrum of an intact cell suspension (-19.65
ppm). This indicates that in intact sycamore cells ATP is
predominantly bound to a divalent metal ion, most probably Mg^{2+},
which was eliminated in the cell extract by the addition of CDTA
as has been observed in Ehrlich ascites tumor cells (Navon et al.,
1977).

^{31}P NMR spectrum of a cell extract of 70 h sucrose-starved
sycamore cells (9 g wet weight) (Fig. 5, spectrum 2) indicated
that although the total amount of NTP, glucose-6-P and sugar-P
resonances have dropped considerably (the 70-h sucrose-starved
cell extract exhibited small vestigial glucose 6-P and sugar-P
resonances), the adenylate energy charge (Pradet and Raymond, 1983)
was maintained at a high value. It is possible therefore, that
the total amount of NTP per unit volume of cytoplasm may be
maintained while the total cytoplasmic and mitochondrial volumes
drop sharply during the starvation period. Furthermore when the
endogenous fatty acids started to decline (see Fig. 1), several
noticeable changes occurred over time in the perchloric extract
from cells (Fig. 5, spectrum 2). Of particular interest was the
marked increase in the amount of glycerylphosphorylcholine,
glycerylphosphorylethanolamine and P-choline [a compound which
has been characterized in plants for the first time by Maizel et
al. (1956)]. These compounds were also found in Ehrlich ascite
tumor (Navon et al., 1977) and Hela cells (Evans and Kaplan, 1977).
Interestingly after a long period of sucrose starvation the most
intense resonances in the spectral region to low field of Pi emanated
from P-choline at 3.25 ppm. P-choline was determined in spectra
obtained after addition of known amounts of P-choline to the PCA
extracts ; it could thus be demonstrated that the total amount
of P-choline that appeared after a long period of sucrose deprivation
corresponded exactly to the total amount of phosphatidylcholine
that disappeared within the same period of time (not shown). For
example after 70 h of sucrose starvation the total amount of P-choline present in intact sycamore cells was approx. 0.7-0.8 μmol
g^{-1} wet weight and the total amount of phosphatidylcholine that
had been hydrolyzed was approx. 0.8-0.9 μmol g^{-1} wet weight. From
these results we can conclude that P-choline accumulated in sycamore
cells after a long period of sucrose starvation derived from phosphatidylcholine catabolism and that P-choline was not significantly
hydrolized.

Effect of sucrose replenishment

Addition of 50 mM sucrose to the medium at 70 h of sucrose
starvation resulted in a marked increase in the cell dry weight
and total cell fatty acids (Fig. 1). Likewise the full rate of
O_2 consumption by 70 h sucrose-starved cells was recovered a few

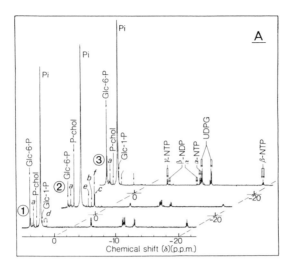

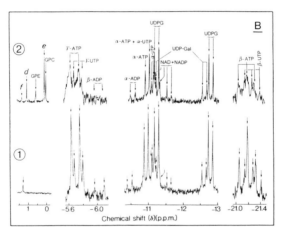

Fig. 5. Proton-decoupled ^{31}P-NMR spectra of PCA extract of sycamore cells

All the interesting regions are shown on an expanded scale (**B**). PCA extracts were prepared from normal aerobic cells (Spectrum 1), 70h-sucrose starved cells (Spectrum 2) and 70 h-sucrose starved cells followed by 10 h recovery after addition of 50 mM sucrose to the culture medium (Spectrum 3) (see Fig. 1). Freeze dried extracts corresponding to a total of $3 \cdot 10^8$ cells (9 g wet weight) were solubilized and supplemented with CDTA as described under Materials and Methods. The samples (2.5 ml) containing 300 µl D_2O were put in a 10 mm diameter NMR tube and spectra recorded at 162 MHz on a Bruker AM400 spectrometer. Spectra were acquired over a period of 2 h with a 50° pulse angle, a sweep-width of 6,000 Hz and a recycling time of 4 s. Free induction decays were accumulated using 8 k data points and zero-filled to 32 k prior Fourier transformation. A 0.1 Hz line broadening was applied. Two levels of broadband H decoupling were employed: 2.5 W during the data acquisition (0.68 s) and 0.5 W during the delay period. Peak assignments: a, position of fructose 6-P, ribose 5-P and P-ethanolamine, Glc-6-P, glucose 6-P ; P-chol, phosphorylcholine ; Glc-1-P, glucose 1-P ; b, glycerylphosphorylethanolamine ; c,

glycerylphosphorylcholine ;   d,e,f,   myo-inositol   hexakisphosphate.
For   phosphoglucomutase   the   similarity   between   the   apparent
equilibrium   constant   (19)   and   the   mass-action   ratio
([Glc-6-P]:[Glc-1-P]=15)   is   good   evidence   that   this   reaction   is
close to equilibrium *in vivo*.

----

hours after the addition of sucrose (Fig. 2). The increase in the
cell dry weight was attributable to a rapid accumulation of sucrose
in the vacuolar reservoir and starch in plastids (not shown) whereas
the increase in total cell fatty acids and in O_2 consumption rates
were attributable to the synghesis of new cytoplasmic material
such as mitochondria (Journet et al., 1986). During this time the
cell wet weight per ml of culture medium remained constant. When
the total fatty acids and the O_2 uptake of sycamore cells returned
to or near the level of normal cells (Fig. 1), growth began and
this lag phase represented the time required to synthesize new
cytoplasmic material such as mitochondria. The ^{31}P-NMR spectrum
of an extract from 70 h-sucrose starved cells 10 h after addition
of sucrose to the medium indicated that glucose-6-P level had increased up to its original level (Fig. 5, spectrum 3). The nucleotide
levels in this system also increased markedly upon refeeding (Fig. 5,
spectrum 3) (see Journet et al., 1986). Furthermore, 10 h after
replenishment of sucrose other substantial changes were evident
in the perchloric extract from cells (Fig. 5, spectrum 3). Of particular interest was the marked decrease in the amount of P-choline
indicating that this compound is probably reused for phosphatidylcholine synthesis. In support of this suggestion we have observed
that, in contrast with fatty acids and glycerol moieties, the polar
head of phosphatidylcholine isolated from 70 h-sucrose starved
cells 10 h after addition of [^{14}C]-sucrose was not labelled. Furthermore we have also shown that, during the course of sucrose replenishment, the linear decrease in cytosolic P-choline previously
accumulated during sucrose starvation, closely correlates with
the linear increase in phosphatidylcholine amount.

DISCUSSION

        These results demonstrate that transfer of sycamore cells
into flasks containing sucrose-free culture medium triggers the
following cascade of reactions. (a) Vacuolar sucrose and starch
present in plastids are consumed. During this time the cell dry
weight decline pregressively. (b) When almost all the intracellular
carbohydrate pools have disappeared the cell fatty acids decline
progressively (the same holds true for the respiration rate) with
a parallel increase in polar lipid deacylation products such as
glycerylphosphorylethanolamine, glycerylphosphorylcholine and P-choline (it is possible that glycerylphosphorylcholine is ultimately
hydrolyzed to glycerol and P-choline). Furthermore decreased levels
of P-choline and glycerylphosphorylcholine were noted in response
to sucrose addition. Thus, the rapidly increased levels of glycerylphosphorylcholine arising in response to sucrose deprivation appear
to be derived from phosphatidylcholine by a lipolytic acyl hydro-
lase-mediated mechanism. These results give a logical explanation
for the fact that all of the plant cells studied so far contain
very powerful lipolytic acyl hydrolase activities (Galliard, 1980)
and possess in their peroxisomes all the enzyme equipment capable

to oxidize free fatty acids (Gerhardt, 1983). The fatty acids thus released are probably oxidized in peroxisomes and mitochondria operating in a concerted manner. The presence of glycerylphosphoryl-choline and P-choline in plant cells in large excess should be considered, therefore, as good markers of membrane utilization after a long period of sucrose starvation and is very likely related to stress.

The results presented here also demonstrate that the slow decrease in the uncoupled rate of O_2 consumption observed in sycamore cells after a long period of sucrose starvation is attributable to a progressive diminution of the number of mitochondria/cell. The arguments in favor of this autophagic activity are (a) the decrease in the rate of respiration triggered by sucrose starvation is not attributable to the availability of substrate for mitochon-drial respiration (Journet et al., 1986) ; (b) on a protein basis, the rates of O_2 uptake in state 3 are about the same for normal and sucrose-starved mitochondria. Likewise, the cardiolipin and cytochrome aa₃ content of sycamore mitochondria (two typical mito-chondrial markers) is invariant to sucrose depletion ; (c) the uncoupled respiration rates decrease in the same ratio as the de-crease in total intracellular cardiolipin or cytochrome aa₃. The same things probably hold true with the cytosolic and plastidial fractions because glycolytic enzymes in the cytosol (Journet et al., 1986) and galactolipids in plastids (Journet et al., 1986) progressively disappear within the same period of time. It is possible, therefore, that the number of mitochondria/unit volume of cytoplasm may be maintained while the total protein of the cell drops sharply during the starvation period.
The data reported here also raise the problem of the mechanisms involved in the slow disappearance of mitochondria in sycamore cells during their ageing in sucrose-free culture medium. Following the work of Matile's group (for review, see Matile and Wiemken, 1976) it is possible that cytoplasmic material containing various cell organelles may be engulfed by the tonoplast membrane. Under these conditions, the products of protein, nucleic acid, and lipid breakdown formed in the vacuoles could be metabolized in the cyto-plasm to feed the remaining mitochondria with respiratory substrates. In support of this suggestion electron micrographs of numerous plant cells suggest that phagocytosis-like invaginations of the tonoplast result in the formation of intravacuolar vesicles contai-ning a portion of cytoplasm (for review, see Matile and Wiemken, 1976). For example, within senescent protoplasts, Wittenbach et al. (1982) have shown that chloroplasts appeared either to move into invaginations of the vacuole or to be taken up into the vacuole. Furthermore, Blum and Buetow (1963) have demonstrated that starvation of *Euglena* in the dark on an inorganic medium triggers an intracellu-lar digestion of cytoplasmic constituents such as RNA, DNA, and protein (autophagic metabolism).
It is obvious that these results emphasize the very flexibility and complexity of plant cell metabolism. This flexibility is com-pounded by the fact that cytoplasm, in particular, can be utilized as a carbon source after a long period of sucrose starvation without significantly affecting the survival of these cells. Under these conditions, plant cells, owing to the presence of intracellular pools of carbohydrate and to their ability to control an autophagic process, can survive adequately for a long period of time without receiving any external supply of organic carbon.
Finally, we have also shown that the total protein of the cell also decreased sharply during the starvation period. Conse-

quently, amino acids released during the course of protein breakdown could be metabolized to provide the remaining mitochondria with respiratory substrates. In general, the catabolic sequence for each amino acid leads to a compound which is capable of entering the tricarboxylic acid cycle (Mazelis, 1980). The fate of NH_3 thus released should be considered.Preliminary experiments carried out in our laboratory have indicated that part of NH_3 released is engaged in the form of asparagine. Under these conditions catabolism of fatty acids and amino acids must be sufficiently intense to supply mitochondria with respiratory substrates in order to maintain a high nucleotide energy charge (Pradet and Raymond, 1983) in the remaining cytoplasmic fraction (Fig. 5).

REFERENCES

Bligny, R. (1977) Plant Physiol. 59, 502-505

Bligny, R. & Douce, R. (1980) Biochim. Biophys. Acta 617, 254-263

Blum, J.J. & Buetow, D.E. (1963) Exp. Cell Res. 29, 407-421

Douce, R., Christensen, E.L. & Bonner, W.D. (1972) Biochim. Biophys. Acta 275, 148-160

Evans, F.E. & Kaplan, N.O. (1977) Proc. Natl. Acad. Sci. U.S.A 74, 4909-4913

Galliard, T. (1980) in The Biochemistry of Plants. A comprehensive treatise (Stumpf, P.K. ed.), vol. 4, pp. 85-116, Academic Press, New York

Gerhardt, B. (1986) Physiol. Veg. 24, 397-410

Jacobus, W.E., Moreadith, R.W. & Vandegaer, K.M. (1982) J. Biol. Chem. 257, 2397-2402

Journet, E., Bligny, R. & Douce R. (1986) J. Biol. Chem. 261, 3193-3199

Maizel, J.V., Benson, A.A. & Tolbert, N.E. (1956) Plant Physiol. 31, 407-408

Matile, Ph. & Wiemken, A. (1976) in Encyclopedia of Plant Physiology (Stocking, C.R. & Heber, U. eds.) vol. 3, pp. 255-287, Springer-Verlag, Heidelberg

Mazelis, M. (1980) in The Biochemistry of Plants. A comprehensive treatise vol. 5 (Miflin, B.J. ed.) pp. 541-567, Academic Press, New York

Navon, G., Ogawa, S., Shulman, R.G. & Yamane, T. (1977) Proc. Natl. Acad. Sci. U.S.A. 74, 87-91

Navon, G., Shulman, R.G., Yamane, T., Eccleshall, T.R., Lam, K.B., Baronofsky, J.J. & Marmur, J. (1979) Biochemistry 18, 4487-4499

Nishimura, M., Douce, R. & Akazawa, T. (1982) Plant Physiol. 69, 916-920

Pradet, A. & Raymond, P. (1983) Annu. Rev. Plant Physiol. 34, 199-224

Rébeillé, F., Bligny, R., Martin, J.B. & Douce, R. (1985) Biochem. J. 226, 679-684

ap Rees, T. (1985) in Encyclopedia of Plant Physiology vol. 18(Douce, R. & Day, D.A. eds.) pp. 391-417, Springer-Verlag, Heidelberg

Roberts, J.K.M. & Jardetzky, O. (1981) Biochim. Biophys. Acta 639, 57-76

Roby, C., Martin, J.B., Bligny, R. & Douce, R. (1987) J. Biol. Chem. 262, 5000-5007

Satre, M., Klein, G. & Martin, J.B. (1986) Biochimie 88, 1253-1261

Wittenbach, V.A., Lin, W. & Hebert, R.R. (1982) Plant Physiol. 69, 98-102

Herbicide action on photosynthetic membranes

Achim TREBST

Lehrstuhl für Biochemie der Pflanzen, Ruhr-Universität Bochum, Postfach 10 21 48, D-4630 Bochum 1, Federal Republic of Germany

SYNOPSIS

The mode of action of herbicides inhibiting photosystem II is in displacing plastoquinone Q_B from its specific binding niche on the D-1 protein subunit of the photosystem II complex. The DNA sequence of the psbA gene encoding the herbicide binding protein is known. The secondary structure of the herbicide or Q_B binding protein D-1 can be predicted and the dimensions and topology of the binding niche described already in great detail. DNA sequencing of the psbA gene from herbicide tolerant mutants, sequencing of the protein with covalently attatched herbicides, site specific antibodies and other approaches for probing membrane protein structures contributed to the present state of knowledge on the folding of the herbicide binding protein. The herbicides can now be modelled into their binding niche. The interaction of essential chemical elements in the structure of the herbicides with and their binding affinities to specific amino acid residues in the target can be evaluated.

INTRODUCTION

Numerous herbicides are known to interfere with photosynthesis inhibiting either electron transport, ATP synthesis or the biogenesis of pigments, lipids or proteins (for review see Fedtke, 1982; Corbett et al., 1984; Sandmann & Böger, 1986). Most advanced is the understanding of the inhibition by herbicides of photosynthetic oxygen evolution, by blocking the acceptor site of photosystem II. These inhibitors interfere with photosynthetic electron flow by displacing plastoquinone from one of its binding sites in the membrane, in this case from the Q_B site on the D-1 subunit of photosystem II. This has been reviewed often, (for example Pfister & Urbach, 1983; van Rensen, 1985; Trebst & Draber, 1986; Renger, 1986). Research on the mode of action of herbicides on photosystem II has greatly aided both photosynthesis research in general and on photosystem II in particular, as well as the molecular understanding of herbicide mode of action and of their structure function correlationships.

The compounds

        Diuron was the first herbicide to be known to inhibit specifi-
cally photosynthetic electron flow. It was followed by analogues
that had the same basic urea structure. But then many compounds
with - seemingly - quite different chemical structures were introduced as photosystem II herbicides: anilides, triazines, triazinones, pyridazinones, benzimidazoles, (bis)-carbamates and many
others (see Hansch, 1969; Moreland, 1970; Büchel, 1979; Fedtke,
1982; Corbett et al., 1984; Sandmann & Böger, 1986, Trebst &
Draber, 1986), more recently cyanoacrylates (Phillips & Huppatz,
1984), pyrones and cyclohexanediones (Asami et al., 1986). Quantitative structure activity correlations (QSAR) were made early (see
Hansch, 1969; Büchel, 1979) that are being continuously improved
(Kakkis et al., 1984; Mitsutake et al., 1986; Draber, 1987). Common
basic elements in the chemical structure of the herbicides were
defined from there (Hansch, 1969; Moreland, 1970; Büchel, 1979).
Interesting historically, an interaction of this chemical element
with peptide bonds in the membrane was discussed early (Hansch,
1969), later dismissed, but is now shown to be correct (see below).
Also an analogy of the inhibitors to plastoquinone was discussed
(van Rensen, 1971), dismissed then by most, but now accepted (see
below). Molecular orbital calculations (for example Trebst et al.,
1984; Bühmann et al., 1987) refined the physico-chemistry of the
herbicides, as it is now particular relevant for the molecular
modelling of the interaction of the compounds with their target.
The dimensions of the binding pocket in the target were deduced
from QSAR of triazinones (Draber & Fedtke, 1979). The rational
design of new compounds improved this way continuously from rules
of thumbs to computer aided design and now to docking of specific
substituents of the inhibitors onto specific amino acids in the
binding niche, revealed by the X-ray structure of the homologous
target protein from photosynthetic bacteria (see below).
        Another group of inhibitors of photosystem II developed some
years ago from the observation that substituted phenols not only
uncouple ATP synthesis from respiratory and photosynthetic electron
transport, but are also effective electron flow inhibitors (Trebst
& Draber, 1979; van Rensen & Hobe, 1979). Herbicides like dinoseb
or ioxynil and many other phenol derivatives block the Q_B site of
photosystem II as well as DCMU/triazines do. The structure activity
correlationships, however, do not follow those of the urea/triazine
type inhibitors where electronic and lipophilic parameters are
governing the inhibitory potency (Hansch, 1969; Büchel, 1979).
Instead, steric parameters, but also again hydrophobic interactions
contribute to the potency of phenol type inhibitors (Trebst &
Draber, 1979). Other compounds of this class followed: hydroxypyri-
dines, chinolones, ketonitriles, quinones, pyrones and chromones
(see Trebst & Draber, 1986 for references). A concept of overlapping binding sites in a common binding domain in the herbicide
target was formulated in order to accomodate all classes of herbicides on the same area in the protein in spite of the different
chemistry of the compounds (Trebst & Draber, 1979; Pfister &
Arntzen, 1979; Renger, 1979).
        With the increased knowledge of the chemistry and structure of
the target, as discussed below, the fitting of the herbicides into
their binding niche as well as their cross resistance in herbicide
tolerant mutants of higher plants and algae allowed to specify even
more detailed the essential chemical elements in the herbicides,
subgroup them and show on molecular terms the role of substituents

in their interaction with the sidechains of the amino acids in the target. From this, for example, a serine family of herbicides (the urea/triazine group) was contrasted to a histidine family (the phenol-type group) (Trebst, 1987). MO and forcefield calculation showed that indeed the first group has the ability to undergo hydrogen bridges to carbonyl-oxygen or to nitrogen from a peptide or serine OH bond in the target, whereas the second inhibitor class has a hydrogen group at the equivalent position and therefore will be pushed away from serine 264 towards histidine 215 in the binding niche (Trebst, 1987; Bühmann et al., 1987).

The target on the photosystem II complex

Functional studies with isolated photosynthetic membrane systems had clearly established that herbicides of the diuron-type inhibit specifically photosystem II. The details of the inhibitory pattern showed that the precise target is the acceptor side of PS II at a protein that "shields" the reduction site of plastoquinone (Renger, 1976), called now the Q_B site. By binding to that protein diuron would inhibit electron transfer from Q_A to Q_B, likely by displacing Q_B from its site (Velthuys, 1981). The "shield" protein or Q_B^- or herbicide-binding protein was then identified as the "rapidly turning over" D-1 subunit of PS II encoded by the psbA gene of the chloroplast genome (Mattoo et al., 1981). Essential for this identification was the photoaffinity labeling technique, as a radioacitve azido-triazine derivative would upon irridation bind covalently to a 32 kDa protein (Pfister et. al., 1981). This has now been shown also for other azido-substituted herbicides (Oett-meier et al., 1984; Boschetti et al., 1985; Bühmann et al., 1987). With the identification of the nucleotide sequence of the psbA gene also of herbicide tolerant mutants of higher plants (Hirschberg & McIntosh, 1983) and algae (see Rochaix & Erickson, 1988) and finally with the transformation of both higher plants (Cheung et al., 1988) and blue-green algae with the mutated psbA gene to yield herbicide tolerant organisms it is now beyond doubt that the D-1 subunit of PS II is the herbicide target protein. Secondary effect in the physiological response of a herbicide tolerant organism to the changed photosynthetic capacity, for example in a changed lipid composition of the membrane (St.John, 1982), should not be confused with the primary effect. But there are other protein targets in the thylakoid membrane as well, though of lower binding affinity. The binding curves of all herbicides to the thylakoid membrane indicate several, in case of phenol herbicides even many, additional "unspe-cific" binding sites (see Oettmeier & Trebst, 1983). These are not changed in the point mutations of herbicide tolerant organisms.
The structure of the D-1 protein - the three-dimensional folding of its primary amino acid sequence - should reflect as mirror image the inhibitors as they dock with their essential ele-ments and substituents onto specific amino acid residues of the protein target.
The primary amino acid sequence of the D-1 polypeptide is derived from the nucleotide sequence of the psbA gene (Zurawski et al., 1982; Hirschberg & McIntosh, 1983). The sequence can be ana-lyzed with algorithms for - for example - identification of hydro-phobic α-helices (Kyte & Doolittle, 1982). The comparison with homologous sequences of the D-1 polypeptide and the L and M subunit of bacterial reaction centers (Deisenhofer et al., 1985; Hearst, 1986; Trebst, 1986) yielded the interpretation that the D-1 poly-

peptide chain contains five domains long and hydrophobic enough to span the membrane in α-helices, i.e. there are five transmembrane helices and the polypeptide is exposed on either side of the membrane. The first protein sequences are available now also (Dostatni et al., 1988; Michel,et al., 1988). Accordingly the N-terminus starts with a phosphorylated and acetylated threonine (Michel et al., 1988) the second amino acid in the open reading frame of the nucleotide sequence. This indicates that the N-terminus is on the (for ATP) accessible matrix side of the thylakoid membrane and therefore the carboxyl terminus on the lumen side, as predicted (Trebst, 1986).

Further information of the folding of the amino acid sequence of the D-1 polypeptide and in particular of the amino acids involved in the binding niche were derived from
1. homology in amino acid sequences and in conservation of essential amino acids of the D-1 and D-2 subunit of PS II to the L and M subunit of the bacterial reaction center (Hearst, 1986; Deisenhofer et al., 1985; Trebst, 1986; Trebst, 1987; Michel & Deisenhofer, 1988) and in particular to the X-ray structure of the latter (Deisenhofer et al., 1985; Michel et al., 1986a; Michel & Deisenhofer, 1986),
2. site directed antibodies against the D-1 and the homologous D-2 subunit (Sayre et al., 1986; Geiger et al., 1987),
3. amino acid changes in herbicide tolerant mutants (for review see Rochaix & Erickson, 1988; Michel & Deisenhofer, 1988) and in site directed mutagenesis of the primary donor for PS II (Debus et al., 1988),
4. trypsination experiments - together with the antibodies (Mattoo et al., 1981; Trebst et al., 1988),
5. protein sequencing of the D-1 polypeptide with covalently bound herbicides, attached to the protein by photoaffinity labeling (Wolber et al., 1986; Dostatni et al., 1988).

Table 1. Amino acids in the Q_B site of the D-1 polypeptide subunit of PS II

Identified by:

Mutations in herbicide tolerance:	Phe 211, Val 219, Ala 251, Phe 255, Gly 256, Ser 264, Leu 275
Photoaffinity labelling with azido-triazine/monuron:	Met 214, Tyr 237, Tyr 254
Protection of PEST and cleavage site in rapid turnover and photoinhibition:	Glu 228 to Arg 238 and on to Glu 244
Protection of trypsin cut:	Arg 238
Fe-binding:	His 215, His 272
Quinone binding:	Tyr 262 to Phe 265 and His 215

---

6. the identification of an endogenous cleavage site in the D-1
   polypeptide that is controlled by the herbicides (Greenberg et
   al., 1987), see below.
This yields individual amino acids and the sequences that are part
of the Q_B and herbicide binding site on the D-1 polypeptide,
summarized in Table 1.

   This evidence predicted that the D-1 polypeptide is not only
the Q_B and herbicide binding protein of photosystem II, but actually carries the reaction center of PS II (Deisenhofer et al., 1985;
Trebst, 1986; Michel & Deisenhofer, 1988). Together with the
homologous D-2 polypeptide (the Q_A binding polypeptide), the D-1
polypeptide binds the reaction center chlorophyll (probably a dimer
on his 198), pheophytin, carotene, Fe and Mn and the primary
electron donor(s) for PS II, called Y or Z and D. New reaction
center preparations of PS II proved this predictions correct (Nanba
et al., 1987; Barber, 1987).

   The folding model of the D-1 polypeptide (Trebst & Draber,
1986; Trebst, 1986; Trebst, 1987) indicates that all amino acid
changes in the herbicide tolerant mutants are in a relatively small
area on the matrix side of the membrane. These changes correspond
well with mutations in the L subunit of purple bacteria (Gilbert et
al., 1985; Michel et al., 1986; Paddock et al., 1987) and in
particular with the dimensions of terbutryn binding in the X-ray
structure of a crystallized bacterial reaction center (Michel et
al., 1986a; Michel & Deisenhofer, 1988). Therefore quite detailed
information on the dimensions of the herbicide binding niche
becomes available. Accordingly (Trebst, 1987) the urea/triazine
type of inhibitors and plastoquinone Q_B may interact with the
peptide bond close to and the hydroxy group of serine 264 (counted
from met 1 from the nucleotide sequence; it would be ser 263, if
counted from thr 1 from the protein sequencing (Michel et al.,
1988)). The binding niche would extend towards his 215. The smaller
sidechain would extend towards ser 264 with phe 255 below and ala
251 above the ring. The hydrophobic sidechains would extend towards
the other side. Longer sidechains in the herbicides would pass by
tyr 254. On top of the binding niche would be the sequence around
arg 238 that contains both the trypsin (at arg 238) and an endogenous cleavage site, as well as tyrosine 237 observed in the photoaffinity labeling site (discussed below).

   The cross-resistance of herbicide tolerant *Chlamydomonas*
mutants, isolated via triazine, triazinone or urea screening (see
Rochaix & Erickson, 1988) tells that the binding affinities of
inhibitors towards specific amino acids is not the same for all
herbicides. Reactive substituents of the inhibitors interact with
certain amino acid residues. As the substituents are not the same
in the different inhibitors, they react also with different amino
acids. This is particularly well seen in the binding of azido-
triazine vs azido-monuron at met 214 vs tyr 254 (Wolber et al.,
1986; Dostatni et al., 1988). Each interaction varies in their
contribution to the total binding energy of a herbicide onto the
protein. This is rephrasing in molecular terms the concept of
"specific binding in a common binding domain" developed at the
beginning for the different groups of inhibitors (see above and
Pfister & Arntzen, 1979; Renger, 1979; Trebst & Draber, 1979;
Trebst & Draber, 1986). MO and forcefield calculations and the
molecular modelling of inhibitors into the target protein are
already about to provide further details.

   Further important information on amino acid sequences in the
herbicide binding niche comes from the special property of the D-1

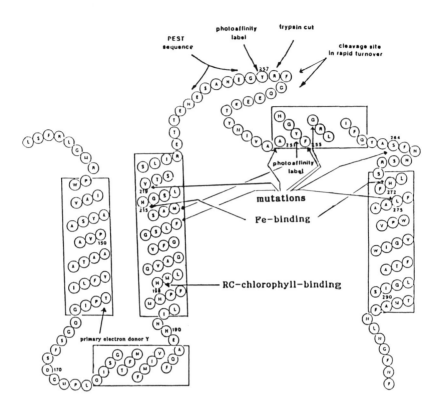

Fig. 1. The predicted folding of part of the amino acid sequence of the Q_B and herbicide binding protein subunit D-1 of photosystem II Indicated are amino acids that are functionally involved in pigment and carrier binding. In quinone and herbicide binding those amino acids participate that are indicated by mutations, photoaffinity labeling and protease attack.

----

polypeptide: its "rapid turnover" (Mattoo et al., 1984). The Q_B protein is in the light much faster synthesized, but also degraded than all the other protein subunits of the thylakoid membrane. This is a long known phenomenon and it has led – as stated above – in the identification of the psbA gene coding for the herbicide binding polypeptide D-1 (Mattoo et al., 1981). Obviously the D-1 polypeptide is cleaved by a specific protease. Although that enzyme has not been characterized yet, the specific cleavage site area in the protein sequence has been identified (Greenberg et al., 1987). It is the sequence around arg 238. The cleavage is prevented by diuron (Mattoo et al., 1984). Diuron and other herbicides also block a specific trypsin cut at also arg 238 of the D-1 polyppetide (Trebst et al., 1988). Furthermore – as already discussed – azido-monuron interacts with tyr 237 (Dostatni et al., 1988). From these three observations it can be concluded that the amino acid sequence around arg 238 of the D-1 polypeptide is part of the herbicide and Q_B binding site in addition to those amino acid sequences indicated by the mutations. That sequence folds above and sideways on those amino acids identified by the mutations. It is this sequence that is mainly responsible for the difference between the bacterial and the plant photosystem and the L and D-1 subunit respectively. It is

absent in the L subunit of the bacteria and this may be part of the
reason why bacterial photosynthesis is inhibited by very few of the
herbicides that inhibit photosystem II.

REFERENCES

Asami, T., Takahashi, N. & Yoshida, S. (1986) Z. Naturforsch. 41c,
751-757
Barber, J. (1987) Trends Biochem. Sci. 12, 321-326
Boschetti, A., Tellenbach, M. & Gerber, A. (1985) Biochim. Biophys.
Acta 810, 12-19
Büchel, K.-H. (1979) Pestid. Science 3, 89-110
Bühmann, U., Herrmann, E.C., Kötter, C., Trebst, A., Depka, B. &
Wietoska, H. (1987) Z. Naturforsch. 42c, 704-712
Cheung, A.Y., Bogorad, L., van Montagu, M. & Schell, J. (1988)
Proc. Natl. Acad. Sci. USA 85, 391-395
Corbett, J.R., Wright, K. & Baillie, A.C. (1984) The Biochemical
Mode of Action of Pesticides. Second edition. Academic Press, New
York
Debus, R.J., Barry, B.A., Babcock, G.T. & McIntosh, L. (1988) Proc.
Natl. Acad. Sci. USA 85, 427-430
Deisenhofer, J., Epp, O., Miki, K., Huber, R. & Michel, H. (1985)
Nature 318, 618-624
Dostatni, R., Meyer, H.E. & Oettmeier, W. (1988) submitted
Draber, W. & Fedtke, C. (1979) in Advances in Pesticide Science
(Geissbühler, H., ed.), Part 3, pp. 475-486, Pergamon Press,
Oxford, New York
Draber, W. (1987) Z. Naturforsch. 42c, 713-717
Fedtke, C. (1982) Biochemistry and Physiology of Herbicide Action,
Springer-Verlag, Berlin, Heidelberg, New York
Geiger, R., Berzborn, R.J., Depka, B., Oettmeier, W. & Trebst, A.
(1987) Z. Naturforsch. 42c, 491-498
Gilbert, C.W., Williams, J.G.K., Williams, K.A.L. & Arntzen, C.J.
(1985) in Molecular Biology of the Photosynthetic Apparatus, pp.
67-71, Cold Spring Harbor Laboratory, Cold Spring Harbor
Greenberg, B.M., Gaba, V. & Mattoo, A.K. (1987) EMBO J. 6, 2865-
2869
Hansch, C. (1969) in Progress in Photosynthesis Research, Vol. III.
Photophosphorylation , CO_2 Fixation, Action Mechanisms of Herbi-
cides (Metzner, H., ed.), pp. 1685-1692, Metzner, Tübingen
Hearst, J.E. (1986) in Encyclopedia of Plant Physiology, New Series
Vol. 19. Photosynthesis III. Photosynthetic Membranes and Light
Harvesting Systems (Staehelin, L.A. & Arntzen), pp. 382-389
Springer-Verlag, Berlin, Heidelberg, New York, Tokyo
Hirschberg, J. & McIntosh, L. (1983) Science 222, 1346-1348
St.John, J.B. (1982) in Biochemical Responses Induced by Herbi-
cides. ACS Symposium Series 181 (Moreland, D.E., St.John, J.B. &
Hess, F.D., eds.), pp. 97-109, American Chemical Society, Washing-
ton
Kakkis, E., Palmire, V.C., Strong, C.D., Bertsch, W., Hansch, C. &
Schirmer, U. (1984) J. Agric. Food Chem. 32, 133-144
Kyte, J. & Doolittle, R.F. (1982) J. Mol. Biol. 157, 105-132
Mattoo, A.K., Pick, U., Hoffmann-Falk, H. & Edelman, M. (1981)
Proc. Natl. Acad. Sci. USA 78, 1572-1576
Mattoo, A.K., Hoffmann-Falk, H., Marder, J.B. & Edelman, M. (1984)
Proc. Natl. Acad. Sci. USA 81, 1380-1384
Michel, H. & Deisenhofer, J. (1986) in Encyclopedia of Plant
Physiology, New Series Vol. 19. Photosynthesis III. Photosynthetic

Membranes and Light Harvesting Systems (Staehelin, L.A. & Arntzen), pp. 371-381, Springer-Verlag, Berlin, Heidelberg, New York, Tokyo

Michel, H.P., Weyer, K.A., Gruenberg, H., Dunger, I., Oesterhelt, D. & Lottspeich, F. (1986) EMBO J. 5, 1149-1158

Michel, H., Epp, O. & Deisenhofer, J. (1986a) EMBO J. 5, 2445-2451

Michel, H. & Deisenhofer, J. (1988) Biochemistry 27, 1-7

Michel, H., Hunt, D.F., Shabanowitz, J. & Bennett, J. (1988) J. Biol. Chem. 263, 1123-1130

Mitsutake, K., Iwamura, H., Shimizu, R. & Fujita, T. (1986) J. Agric. Food Chem. 34, 725-732

Moreland, D.E. (1970) Ann. Rev. Plant Physiol. 35, 659-593

Nanba, O. & Satoh, K. (1987) Proc. Natl. Acad. Sci. USA 84, 109-112

Oettmeier, W. & Trebst, A. (1983) in The Oxygen Evolving System to Photosystem II (Inoue, Y., ed.), pp. 411-420, Academic Press Japan, Tokyo

Oettmeier, W., Masson, K., Soll, H.-J. & Draber, W. (1984) Biochim. Biophys. Acta 767, 590-595

Paddock, M.L., Williams, J.C., Rongey, S.H., Albresch, E.C., Feher, G. & Okamura, M.Y. (1987) in Progress in Photosynthesis Research, Vol. III (Biggins, ed.), pp. 11.775-11.778, Martinus Nijhoff Publishers, Dordrecht

Pfister, K. & Arntzen, C.J. (1979) Z. Naturforsch. 34c, 996-1009

Pfister, K., Steinback, K.E., Gardner, G. & Arntzen, C.J. (1981) Proc. Natl. Acad. Sci. USA 78, 981-985

Pfister K. & Urbach W. (1983) in Encyclopedia of Plant Ecology IV. (Lange, O.L., Nobel, P.S., Osmond, C.B. & Ziegler, H., eds.), Springer-Verlag, Berlin, Heidelberg

Phillips, J. & Huppatz, J. (1984) Z. Naturforsch. 39c, 335-337

Renger, G. (1976) Biochim. Biophys. Acta 635, 236-248

Renger, G. (1979) Z. Naturforsch. 34c, 1010-1014

Renger, G. (1986) Physiol. Veg. 24, 509-521

van Rensen, J.J.S. (1971) Thesis, Wageningen

van Rensen, J.J.S. & Hobe, J.H. (1979) Z. Naturforsch. 34c, 1021-1023

van Rensen, J.J.S. (1985) in Weed Science Advances (Turner, R.G., ed.), Vol. 1, pp., Butterworth, London

Rochaix, J.-D. & Erickson, J. (1988) Trends Biochem. Sci. 13, 56-59

Sandmann, G. & Böger, P. (1986) in Encyclopedia of Plant Physiology, New Series Vol. 19. Photosynthesis II (Staehelin, L.A. & Arntzen, C.J., eds.), pp. 595-602, Springer-Verlag, Berlin, Heidelberg, New York, Tokyo

Sayre, R.T., Andersson, B. & Bogorad, L. (1986) Cell 47, 601-608

Trebst, A. & Draber, W. (1979) in Advances in Pesticide Science (Geissbühler, H., ed.), Part 2, pp. 223-234, Pergamon Press, Oxford, New York

Trebst, A., Donner, W. & Draber, W.T. (1984) Z. Naturforsch. 39c, 405-411

Trebst, A. (1986) Z. Naturforsch. 40c, 237-241

Trebst, A. & Draber, W. (1986) Photosynthesis Res. 10, 381-392

Trebst, A. (1987) Z. Naturforsch. 42c, 742-750

Trebst, A., Depka, B., Kraft, B. & Johanningmeier, U. (1988) Photosynthesis Res., in press

Velthuys, B.R. (1981) FEBS Lett. 126, 277-281

Wolber, P.K., Eilmann, M. & Steinback, K.E. (1986) Arch. Biochem. Biophys. 248, 224-233

Zurawski, G., Bohnert, H.J. Whitfeld, P.R. & Bottomley, W. (1982) Proc. Natl. Acad. Sci. USA 79, 7699-7703

Effects of water stress on photosynthesis and related processes

Michael SPEER, Jutta E. SCHMIDT and Werner M. KAISER

Botanisches   Institut   der   Universität,   Mittlerer
Dallenbergweg 64, D-8700 Würzburg

SYNOPSIS

Direct effects of rapid or slow dehydration on leaves
or leaf tissues from various plant species were compared.
Photosynthetic capacity and leaf nitrate reduction were
measured as a function of the relative water content of
leaves. Solute efflux from dehydrated or rehydrated leaf
tissue was used as an indicator for membrane damage during
dehydration and subsequent rehydration.

INTRODUCTION

Plants have developed different strategies to cope
with drought. Poikilohydrous plants change their tissue
water potential in parallel with the water potential of
air and soil, and recover rapidly even after severe
desiccation. Higher plants can buffer such variations to
different degrees e.g. by slowing down transpirational
water loss or by storing water in organs with a low
surface/volume ratio or by developing deep root systems.
As a first response to drought, higher plants usually
close their stomata in order to prevent water loss.
Inevitably this slows down CO_2 uptake, which in turn will
affect not only photosynthesis, but sooner or later
primary metabolism in general. When dehydration proceeds,
all organisms - and not only higher plants - have to face
similar problems at the cellular level: With increasing
water deficit, cells shrink; as a direct consequence,
internal solutes are concentrated to such an extent that
they eventually reach toxic concentrations for certain
enzymes. In addition, changes in cell size and form may
exert "mechanical" stress on membranes leading to membrane
rupture. Productivity and growth under mild dehydration

ABBREVIATIONS: DHAP, dihydroxyacetone   phosphate;   OAA,
oxaloacetate;   PEP,   phosphoenolpyruvate;   PGA,   3-
phopsphoglycerate; RubP, ribulose-1,5-bisphosphate; RWC,
relative water content.
Financial support by the DFG is gratefully acknowledged.

need unimpaired enzymatic activity. Survival of severe
dehydration needs avoidance of membrane damage or repair
mechanisms.

In the following we show that moderate dehydration
down to relative water contents of 70 % often inhibited
photosynthesis only via stomatal closure, without a change
in mesophyll capacity. However, as a consequence of
decreased CO_2 availability, other basic processes such as
nitrate reduction were also affected. Direct impairment
of chloroplast functions was observed only when
dehydration becomes more severe. It appeared to be due
mainly to inhibition of water soluble enzymes. Membrane
damage was observed only during extreme desiccation and
even more during rapid rehydration.

RESULTS AND DISCUSSION

Photosynthetic capacity as a function of the relative
water content of leaves: a comparison of different species
and stress conditions.

Among all metabolic processes, the response of
photosynthesis to drought has gained most attention. One
reason is certainly that plant growth depends ultimately
on photosynthesis. Another reason is that photosynthesis
is often limited by stomatal resistance which increases
under drought. However, in addition to stomatal effects on
photosynthesis, a non-stomatal component has been proposed
to contribute to the overall inhibition of photosynthesis
under drought (for review see Hsiao 1973, Bradford & Hsiao
1982, Schulze 1986).
Evidence for a non-stomatal inhibition of photosynthesis
by drought has been mainly derived from gas exchange data,
where water stress caused a decrease in net
photosynthesis, with little or no change in calculated
intercellular CO_2 concentrations (for review see Hsiao
1973, Kriedemann & Downton 1981, Bradford and Hsiao 1982,
Schulze 1986). The validity of such calculations has been
questioned recently at least for some conditions (Laisk
1983, Sharkey 1984). Therefore it seems desirable to
measure photosynthetic capacity without limitation by
stomatal resistance. This is achieved e.g. by applying
very high external CO_2 concentrations up to 15 % in a
leaf disc oxygen electrode chamber.
Fig. 1 compares photosynthetic capacity of various
plant species under two different conditions: detached
leaves were rapidly (over periods of up to 30 hours)
wilted under a stream of air at very low photon flux
density (10 µE m^{-2} sec^{-1}), or whole potted plants were
wilted slowly (i.e. over periods of days or weeks or even
months) under their standard growth conditions.
Photosynthetic capacity was measured at saturating photon
flux density. When calculated on a chlorophyll basis, it
decreased as a function of the relative water content
(RWC), and the response of the different species examined
was quite similar when wilting was done rapidly and at low
photon flux density (Fig.1). Basically, an effect of
dehydration on photosynthesis became visible only after
RWC had decreased by at least 30%. For some species such
as Peperomia magnoliaefolia, up to 50 % water loss had

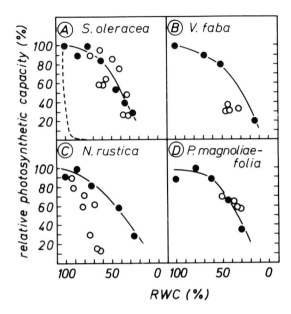

Fig. 1: Photosynthetic capacity (on a chlorophyll basis)
of leaves from various plant species in response to rapid
(●) or slow (○) wilting.
Photosynthetic capacity was measured in a leaf disc oxygen
electrode chamber (Hansatech, Bachofer, Reutlingen, FRG)
at 15 % external CO_2, at a photon flux density of 400 µE
m^{-2} sec^{-1}. Leaf discs (40 mm diameter) were cut out of
turgid leaves, and their fresh weight and photosynthetic
capacity were determined after sealing the wounded rim
with silicone paste. Subsequently, discs were exposed to
"rapid wilting" in a stream of air at room light (about
10µE $m^{-2}sec^{-1}$), and the actual fresh weight and
photosynthetic capacity were determined at various time
intervals. At the end of the experiment, part of the discs
were used for determination of dry weight and chlorophyll
content. The time to reach the final desired RWC varied
from about 12 h (N. rustica) to about 40 h (Peperomia
magnoliaefolia). For "slow wilting", water was withhold
from whole plants under their normal growth conditions.
Leaves were harvested at various times, and fresh weight,
photosynthetic capacity, dry weight and chlorophyll
content were determined. The relative water content (RWC)
was determined from the initial and the actual ratio of
fresh weight and dry weight of individual leaves.
For the different species, these " slow wilting"
conditions were: a) Spinach was grown hydroponically in a
growth chamber at about 400 µE m^{-2} sec^{-1}. Plants were
exposed to water stress by removing the culture solution
during the day; roots were sprayed intermittendly with
culture solution; during the night, plants were brought
back to the culture medium and were allowed to recover.
Total time to bring the plants to 25% RWC was 8 days.
b) All other plants were grown in pots in a green house
under additional artificial illumination at a photon flux
density of 150 µE $m^{-2}sec^{-1}$ (N. rustica) to 500 µE $m^{-2}sec^{-1}$
(P. magnoliaefolia). Stress was imposed by withholding

water. The time to reach the final RWC was: for V. faba 7
to 10 days; for N. rustica   7 days; for P. magnoliaefolia
5 months. 100 % rates were (in μmoles O_2 mg^{-1} chlorophyll
h^{-1}): S. oleracea 350; V. faba 312; N. rustica 356; P.
magnoliaefolia 159. In A, the dashed line indicates the
response of rapidly wilted, detached spinach leaves, when
photosynthesis was measured as CO_2 uptake in ambient air.

---

very little effect on photosynthetic capacity. The
experiments indicate that the photosynthetic machinery is
surprisingly insensitive to dehydration, at least at the
high CO_2 concentrations applied in our experiments. It has
to be mentioned here that this does not necessarily hold
for normal CO_2 concentrations. Loske and Raschke (1988)
recently described a partial inhibition of RuBP-
carboxylation in leaves treated with absisic acid or
exposed to dry air. If the inhibition were due to a
decreased CO_2 affinity of Rubisco, oxygen electrode
measurements at extremely high external CO_2 would lead to
an overestimation of photosynthetic capacity under
dehydration. In Fig. 1 A, (dashed line), photosynthesis of
spinach leaves which were rapidly wilted at low photon
flux density was additionally measured at ambient CO_2
concentrations. Here, inhibition was complete at water
deficits which caused no impairement of photosynthesis at
the mesophyll level at high external CO_2. This
demonstrates quite impressively the role of stomatal
resistance at water deficits from 0 to 30%.

Involvement of photoinhibiton

    When dehydration developed more slowly and at higher
photon flux density, the relation between RWC and
photosynthetic capacity remained much the same for some
species, but not for others (Fig.1). In spinach, effects
of dehydration on photosynthetic capacity were completely
independent of whether leaves were exposed to low or
normal light conditions (Fig. 1). Even a 24 h exposure to
full sunlight during rapid wilting of detached leaves
caused no additional inhibition of photosynthesis (Kaiser
1987b), indicating a remarkable insensitivity of spinach
towards photoinhibition. Absence of photoinhibition in
water stressed leaves from Helianthus annuus was
recently reported by Sharp & Boyer (1986). In some
species, however, (e.g. Nicotiana rustica, Fig. 1),
photosynthetic capacity was strongly impaired at rather
low water deficits by slow wilting at moderate photon
flux density. But even in such cases where light caused
additional impairment of photosynthetic capacity, this
might reflect a downregulation of biochemical energy
conversion to the level pretended by stomatal resistance
(in favour of energy dissipation as heat), rather than a
new limitation (for an ample discussion of the interaction
of light and water stress see Powles 1984, Osmond et al.,
1987). In leaves with closed stomata, photooxidative
damage by excess absorbed energy is obviously prevented by
photorespiration. By measuring modulated chlorophyll

fluorescence of leaves at various oxygen – and CO_2 –
concentrations, Heber et al. (in press) showed that
photorespiratory CO_2 cycling behind closed stomata
supported electron transport rates of about 100
μequivalents $mg^{-1} h^{-1}$ in leaves from <u>Arbutus unedo</u> which
had lost about 15% of their water. They showed further
that photoinhibition was prevented in this species when
photorespiration was allowed to proceed in 21% oxygen. In
1% oxygen, which does not support photorespiration,
photoinhibition (measured as decreased quantum yield)
occured. Inspite of such protective mechanisms, long-
lasting and strong water stress in some cases caused
partial bleaching of chlorophyll and leaf senescence even
at moderate photon flux densities. Spinach leaves lost
about 10% chlorophyll/ leaf area during a 8 day wilting
period at 350 μE $m^{-2} sec^{-1}$, and leaves from <u>Peperomia</u>
<u>magnoliaefolia</u> up to 40 % chlorophyll during a period of 5
months without water at about 500 μE $m^{-2} sec^{-1}$ (not shown).
As discussed below, strong dehydration affects soluble
enzymes in the water phase of subcellular compartments.
Therefore photorespiration is expected to be impaired by
severe dehydration to a similar extent as photosynthesis,
and this will inevitably increase the danger for
photoinhibition.

Effect of mild water stress on nitrate reduction

When stomata closed under mild dehydration (RWC 90 to
95%) of leaves, CO_2 assimilation was not the only process
which suffered directly and immediately. Nitrate reduction
in spinach leaves was also inhibited. When detached turgid
leaves were illuminated in air, their endogenous nitrate
level (mainly vacuolar) decreased, indicating reduction
(Fig.2). Initial rates were 5 to 10 μmoles mg^{-1} chlorophyll
h^{-1}. When leaves were slightly wilted prior to
illumination, nitrate reduction slowed down drastically
(Fig.3). When wilted leaves were illuminated in 15% CO_2,
nitrate reduction occured at a normal rate (Fig. 3). It is

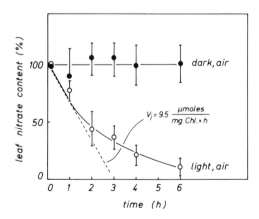

Fig. 2: Changes of the nitrate content of fully turgid,
detached spinach leaves in the dark or in the light (400
μE $m^{-2} sec^{-1}$), in air.

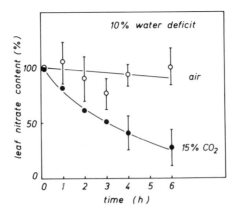

Fig.3: Effect of mild water stress on the light induced
changes of the nitrate content of spinach leaves in air or
at 15% CO_2 in air. Leaves were rapidly wilted within 1 h
to about 90% RWC in the dark prior to the experiment.

thus   obvious that: i) nitrate reduction and CO_2 reduction
are   strictly   coupled,   and ii) due to   that   , stomatal
closure   under   stress   will   affect   not   only CO_2
assimilation,   but also nitrate assimilation.   The mode of
coupling of the two processes is not yet known.   In algae,
the CO_2-dependence   of   nitrate   assimilation   has   been
attributed   to   a lack of export of   reducing   equivalents
from   chloroplasts   to cytosol under CO_2 deficiency   (For
review see Ullrich 1983,   Guerrero et al. 1981). According
to   our present knowledge on the transport systems in   the
inner chloroplast envelope, export of reducing equivalents
should   work   even   in the absence of   net CO_2 fixation,
either by malate/OAA or DHAP/PGA exchange. Thus it is hard
to believe that inhibition of nitrate reduction by low CO_2
is   due to a lack of reducing equivalents in   the   cytosol
(Guerrero et al., 1981).

Inhibition of metabolism by severe dehydration

When   dehydration   proceeded,   various   metabolic
reactions were directly and increasingly affected.   Dark-
CO_2-fixation   by   cytosolic   PEP-carboxylase   was   as
sensitive as the dark reactions of photosynthesis (Kaiser
and Heber 1981,   Kaiser et al. 1983) . The conclusion was
that   a   first consequence of   severe   dehydration   at   the
cellular level was inhibition of soluble enzymes, not only
in   the cytosol but also in chloroplasts(Kaiser and Heber
1981). This view is established by the observation that by
rapidly   wilting   leaves   at   low   light   intensities,
photosynthetic capacity at high external CO_2 was inhibited
at   a   degree of dehydration were   electron   transport   or
thylakoid   membrane   energization were   not   yet   affected
(Dietz & Heber 1983,   Kaiser 1987b,   Schreiber & Bilger
1987).   This not only true for rapidly wilted leaves,   but
also for plants stressed under more natural conditions: In

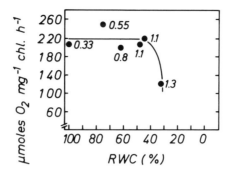

Fig. 4: Rates of uncoupled electron transport (H_2O -- Ferricanid, measured as oxygen evolution) of thylakoids isolated from leaves of slowly wilted spinach plants. Wilting conditions were as in Fig. 1. On subsequent days, thylakoids were isolated in buffer solution (containing 20 mM HEPES-KOH pH 7,8 and 1 mM $MgCl_2$), brought to the actual osmolality by addition of sorbitol. Numbers at the symbols give the sorbitol concentrations in isolation and reaction media.

the experiment shown in fig. 4, spinach leaves were wilted slowly over a period of 1 week under normal light conditions as in fig.1. On subsequent days, leaf sap osmolality was measured and thylakoids were rapidly isolated in isotonic sorbitol media. Open chain electron transport (H_2O -- Ferricyanide) in presence of ammonium chloride as uncoupler was not affected by slow dehydration of leaves until their RWC decreased below 40%. It is important to be aware that in the experiment given in fig. 4, *all* thylakoids from a small leaf piece were isolated and assayed, thus avoiding *selection* of a certain, undamaged membrane population (for experimental methods see legend of fig. 4.)
Unimpaired electron transport rates from water to ferricyanide and maintenance of the transthylakoid proton gradient in stressed leaves does not exclude an impairment of both, ATP formation and NADP reduction. In fact Younis et al. (1979) found an inhibition of the chloroplast coupling factor (CF_1) isolated from water stressed leaves. Recently, Boyer et al. (1987) have reported that in water stressed sunflower leaves, the relaxation time constant of the flash- induced absorbance change ($\Delta A_{518-540}$), which is indicative for thylakoid membrane depolarisation by proton movement through CF_1, increased considerably. Their interpretation was again that the coupling factor activity was reduced. It remained unclear, however, whether this was really caused by an inhibition of CF_1. or by decreased ADP-regeneration (with the Calvin cycle being partially inhibited).
      The postulated inhibition of soluble enzymes or of enzymes at the membrane/water interphase in dehydrated cells has been studied in more detail with the help of an "artificial stroma medium" (Kaiser et al. 1986). The composition of that medium was designed by analyzing all

major osmotically relevant solute concentrations in isolated intact spinach chloroplasts. Enzyme activities in dehydrated cells were mimicked by increasing the concentration of the medium. A major result was that the only solutes which are critical for enzyme activities were the divalent anions phosphate and sulfate. Both anions competitively inhibit enzymes catalyzing reactions with anionic substrates. E.g., the activity of Rubisco was inhibited by phosphate with an apparent K_I of 0.52 mM (with respect to RuBP). The inhibition was non-competitive with respect to CO_2, and in that respect it resembled the

Table I: Capability to recover from osmotic (intact chloroplasts, leaf slices) or evaporative dehydration (leaves).
a) Intact spinach chloroplasts were isolated from turgid leaves. Intactness was determined by measuring ferricyanide dependent oxygen evolution in presence of NH_4Cl (10 mM) as an uncoupler. Intactness was 75-98 %. Photosynthesis was measured as CO_2-dependent oxygen evolution. 100% rates were 120 µmoles mg^{-1} chlorophyll $^{-1}$. Part of the chloroplasts was suspended in a medium containing 1 M sorbitol, and intactness and photosynthesis were measured (= dehydrated). Another part was also suspended in 1 M sorbitol for 10 min, and then resuspended in isotonic medium (= rehydrated).
b) Spinach leaf slices were preincubated in a medium containing either 0.22 M sorbitol (= hydrated) or 1,5 M sorbitol (= dehydrated). After 10 min., part of the hypertonically treated slices was brought back to an isotonic medium (= rehydrated). Photosynthesis was measured as fixation of $NaH^{14}CO_3$ during a 10 min incubation at saturating light intensity. Rates are given as percentage of the hydrated control; 100% were 240 µmoles mg^{-1} chlorophyll h^{-1}.
c) Photosynthesis of discs from turgid spinach leaves, or from discs wich were rapidly wilted in dim room light down to relative water contents of 27-35% was measured in a leaf oxygen electrode chamber at 15% CO_2 and at saturating light intensity. Subsequently, discs were rehydrated by floating on water for about 12 hours, followed by a 12 h period in a moist chamber, and photosynthesis was measured again. 100 % rates were initially 350 µmoles mg^{-1} chlorophyll h^{-1}, but rates of control leaves decreased over the duration of the experiment by about 1 third.

	dehydrated	rehydrated
	(% of control)	
a) intact chloroplasts		
intactness	100	25
photosynthesis	30	0
b) photosynthesis of		
leaf slices in solution	20	100
c) photosynthesis of		
wilted leaves	30	85

response of photosynthesis in vivo, which was characterized by a decrease of both, the initial slope of the CO_2-response curve and of V_{max} (Kaiser 1984).

Heavy solute loss from cells or membrane damage during dehydration should cause a more or less *irreversible* inhibition of metabolism. Inhibition of photosynthesis of osmotically dehydrated chloroplasts was absolutely irreversible following rehydration , and in fact the intactness of the chloroplast envelope (as indicated by its permeability to ferricyanide) was worse after rehydration than in the dehydrated state (Table I). A possible explanation is that in strongly hypertonic sorbitol solutions, sorbitol influx causes excessive swelling after rehydration. In contrast, inhibition of photosynthesis of osmotically dehydrated leaf tissue slices was near perfect when slices were transferred from strongly hypertonic to *isotonic* solutions (Table I). Effects of wilting on leaves were also more or less reversible (Table I) . The degree of reversibility was much more species specific than the effect of dehydration itself (not shown). This might simply reflect differences in leaf architecture rather than biochemical differences among species. Photosynthesis of spinach leaves was irreversibly damaged by wilting only when RWC decreased below 20% (not shown).

In all species examined, solute efflux from leaf slices into a surrounding medium was only slightly

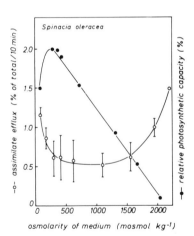

osmolarity of medium (mosmol kg⁻¹)

Fig. 5: Photosynthesis of spinach leaf slices, and efflux of ^{14}C-labelled assimilates from prelabelled leaf slices in response to osmotic dehydration. Photosynthesis was measured as $NaH^{14}CO_3$ fixation. For the efflux experiments, slices were prelabelled with $NaH^{14}CO_3$ for 20 min in a sorbitol-free medium. Subsequently, they were rinsed with non-radiaoactive medium and transferred into media with various osmolarities (sorbitol).Time course of efflux was measured for 20 min, but the figure gives efflux during the first 10 min only. All media contained 2 mM $CaCl_2$ in order to keep plasmodesmata closed. Other conditions as in fig. 6.

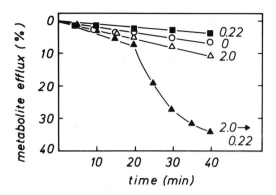

Fig. 6: Effect of osmotic dehydration and rehydration on
on efflux of metabolites from spinach leaf slices into a
surrounding medium. Slices from turgid leaves were
prelabelled with $^{14}CO_2$ in sorbitol -free buffer solution
(20 mM HEPES-KOH pH 7.6 , 1mM $CaCl_2$) for 20 min .
Subsequently they were briefly rinsed with unlabelled
medium, and transferred to media containing various
sorbitol concentrations. Part of the slices was first
suspended in hypertonic buffer solution (2 M sorbitol) for
20 min, and then brought back into a slightly hypotonic
(0.22 M sorbitol) medium . The media were exchanged at
various time intervals, and radioactivity was determined
by liquid scintillation counting after removal of non-
fixed $^{14}CO_2$ by acidification with glacial acetic acid.
Efflux is expressed as percentage of total labelled
material.

increased by hypertonic dehydration. The plasma membrane
remained intact at a stage of dehydration which caused
significant inhibition of photosynthetic capacity . This
is shown in Fig. 5. Here, spinach leaf slices were
dehydrated in hypertonic sorbitol solutions.
Photosynthesis was measured as $^{14}CO_2$-fixation, and
integrity of the plasma membrane was followed by measuring
efflux of ^{14}C-labeled assimilates from slices prelabelled
under isotonic conditions. Inhibition of photosynthesis
was obvious at sorbitol concentrations which caused no
increase in assimilate efflux. However, efflux increased
transiently upon transfer of slices from hypertonic to
isotonic or hypotonic media (Fig. 5). Rapid rehydration
appeared to be more of a problem to membranes than
dehyration. Here again, influx of osmoticum during
dehydration might produce artificial cell rupture after
rehydration without reflecting a natural situation .
    Therefore, solute efflux was also measured from leaf
tissue slices obtained from wilted or rehydrated whole
leaves. Potassium efflux into a hypotonic medium
increased with decreasing RWC of leaves (Fig. 6). It
was non-linear with time, reaching a new steady state
after about 60 to 90 min. Efflux from heavily wilted leaf
tissue into hypotonic solutions was faster than efflux
into isotonic solutions (Fig. 7), indicating again
deleterious effects of a rapid "upshock" .

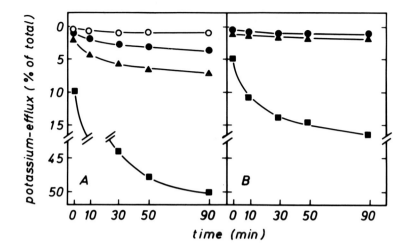

Fig. 7: Efflux of potassium from spinach leaf slices
obtained from rapidly wilted (A) or wilted and rehydrated
leaves. Slices were suspended in a potassium-free medium
(as in fig. 6), containing 0.4 M sorbitol (isotonic with
respect to turgid leaves). The medium was exchanged at the
times indicated and replaced by new , potassium -free
solution. The relative water content (RWC) of leaves was:
O 100%, ● 60 %, ▲ 25%, ■ 10%.

        When leaf slices from wilted and subsequently
rehydrated leaf tissue were suspended in isotonic
solutions, potassium efflux was again as small as from
turgid leaves. Only when the leaves experienced water
deficits below 25 %, potassium efflux after rewatering
remained high, indicating presumably irreversible
membrane damage. A comparison of data from fig.4 and
fig.7 points to the possibility that photosynthetic
membranes from leaves wilted slowly under natural
conditions were damaged earlier than the plasmamembrane.
It is not known yet wether this is a specific light
effect. When isolated intact chloroplasts were osmotically
dehydrated in sorbitol solutions mimicking a relativ water
content of 20% (about 40 bars), open chain electron
transport from water to methylviologen measured during a 2
min light period was not at all affected (Kaiser et al.
1981).
        The above described efflux experiments indicate to
what extent dehydration/rehydration cycles might effect
membrane properties. Since the apoplastic volume in
leaves is rather small (10 to 20% of the total aqueous
phase in leaf tissues, solute loss from the cytosol in
situ is presumably much smaller than solute efflux from
leaf slices into a surrounding medium, where volume ratios
medium /cytoplasm are at least 100 . However, even
transient changes in the permeability properties of
intracellular compartments might cause drastic
perturbations of metabolism. Together with a direct

inhibition   of   enzymes   by   high   internal   salt
concentrations,   this might well contribute to the overall
impairment of the functioning of the dehydrated cell, and
to a retardation of recovery after rehydration.

CONCLUSIONS

    1. Mild or moderate, rapidly imposed water stress
(i.e. RWC ≥ 70%) inhibits photosynthesis via stomatal
closure,   without   direct   effects   of dehydration   at   the
chloroplast level. In many fast growing herbaceous plants,
this holds also when stress developes slowly and at more
natural photon flux densities.   However, as a consequence
of stomatal closure,   other metabolic key processes like
nitrate reduction ar also affected.
    2. Direct inhibition of photosynthetic capacity at
RWC from 70 to about 40 % is caused primarily by
inhibition of enzymes in the water phase of cytosol and
organelles. The inhibition is not due to membrane damage
and impaired compartmentation,   but rather to increased
salt concentrations in dehydrated cells. Accordingly,
electron transport or thylakoid membrane energization are
rather insensitive, at least in the absence of high photon
flux densities. The inhibition is largely reversible upon
rehydration.
    3. Membranes become transiently permeable mainly
during rapid rehydration. Irreversible membrane rupture is
observed only when RWC decreases below 20 %.

REFERENCES

Boyer, J.S., Ort, D.R. & Ortiz-Lopez, A.O. (1987) in
    Current Topics in Plant Biochemistry and Physiology
    (Randall, D.D., Sharp, R.E., Novacky, A.J. & Blevins,
    D.G., eds.) vol. 6, University of Missouri, Columbia,
    pp.   69-73
Bradford, K.J. & Hsiao, T.C. (1982) in Encyclopedia of
    Plant Physiology (Lange O.L., Nobel, P.S., Osmond, C.B.
    & Ziegler, _H. , eds.) Vol 12 B, Springer, Berlin,
    Heidelberg, New York, pp. 264-324
Dietz, K.J. & Heber, U. (1983) Planta 158, 349-356
Guerrero, M.G., Vega, J.M. & Losada, M. (1981) Annu Rev.
    Plant Physiol. 32, 169-204
Heber, U., Neimanis, S., Setlikova, E, & Schreiber, U. (in
    press). Proceedings of the International Congress on
    Plant Physiology, New Delhi 1988
Hsiao, T.C. (1973) Annu. Rev. Plant Physiol. 24, 519-570
Kaiser, W.M. & Heber, U. (1981) Planta 153, 423-429
Kaiser, W.M., Kaiser, G., Schöner, S. & Neimanis, S.
    (1981) Planta 153, 430-435
Kaiser, W.M., Schwitulla, M. & Wirth, E. (1983) Planta
    158, 302-308
Kaiser, W.M. (1984) J. Exp. Bot. 35, 1-11
Kaiser, W.M., Schröppel-Meier, G. & Wirth, E. (1986)
    Planta 167, 292-299
Kaiser, W.M. (1987) Physiol. Plantarum 71, 142-149
Kaiser, W.M. (1987) in Current Topics in Plant
    Biochemistry and Physiology, (Randall, D.D., Sharp,

R.E.,   Novacky, A.J., & Blevins, D.G., eds.), University
of Missouri, Columbia, pp.119-133

Kriedemann,   P.O. & Downton, J.S. (1981) in The Physiology
and   Biochemistry of Drought Resistance in   Plants,
(Paleg,   L.G.   & Aspinall,   D.   eds.) , Academic Press,
Sydney, New York, London, San Francisco, pp. 71-95

Laisk, A. (1983), J. Exp. Botany 149, 1627-1635

Loske, D. & Raschke, K. (1988) Planta 173, 275-281

Osmond,   B., Ben, G.Y., Huang, L.K. & Sharkey, T.D. (1987)
in Current Topics in Plant Biochemistry and Physiology (
Randall,   D.D.,   Sharp,   R.E.,   Novacky, A.J. & Blevins,
D.G.,   eds.),   Vol. 6, University of Missouri, Columbia,
pp.134-146

Powles, S.B. (1984) Annu. Rev. Plant Physiology 35, 15-44

Schreiber,   U.   & Bilger,   W. (1987)   in Plant Response to
Stress (Tenhunen,J.D.,   Catarino,   F.M.,   Lange,   O.L. &
Oechel,   W.C.,   eds.),   NATO ASI   Series G,   Vol.   15,
Springer Verlag,   Berlin,   heidelberg, New York, London,
Paris, Tokyo, pp. 27-54

Schulze,   E.D. (1986) Annu. Rev. Plant Physiology 37, 247-
274

Sharp,   R.E. & Boyer, J.S. (1986) Plant Physiol. 82, 90-95

Sharkey, T.D. (1984), Planta 160, 143-150

Ullrich,   W.R.   (1983) in Encyclopedia of Plant Physiology
(Läuchli,   A.,   Bieleski,   R.L.,   eds.),   vol. 15 A,
Springer,   Berlin, Heidelberg, New York, Tokyo, pp. 376-
397 Younis, H.M., Boyer,J.S. & Govindjee (1979) Biochim.
Biophys. Acta 548, 328-340

Chilling sensitivity and phosphatidylglycerol biosynthesis

Norio MURATA, Osamu ISHIZAKI and Ikuo NISHIDA

National Institute for Basic Biology, Myodaiji, Okazaki 444, Japan

SYNOPSIS

Chilling sensitivity of higher plants is correlated with the proportion of saturated and trans-monounsaturated molecular species of phosphatidylglycerol. Fatty acid selectivity of acyl-(acyl-carrier-protein):glycerol-3-phosphate acyltransferase in chloroplasts can interpret the biosynthesis of these molecular species in chilling-sensitive plants. Complementary DNA for this enzyme from squash has been cloned.

INTRODUCTION

In the mechanism proposed by Lyons (1973) and Raison (1973) for the chilling sensitivity of plants, the primary event in chilling injury is the formation of a lipid gel phase in cellular membranes at low temperature. When a membrane goes into the phase separation state in which both gel and liquid crystalline phases co-exist, the membrane becomes leaky to small electrolytes, diminishing ion gradients across the membrane that is essential for the maintenance of physiological activities of the plant cell. We have shown that this mechanism operates in the chilling injury of the blue-green alga, Anacystis nidulans, in which the electrolytes leak out from the cytoplasm to the outer medium when the cytoplasmic membranes enter the phase separation state at low temperature (Ono & Murata

---

ABBREVIATIONS: ACP, acyl-carrier protein; AT1, AT2 and AT3, iso-meric forms of G3PATase from squash cotyledons; DGDG, digalactosyl-diacylglycerol; DPG, diphosphatidylglycerol; LPA, lysophosphatidic acid; MGDG, monogalactosyldiacylglycerol; PA, phosphatidic acid; PC, phosphatidylcholine; PE, phosphatidylethanolamine; PG, phos-phatidylglycerol; PI, phosphatidylinositol; SQDG, sulfoquinovosyl-diacylglycerol; 16:0, palmitic acid; 16:1t, 3-trans-hexadecenoic acid; 18:1, oleic acid; 18:2, linoleic acid; 18:3, α-linolenic acid; the molecular species are represented as, e.g., 16:0/16:1t for sn-1-palmitoyl-2-(3-trans)-hexadecenoylglycerol moiety.

This work was supported in part by a grant from the NIBB Program for Biomembrane Research and Grants-in-Aid for Scientific Research (61440002) and Cooperative Research (62304004) from the Ministry of Education, Science and Culture, Japan.

Table 1. Phase transition temperature of lipid molecular species

Molecular species	Phase transition temperature ($^{\circ}$C)	Reference
18:0/18:0-PC	58	Phillips et al. (1972)
18:0/18:1-PC	3	Phillips et al. (1972)
18:1/18:1-PC	−20	Phillips et al. (1972)
16:0/18:0-PC	47	Silvius (1982)
16:0/18:1-PC	−5	Silvius (1982)
16:0/16:0-PG	42	Murata & Yamaya (1984)
16:0/16:1t-PG	32	Bishop & Kenrick (1987)
18:0/18:0-PG	55	Silvius (1982)
18:1/18:1-PG	−18	Silvius (1982)
16:0/16:0-SQDG	43	Bishop et al. (1986)

1981a,b, 1982; Murata et al. 1984; Murata & Nishida 1987). However, the lipid phase transition in higher-plant membranes has not been well demonstrated, and the validity of the proposed mechanism for chilling injury is still in question.

Higher-plant cells contain MGDG, DGDG, SQDG and PG in plastid membranes, PC, PE, DPG and PG in mitochondrial membranes and PC, PE and PI in endoplasmic reticulum and plasma membranes (Harwood 1980). In addition to these glycerolipids, the plasma and tonoplast membranes contain high proportions of sterols and glycosphingolipids (Yoshida & Uemura 1986). Temperature for transition between the gel and liquid crystalline phases of glycerolipid molecular species varies markedly with a degree of unsaturation in their fatty acyl chains (Table 1). The molecular species containing only saturated fatty acids such as 16:0 and 18:0 reveal phase transition temperatures above 40°C. The molecular species of PC containing a cis-unsaturation bond reveal the phase transition near 0°C (Phillips et al. 1972) and introduction of the second cis-unsaturation bond decreases the phase transition temperature to about -20°C (Phillips et al. 1972). In contrast, the substitution of 16:0 by trans-unsaturated fatty acid, 16:1t, at the C-2 position of PG shifted

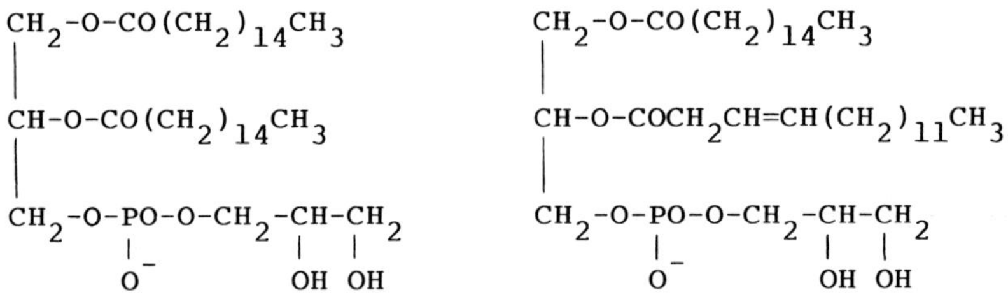

```
CH₂-O-CO(CH₂)₁₄CH₃              CH₂-O-CO(CH₂)₁₄CH₃
|                              |
CH-O-CO(CH₂)₁₄CH₃              CH-O-COCH₂CH=CH(CH₂)₁₁CH₃
|                              |
CH₂-O-PO-O-CH₂-CH-CH₂          CH₂-O-PO-O-CH₂-CH-CH₂
   |_      | |                    |_      | |
   O      OH OH                   O      OH OH


        16:0/16:0                      16:0/16:1t
```

Fig. 1. Chemical structure of 16:0/16:0 and 16:0/16:1t molecular species of PG

These molecular species are at high levels in the chilling-sensitive plants, whereas the PG molecular species containing cis-unsaturated fatty acids are dominant in the chilling-resistant plants.

the phase transition temperature only by 10°C. These findings suggest
that if there are lipid molecular species which induce the membrane
phase transition above 0°C, they should be fully saturated or trans-
monounsaturated ones. In all glycerolipids from leaf cells, only
PG contains high levels of these molecular species (Murata et al.
1982; Murata 1983; Raison & Wright 1983; Murata & Yamaya 1984). Low
levels of saturated molecular species are also found in SQDG from
some plants (Murata & Hoshi 1984; Kenrick & Bishop 1986a,b).

MOLECULAR SPECIES OF PHOSPHATIDYLGLYCEROL AND THEIR BIOSYNTHESIS

    Chilling-sensitive plants contain much higher proportions of
16:0/16:0 plus 16:0/16:1t species of PG (Fig. 1) than chilling-
resistant plants. In about 20 plants examined in our laboratory the
sum of the contents of these molecular species ranges from 3 to 19%
of the total PG in the chilling-resistant plants, and from 26 to
65% in the chilling-sensitive plants (Murata et al. 1982; Murata
1983). This may suggest that these two molecular species are closely
associated with the chilling sensitivity of plants. Roughan (1985a)
has surveyed the fatty acid composition of PG in 74 plants to confirm
the correlation between chilling sensitivity and saturated plus
trans-monounsaturated PG molecular species with exceptional cases
occurring in solanaceous plants and C_4 grasses. In a recent work
by Li et al. (1987), the correlation also exists among rice varieties
having different sensitivities toward chilling. In addition to these
two species, 18:0/16:0 and 18:0/16:1t PG species exist at low proportions. Although the phase transition temperatures of these molecular
species have not been measured, they are estimated to be higher than
those of 16:0/16:0 and 16:0/16:1t PG species, respectively, since
the replacement of 16:0 by 18:0 should increase phase transition
temperature (Table 1). These four molecular species are called, in
some occasions, "high-melting point" PG molecular species (Kenrick
& Bishop 1986b), or "disaturated" PG molecular species (Roughan
1985a); in the latter case, the trans-unsaturated fatty acid is regarded as a saturated one because the trans-unsaturation does not
decrease the phase transition temperature as much as cis-unsaturation
(Table I). Another lipid class which contains saturated molecular
species is SQDG (Murata & Hoshi 1984). The content of saturated SQDG
molecular species (16:0/16:0 plus 18:0/16:0) relative to the total
SQDG ranges from 0 to 20%, which is much lower than in the case of
PG. Although there is little correlation between the saturated SQDG
and chilling sensitivity (Murata & Hoshi 1984), the saturated SQDG,
if present, may interact with the saturated and trans-unsaturated
PG molecular species by stimulating the formation of gel phase
domains in the chloroplast membranes.
    The content of saturated plus trans-unsaturated PG molecular
species relative to the total PG in etioplast and amyloplast membranes is almost identical to that in the chloroplasts of the same
plants (Murata & Kurisu 1984). This suggests that the mechanism of
PG synthesis and temperature-dependent behaviour of the membranes
are essentially the same in these types of plastids.
    Sparace & Mudd (1982) as well as Roughan (1985b) demonstrated
that PG is synthesized in chloroplasts, whereas the other chloroplastic lipids, i.e., MGDG, DGDG and SQDG, are synthesized fully
or partially under the cooperation between chloroplasts and endoplasmic reticulum (Roughan and Slack, 1982, 1984). Based on the
result of the positional distribution of fatty acids in PG molecular
species and the demonstration by Sparace & Mudd (1982) that PG is
synthesized from glycerol 3-phosphate and acetate by chloroplasts,

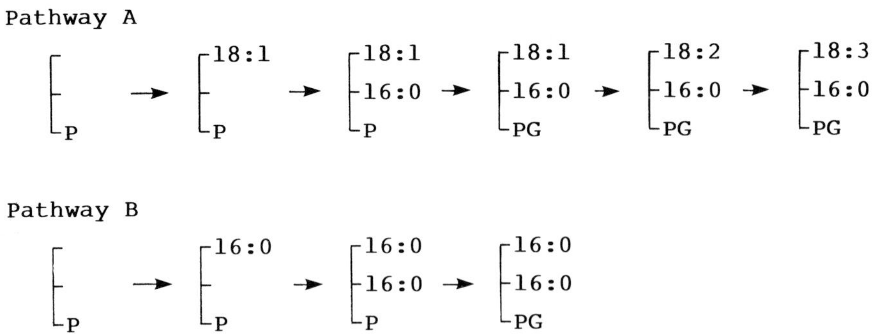

Pathway A

$$\begin{bmatrix} \\ \\ \text{P} \end{bmatrix} \rightarrow \begin{bmatrix} 18{:}1 \\ \\ \text{P} \end{bmatrix} \rightarrow \begin{bmatrix} 18{:}1 \\ 16{:}0 \\ \text{P} \end{bmatrix} \rightarrow \begin{bmatrix} 18{:}1 \\ 16{:}0 \\ \text{PG} \end{bmatrix} \rightarrow \begin{bmatrix} 18{:}2 \\ 16{:}0 \\ \text{PG} \end{bmatrix} \rightarrow \begin{bmatrix} 18{:}3 \\ 16{:}0 \\ \text{PG} \end{bmatrix}$$

Pathway B

$$\begin{bmatrix} \\ \\ \text{P} \end{bmatrix} \rightarrow \begin{bmatrix} 16{:}0 \\ \\ \text{P} \end{bmatrix} \rightarrow \begin{bmatrix} 16{:}0 \\ 16{:}0 \\ \text{P} \end{bmatrix} \rightarrow \begin{bmatrix} 16{:}0 \\ 16{:}0 \\ \text{PG} \end{bmatrix}$$

Fig. 2. **Two pathways proposed for biosynthesis of the PG molecular species in the chloroplast of chilling-resistant and chilling-sensitive plants**
Pathway A is dominant in the chilling-resistant plants, whereas both pathways A and B are of comparable activity in the chilling-sensitive plants. P, phosphate; PG, glycerophosphate.

we have proposed biosynthetic pathways for the PG molecular species (Murata 1983), as presented in Fig. 2. In pathway A, 18:1 is esterified to the C-1 position of glycerol 3-phosphate, and 16:0 to the C-2 position. After the PA thus produced is converted to PG, most of the 18:1 at the C-1 position is desaturated to 18:2 and subsequently to 18:3. A part of 16:0 at the C-2 position is converted to 16:1t. The combination of the fatty acids thus produced forms a variety of cis-unsaturated molecular species which undergo phase transition at about 0°C or below. In pathway B, 16:0 is esterified to both the C-1 and C-2 position of glycerol 3-phosphate. After the conversion of PA into PG, a part of 16:0 at the C-2 position is desaturated to 16:1t, whereas 16:0 at the C-1 position is not desaturated at all, resulting in the formation of only two molecular species, 16:0/16:0 and 16:0/16:1t. In chilling-resistant plants in which the sum of the 16:0/16:0 plus 16:0/16:1t contents is low, pathway A should be favoured over pathway B. In chilling-sensitive plants, on the other hand, pathway A and pathway B should be comparably active, resulting in production of 16:0/16:0 and 16:0/16:1t PG species in addition to cis-unsaturated molecular species.

  If such a scheme is valid, the chilling sensitivity of higher plants, or apparently the variation in the proportion of the 16:0/16:0 plus 16:0/16:1t PG molecular species in higher-plants, should result from the preferential transfer of 16:0 or 18:1 to the C-1 position of glycerol 3-phosphate. Such a preference is possible,

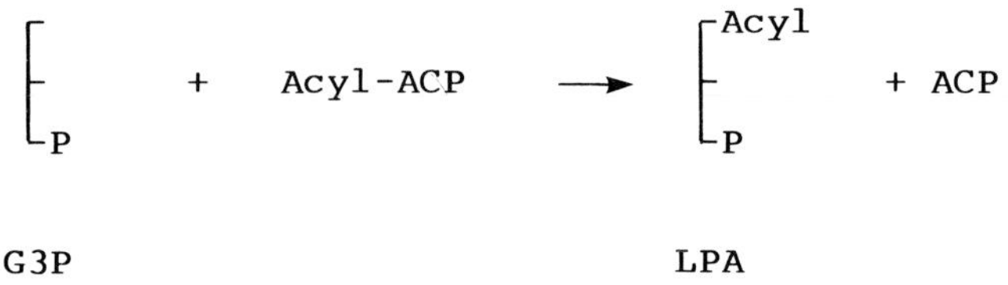

$$\begin{bmatrix} \\ \\ \text{P} \end{bmatrix} + \text{Acyl-ACP} \longrightarrow \begin{bmatrix} \text{Acyl} \\ \\ \text{P} \end{bmatrix} + \text{ACP}$$

    G3P   LPA

Fig. 3. **The reaction catalyzed by acyl-ACP:glycerol-3-phosphate acyltransferase in chloroplasts**
G3P, glycerol 3-phosphate.

if an enzyme which transfers the acyl group to the C-1 position of
glycerol 3-phosphate has a different selectivity toward 18:1 and
16:0 fatty acids between chilling-sensitive and chilling-resistant
plants; i.e. the enzyme in the chilling-resistant plants has a rather
strict specificity for 18:1 over 16:0, whereas that in the chilling-
sensitive plants is unspecific for either of the fatty acids and
transfers both 18:1 and 16:0 at comparable rates.

PLASTIDIAL ACYL-ACP:GLYCEROL-3-PHOSPHATE ACYLTRANSFERASE

    Acyl-(acyl-carrier-protein):glycerol-3-phosphate acyltransferase
(EC 2.3.1.15), designated as G3PATase, in higher-plant chloroplasts
transfers the acyl group from acyl-ACP to the C-1 position of
glycerol 3-phosphate to synthesize lysophosphatidic acid (1-acyl-
glycerol 3-phosphate). This reaction (Fig. 3) is the first step in
the biosynthesis of PG. According to our hypothesis as presented
in Fig. 2, the substrate selectivity of this enzyme toward the acyl
group in the acyl-ACP is very likely to be a determinant of the
chilling sensitivity of higher plants.
    This enzyme was purified first by Bertram & Heinz(1981) in a
soluble form from leaves of two chilling-resistant plants, pea and
spinach, by ion-exchange chromatography, gel-filtration chromatography and isoelectric focusing. Two isomeric forms of the enzyme
are found in pea. Recently, we have purified the enzyme from cotyledons of a chilling-sensitive plant, squash, by ion-exchange
chromatography, gel-filtration chromatography and affinity chromatography with acyl-carrier protein (Nishida et al, 1987). The squash
cotyledons contain three isomeric forms designated as AT1, AT2 and
AT3. AT2 and AT3 have been purified to single components after
40,000- and 32,000-fold purification, respectively. Molecular masses
and isoelectric points of these purified enzymes are summarized in
Table 2. We have also found at least two isomeric forms of G3PATase
having different isoelectric points and molecular masses in all the
plants examined, such as rice, barley, spinach, sunflower and maize.
(Dubacq, Douady, Nishida & Murata, unpublished data).
    To determine whether the G3PATase from chilling-sensitive plants
exhibits fatty acid selectivities different from those of resistant
plants, activities of enzymes from chilling-sensitive plants, squash,
and from chilling-resistant plants, spinach and pea, are compared
in Table 3. When about equal amounts of 18:1-ACP and 16:0-ACP are
mixed as substrate of the enzyme reaction, enzymes from spinach and

Table 2. Molecular mass and isoelectric point of acyl-ACP:glycerol-
          3-phosphate acyltransferase

| Plant | Isomeric form | pI | M_r, kDa | | Reference |
			Gel filtration	SDS-PAGE	
Spinach	---	5.2	42	---	Bertrams & Heinz (1981)
Pea	AT1	6.3	42	---	Bertrams & Heinz (1981)
	AT2	6.6	42	---	Bertrams & Heinz (1981)
Squash	AT1	6.6	30	---	Nishida et al. (1987)
	AT2	5.5	40	39	Nishida et al. (1987)
	AT3	5.6	40	39	Nishida et al. (1987)

**Table 3. Substrate selectivity of acyl-ACP:glycerol-3-phosphate
           acyltransferase**

Origin of enzyme	Glycerol 3-phosphate (mM)	Incorporation into LPA, %		Reference
		18:1	16:0	
Spinach	1.0	94	6	Frentzen et al. (1983)
Pea AT1 and AT2	1.0	90	10	Frentzen et al. (1983)
Squash AT1	0.3	89	11	Frentzen et al. (1987)
AT2	0.3	58	42	Frentzen et al. (1987)
AT3	0.3	55	45	Frentzen et al. (1987)

pea preferentially incorporate 18:1 to the C-1 position of glycerol
3-phosphate. Among the three isomeric forms from squash, AT1 shows
a strong preference to 18:1-ACP as spinach and pea enzymes. AT2 and
AT3, on the other hand, hardly discriminate 18:1-ACP and 16:0-ACP.
However, the observed selectivity of AT1 preferable to 18:1-ACP is
significantly reduced with increase in the pH of the reaction mixture
from pH 7.4 to pH 8.0, the stromal pH of illuminated chloroplasts.
The fatty acid selectivity of G3PATase can therefore explain the
finding that the content of 16:0/16:0 plus 16:0/16:1t PG molecular
species is low in chilling-resistant plants and high in chilling-
sensitive plants.

MOLECULAR CLONING OF ACYL-ACP:GLYCEROL-3-PHOSPHATE ACYLTRANSFERASE

A direct verification of the proposed role of G3PATase in the
chilling sensitivity of higher plants would be obtained by studies
combined with the molecular cloning of G3PATase and subsequent transformation of plants by the cloned gene. For this purpose, we have
produced polyclonal antibodies raised against the AT3 by intra-
peritoneal injections into mice. The antisera thus produced react
with AT2 and AT3.

A λgt11 random-primed cDNA library and a λgt11 oligo(dT)-primed
cDNA library have been constructed from poly(A)$^{+}$RNA from greening
squash cotyledons. The random-primed cDNA library was screened with
the antisera to obtain one positive clone, designated as λAT01, containing a cDNA insert of 400 base pairs (bp). The oligo(dT)-primed
cDNA library was screened with the cDNA insert of λAT01 to yield
one positive clone. This clone, designated as λAT02, contained a
cDNA insert of 718 bp which covered the poly(A) tracks. The random-
primed cDNA library was rescreened with the cDNA insert of λAT01
to yield a clone, designated as λAT03, containing a cDNA insert of
1,426 bp and an open-reading frame of 1188 bp. The nucleotide
sequence determination indicates that the cDNA inserts of the three
clones overlap each other. The amino acid sequence deduced from the
nucleotide sequence of the open-reading frame is compared with the
amino terminal sequence of AT2 and the amino terminal sequence of
a CNBr-fragment of AT2 and AT3, to suggest that the cDNA insert of
λAT03 encodes for either AT2 or AT3 of G3PATase.

When the amino acid sequence deduced from the cDNA sequence is
compared with the amino terminal sequence determined for AT2, it
would appear that G3PATase is synthesized as a precursor protein
with a leader sequence of 28 amino acid residues and processed to

a mature protein of 368 amino-acid residues. The presence of the
leader sequence is consistent with the suggestion (Feierabend 1982)
that the chloroplast G3PATase is one of the nuclear-coded proteins
in which the leader sequence is postulated to be necessary for the
transport of protein from the cytoplasm into the chloroplast stroma.
The relative molecular mass (Mr) calculated for the mature protein
is 40,929 Da, and is consistent with Mr about 40-kDa estimated for
AT2 and AT3 by SDS-polyacrylamide-gel electrophoresis and gel filtra-
tion (Table 2).

The hydropathy profile of the deduced amino acid sequence of
the squash G3PATase does not show any cluster of hydrophobic regions
which may correspond to a transmembrane structure of the protein,
but a homogeneous distribution of hydrophilic regions is evident.
This is consistent with the finding that this enzyme is soluble.
The hydropathy profile of the leader sequence of G3PATase is similar
to that of C-terminal region of transit peptides preceding the small
subunit of ribulose bisphosphate carboxylase/oxygenase (Karlin-Neuman
& Tobin 1986) and also spinach acyl-carrier protein I (Scherer &
Knauf 1987).

G3PATase of Escherichia coli is a membrane-bound protein and
its DNA has been cloned (Lightner et al. 1983). The nucleotide
sequence of E. coli G3PATase contains an open-reading frame of
2418 bp and predicts a polypeptide comprising 806 amino acids. The
homology between the squash and E. coli G3PATases in thier nucleotide sequences and deduced amino-acid sequences was less than 35
and 10%, respectively. Furthermore, no high homology was found even
in partial sequences.

REFERENCES

Bertrams, M. & Heinz, E. (1981) Plant Physiol. 68, 653-657
Bishop, D.G. & Kenrick, J.R. (1987) Phytochemistry 26, 3065-3067
Bishop, D.G., Kenrick, J.R., Kondo, T. & Murata, N. (1986) Plant
  Cell Physiol. 27, 1593-1598
Feierabend, J. (1982) in Methods in Chloroplast Molecular Biology
  (Edelman, M., Hallick, R.B. & Chua, N.H., eds.), pp. 671-680,
  Elsevier Biochemical, Amsterdam
Frentzen, M., Heinz, E., McKeon, T.A. & Stumpf, P.K. (1983) Eur.
  J. Biochem. 129 , 629-636
Frentzen, M., Nishida, I. & Murata, N. (1987) Plant Cell Physiol.
  28, 1195-1201
Harwood, J.L. (1980) in The Biochemistry of Plants, Vol. 4 (Stumpf,
  P.K., ed.), pp. 1-55, Academic Press, New York
Karlin-Neuman, G.A. & Tobin, E.M. (1986) EMBO J. 5, 9-13
Kenrick, J.R. & Bishop, D.G. (1986a) Phytochemistry 25, 1293-1295
Kenrick, J.R. & Bishop, D.G. (1986b) Plant Physiol. 81, 946-949
Li, T., Lynch, D.V. & Steponkus, P.L. (1987) Cryo. Lett. 8, 314-321
Lightner, V.A., Bell, R.M. & Modrich, P. (1983) J. Biol. Chem. 258,
  10856-10861
Lyons, J.M. (1973) Annu. Rev. Plant Physiol. 24, 445-466
Murata, N. (1983) Plant Cell Physiol. 24, 81-86
Murata, N. & Hoshi, H. (1984) Plant Cell Physiol. 25, 1241-1245
Murata, N. & Kurisu, K. (1984) in Structure, Function and Metabolism
  of Plant Lipids (Siegenthaler, P.-A. & Eichenberger, W., eds.),
  pp. 551-554, Elsevier Science Publishers, Amsterdam
Murata, N. & Nishida, I. (1987) in The Biochemistry of Plants, Vol. 9
  (Stumpf, P.K., ed.), pp. 315-347, Academic Press, London.
Murata, N. & Yamaya, J. (1984) Plant Physiol. 74, 1016-1024

Murata, N., Sato, N., Takahashi, N. & Hamazaki, Y. (1982) Plant Cell Physiol. 23, 1071-1079

Murata, N., Wada, H. & Hirasawa, R. (1984) Plant Cell Physiol. 25, 1027-1032

Nishida, I., Frentzen, M., Ishizaki, O. & Murata, N. (1987) Plant Cell Physiol. 28, 1071-1079

Ono, T. & Murata, N. (1981a) Plant Physiol. 67, 176-181

Ono, T. & Murata, N. (1981b) Plant Physiol. 67, 182-187

Ono, T. & Murata, N. (1982) Plant Physiol. 69, 125-129

Phillips, M.C., Hauser, H. & Paltauf, F. (1972) Chem. Phys. Lipids 8, 127-133

Raison, J.K. (1973) J. Bioenerg. 4, 258-309

Raison, J.K. & Wright, L.C. (1983) Biochim. Biophys. Acta 731, 69-78

Roughan, P.G. (1985a) Plant Physiol. 77, 740-746

Roughan, P.G. (1985b) Biochim. Biophys. Acta 835, 527-532

Roughan, P.G. & Slack, C.R. (1982) Annu. Rev. Plant Physiol. 33, 97-132

Roughan, G. & Slack, C.R. (1984) Trends Biochem. Sci. 9, 383-386

Scherer, D.E. & Knauf, V.C. (1987) Plant Mol. Biol. 9, 127-134

Silvius, J.R. (1982) in Lipid Protein Interactions (Jost, P.C. & Griffith, O.H., eds.), pp. 240-281, John Wiley and Sons, New York

Sparace, S.A. & Mudd, J.B. (1982) Plant Physiol. 70, 1260-1264

Yoshida, S. & Uemura, M. (1986) Plant Physiol. 82, 807-812

Calcium, protein kinase and the plasma membrane

Simon GILROY, David BLOWERS, Mark COLLINGE, Helen HARVEY and Tony TREWAVAS

Botany Department, University of Edinburgh, King's Buildings, Mayfield Road, Edinburgh

INTRODUCTION

The last five years has seen an explosion of interest in calcium and signal transduction in plants. In part this has resulted from the molecular dissection of signalling systems in animal tissues but the detection of calmodulin in plants (Anderson & Cormier, 1978), calcium regulated protein kinases (Hetherington & Trewavas, 1982), measurement of higher plant cell cytosol calcium (Gilroy et al., 1986) and putative identification of calcium channels (Hetherington & Trewavas, 1984a) has indicated that the potential for an equivalent signalling system is present.

Calcium has been implicated as a possible regulatory molecule in numerous plant tissues and aspects of development (Allan & Trewavas, 1987; Gilroy et al., 1987; Trewavas, 1985). Few of these have been established in a sufficiently critical manner to warrant confidence that calcium acts necessarily as a pivotal element. It is known that numerous metabolic reactions in plants do seem to be coupled to calcium. An incremental change in cytosol calcium level could then initiate a greater variety of effects than changes in many other metabolites. Thus calcium has the potential in plants as it does in animals for contributing substantially to regulation and control processes.

Whether calcium contributes to regulation solely in an active fashion, i.e. by changing concentration, is as yet largely unknown. Whether calcium contributes in a sensitive fashion (sensitivity defined as a measured incremental change in response induced experimentally by an incremental change in cytosol calcium) is still surmised. It is known that the sensitivity of calcium coupled reactions can be modified by changes in calmodulin level or phosphorylation of other protein components or other enzyme cofactors and probably by a variety of other as yet unknown metabolic changes (Rasmussen et al., 1986). A simple perusal of the article by Rasmussen et al. (1986) indicates how complex controlling reactions and sequences involving calcium can be.

The rather blunt and severe experimental treatments often currently used to determine calcium involvement need to be replaced by more refined and delicate treatments. Many cells may simply require a functional calcium system. Since the metabolic system is a complex, highly interrelated network, inhibition or gross interference of normal calcium functions may simply constrain the

This work was supported by grants from the S.E.R.C. and A.F.R.C.

behaviour of the whole even if normally calcium was contributing to
control in an insensitive fashion.

Reasons for localising calcium studies initially to the plasma membrane

        Our studies so far are limited to two developmental systems.
Firstly, the specification of polarity in the Fucus zygote which is
believed to involve cytosol calcium ions at an early stage
(Trewavas, 1982) and secondly, protoplast fusion (Boss et al.,
1984) which we use as a model system to understand membrane fusion
events. Neither system is ideal for biochemical work and we have
had to resort to using other tractable tissues with the hope of
applying what information we obtain to these two systems.
        Our studies have been largely limited to the calcium relations
of the plasma membrane for the following reasons.
1. Perturbations which increase calcium entry rates often result in
cytosol calcium transients only (Rasmussen et al., 1986).
Increased entry is followed by increased expulsion. Thus the system
settles to an increased rate of calcium cycling around the plasma
membrane. Elevation of cytosol calcium will probably in many cases
be limited to the immediate locale of the plasma membrane.
2. The diffusion coefficient of calcium in cytoplasm is two orders
of magnitude lower than that in free solution (Hodgkin & Keynes,
1957). This may result from binding to cytoskeletal components or
other organelles but again entry through the plasma membrane leads
to a much higher concentration at the membrane and its immediately
attached constituents.
3. The plasma membrane can act as a signal amplifier for many
signal transduction phenomena. The opening of one calcium channel
in the plasma membrane can permit the entry of thousands of calcium
atoms down their electrochemical gradient.
4. Finally it is believed that localised calcium entry can have
very specific effects on the distribution of proteins in the plasma
membrane. This phenomenon is believed to underlie such disparate
but challenging problems as polarity specification, localised growth
and secretion and unequal cell division (Trewavas, 1982).
        Our current research involves four aspects of the plasma
membrane; calcium channel protein isolation, plasma membrane bound
calcium regulated protein kinase, cytosol calcium estimation and
plasma membrane localised calmodulin. Much of the described work is
unpublished.

Calcium channel protein identification in plant cells

        A variety of drugs are available for binding to channels and
inhibiting calcium channel function. Verapamil and dihydroxypyridines have been used for channel protein identification and
purification from animal cells (Reuter, 1983). Both verapamil and
the dihydroxypyridines bind to plant membranes in a saturable manner
and verapamil inhibits Ca uptake into protoplasts (Andrejauskas
et al., 1985; Drakeford & Trewavas, 1985; Graziana et al., 1987;
Hetherington & Trewavas, 1984a). Verapamil also inhibits the uptake
of non-exchangeable calcium into maize roots and inhibits root
extension, phototaxis, bud formation and cell division (Allan &
Trewavas, 1987; Drakeford & Trewavas, 1985).
        Phase partitioned plasma membrane preparations from maize
(Larsson, 1983) have a 10 fold higher specific activity of
verapamil binding than other cellular membranes. The verapamil

binding site can be solubilised by detergents and the solubilised
site has similar binding characteristics to the membrane binding
site. Partial purification (100 fold) has been obtained using a
combination of ion exchange, exclusion and hydrophobic
chromatography. The final preparation has proteins with molecular
weight of about 170,000, 55,000 and 50,000 on SDS gel electrophoresis. There is an additional component at 100,000 which we
believe to be a contaminant. Proteins of 160-170,000, 40-50,000 and
30,000 and in the early papers a contaminant of 100,000 are
characteristic of animal calcium channels (Borsotto et al., 1985;
Glossman et al., 1984). The final preparation intriguingly has
associated protein kinase activity. Our aim in this research is the
eventual production of immune probes which can selectively impair
and label calcium channels.

Estimation of cytosol calcium in higher plant cells

        The three methods currently used in animal cells for cytosol
calcium measurement are microelectrodes, calcium fluorescent dyes
and aequorin, a calcium luminescent protein. None of the methods is
ideal for plant work and all suffer some ambiguity in assessing
cytosol calcium levels.
        Early attempts to use the calcium sensitive fluorescent dyes
Quin 2 and Fura 2 were frustrated by an inability of many but not
all plant cells to permit passage of the acetomethoxy esters across
the plasma membrane (Gilroy et al., 1985). This difficulty was
circumvented by using electropermeabilisation of protoplasts in the
presence of Quin 2 or by microinjection of Fura 2. However in the
case of the latter dye a further problem is the very rapid
accumulation into vacuoles (Bush & Jones, 1987). Even with Quin 2
there is a small ATP dependent accumulation in vacuoles although
this does not appear to be serious (Gilroy et al., 1988). The
values of cytosol calcium estimated by Quin 2 in mung bean root,
carrot or wheat vary between 80-350 nanomolar (Gilroy et al.,
1986-88) comparable with animal cells. Calmodulin binding
inhibitors cause a rapid elevation of cytosol calcium suggesting a
possible inhibition of a calmodulin dependent calcium ATPase perhaps
located in the plasma membrane (Gilroy et al., 1987). Carrot cell
protoplasts loaded with Quin 2 maintain a constant cytosolic calcium
concentration when the external concentration varies from micromolar
to ten millimolar, suggestive of a calcium stat, but incubation in
100mM calcium chloride leads to elevation above micromolar
concentrations (Gilroy et al., 1988).
        Quin 2 is suitable as an indicator of cytosol calcium in
protoplasts but there are features of it which we do not like.
Firstly, a considerable quantity of Quin 2 (a cytosol concentration
of 10^-M) has to be loaded into protoplasts for calcium
estimation. Some 10-15% of this finds its way into a vacuole
fraction (Gilroy et al., 1988). However the binding constant of
Quin 2 and calcium, which is approximately 100 nanomolar, ensures
that variations in vacuolar calcium concentrations (normally
millimolar or higher) will not affect measurement of cytosol
variations. Similar problems of organelle distribution of dyes have
been reported in animal cells. Secondly, we have been able to
demonstrate that protoplasts electroporated in Quin 2 possess lower
ATP levels than simple electroporated controls. This suggests that
Quin 2 is in some way affecting the metabolism of the protoplast.
Even so the Quin 2 loaded protoplast can maintain relatively
constant cytosol calcium with variation in external potassium or

magnesium from one micromolar to 100 millimolar.

These observations have led to the development of an
alternative procedure. Aequorin, a protein of molecular weight
about 20,000, will penetrate the plasma membrane of electro-
permeabilised protoplasts and sufficient enters the short period the
electropores are open for simple detection of luminescence in a
scintillation counter (Gilroy et al., 1988). More importantly
aequorin seems to have no effect on protoplast ATP levels. The
disadvantage of aequorin is that it is not really sensitive to
calcium below about 200 nanomolar and thus is only useful when
calcium increases above this value. However estimated resting
calcium levels are not dissimilar when measured with aequorin or
Quin 2. Quite fortuitiously we have found that silver ions increase
cytosol calcium rapidly when estimated by aequorin. The exact
increase is difficult to estimate because uptake of silver ions will
inevitably discharge the luminescent aequorin without light
emmission, but the increase is substantial. Silver ions have
frequently been used as a supposedly-specific inhibitor of the
developmental influence of ethylene. We have also used aequorin
loaded protoplasts to demonstrate that the inhibition of respiration
by azide causes a rapid elevation of cytosol calcium. This may
reflect a requirement for oxidative respiration (ATP) for a
functioning calcium ATPase to keep cytosol calcium level low.

Using either aequorin or Quin 2 we have not observed growth
regulators or red light treatments to modify cytosol calcium
concentrations detectably. This may be because protoplasts are not
ideal objects for such studies but we feel that it is more likely
that only local cytoplasmic variations may occur and only a limited
number respond. Consequently we have now resorted to fluorescence
ratio imaging (Tsien & Poenie, 1986) as more likely to yield
significant data.

Calcium regulated protein kinase in the plasma membrane

Simple membrane preparations of etiolated pea plants contain
protein kinase activity. Phosphorylating activity can be increased
5-10 fold by addition of micromolar calcium (Hetherington &
Trewavas, 1982, 1984b,c). On a specific activity basis preparations
of phase partitioned plasma membrane contain five times the amount
of calcium regulated kinase as other membranes (Blowers et al.,
1985). Limited hydrolysis of some of the membrane phospholipids by
phospholipase C diminishes activity rather than promotes enzyme
activity (Hetherington & Trewavas, 1984c). Thus the enzyme seems to
be unlike protein kinase C and is a simple calcium regulated protein
kinase.

We have solubilised and further purified the enzyme 500-1000
fold using ATP agarose (Blowers et al., 1987). The final
preparation contains 3 peptides characterised by SDS gel
electrophoresis. One of these, the 18,000 kilodalton peptide we
have previously shown as likely to be the catalytic site since
autophosphorylation takes place on this peptide (Blowers & Trewavas,
1987). Autophosphorylation was demonstrated using dilution kinetics
and reconstitution of enzyme activity after SDS gel electrophoresis
(Blowers & Trewavas, 1987). Autophosphorylation of the purified
enzyme can be increased about 2 fold by adding calcium and
calmodulin. The effect of calcium is a great deal more pronounced
on endogenous proteins in the membrane than on autophosphorylation.

We had difficulty initially in demonstrating calmodulin
dependence of enzyme activity. Calmodulin added to washed membranes

or trifluoperazine failed to promote or inhibit calcium activation. But with solubilisation, calcium activation is inhibitable by trifluoperazine (Hetherington & Trewavas, 1984c). Our membrane preparations possess substantial endogenous calmodulin which is not removed by EGTA washing but can be solubilised by detergents. This calmodulin in membranes may in part account for some of our early difficulties with calmodulin dependence. In an effort to establish more clearly that this protein kinase is calmodulin dependent, we have resorted to using azido iodinated calmodulin as an affinity label (Andreasen et al., 1981). On purified enzyme preparations a complex of about 35,000 kilodaltons is produced upon U.V. illumination with azido calmodulin (Collinge & Trewavas, 1988). This can be accounted for as calmodulin (17,000) linked to the enzyme (18,000). Thus it is likely that the enzyme binds calmodulin and is calmodulin dependent in the plasma membrane.

In the purified state the enzyme appears to preferentially auto-phosphorylate. In short incubations histone H1 receives only one-third the label from $-^{32}$P-ATP compared to the enzyme itself. Furthermore prior autophosphorylation reduces the protein kinase activity towards histone H1 (Blowers & Trewavas, 1987). We have recently found that there are two autophosphorylation sites (one is serine) and that the enzyme has intrinsic phosphatase activity as well. The phosphorylation on serine is in reality a very rapid cycling of phosphate rather than a permanent modification.

The enzyme has the potential to act as a bistable molecular switch as described by Lisman (1985) for membrane bound protein kinase. But demonstration of bistability will not be easy. At present we are attempting to characterise the identity of substrates for the kinase. Membrane bound-tubulin is one exciting possibility.

Calmodulin in the plasma membrane

As assayed by purification and NAD kinase activation the level of calmodulin can vary by an order of magnitude during cell development (Allan & Trewavas, 1985a,b). This establishes a potential for regulating calcium events by sensitivity variation. Approximately half the calmodulin in pea is located in pelletable membrane and the remainder is soluble. Using iodinated calmodulin we have found that there is considerable in vitro binding to the major cell fractions and much of this is independent of added calcium ions. Examination of these fractions by gel overlay procedures has shown that the highest calcium-independent binding is found in nuclear and ribosomal preparations and suggests basic proteins to be responsible for the absence of good calcium dependent binding (Collinge & Trewavas, 1988).

However calcium dependent calmodulin binding of phase partitioned plasma membrane is much easier to detect, being some 20 fold higher for membranes incubated in calcium compared to those incubated in EGTA. Gel overlay procedures reveal several binding proteins. Few of these binding proteins appear to be phos-phorylatable although a peptide in the 18K region also binds as described.

However the most surprising observation is the amount of calmodulin in phase partitioned plasma membrane. We have used 3 procedures, Ca^{45} overlays, phosphodiesterase and Western blots using antibodies to spinach calmodulin to show that about 1-2% of plasma membrane protein is calmodulin. At present we are trying to understand the positioning of this calmodulin in the membrane itself with the exciting possibility of projecting arms from the membrane to bind other molecules of significance to the plasma membrane.

SUMMARY

    The plasma membrane is the first part of the cell upon which
the environment impacts and it is often regarded as an essential
component of any signal transduction system. The ongoing research
described here indicates the presence of some components of a signal
transduction and amplification system involving calcium. The
putative presence of calcium channels and a calcium/calmodulin
regulated protein kinase located in the plasma membrane indicates
the potential for rapid metabolic responses to an increase either in
cytosol calcium or more simply an increased calcium cycling through
the plasma membrane. A major lacuna of knowledge is the substrates
for the protein kinase. There will be no swift solution to that
question but technological improvements using known labelled
proteins for probing Western blots should see an identification of
some of them. Experiments are currently in progress to identify
some of those concerned with the cytoskeleton.

REFERENCES

Allan, E. & Trewavas, A.J. (1985a) Planta 165, 493-501
Allan, E.E.F. & Trewavas, A.J. (1985b) in Molecular and Cellular
  Aspects of Calcium in Plant Development (Trewavas, A.J., ed.),
  pp. 311-312, Plenum Press, New York
Allan, E.F. & Trewavas, A.J. (1987) in The Biochemistry of Plants
  (Davies, D.D., ed.), vol. 12, pp. 117-153, Academic Press, London
Anderson, J.M. & Cormier, M.J. (1978) Biochem. Biophys. Res. Commun.
  84, 595-603
Andreasen, T.J., Keller, C.H., Laporte, D.C., Edelman, M.M. & Storm,
  D.R. (1981) Proc. Nat. Acad. Sci. U.S. 78, 2782-2785
Andrejauskas, E., Hertel, R. & Marme, D. (1985) J. Biol. Chem.
  260, 5411-5416
Blowers, D., Collinge, M., Gilroy, S., Harvey, H. & Trewavas, A.J.
  (1987) in Plant Membranes Structure, Function, Biogenesis (Leaver,
  C.J. & Sze, H., eds.), pp. 371-381, Liss, New York
Blowers, D.P., Hetherington, A. & Trewavas, A.J. (1985) Planta
  166, 208-215
Blowers, D.P. & Trewavas, A.J. (1987) Biochem. Biophys. Res. Commun.
  143, 691-696
Borsotto, M., Barhavin, J., Norman, R.I. & Lazdunski, M. (1985)
  J. Biol. Chem. 260, 14255-14263
Boss, W.F., Grimes, H.D. & Brightman, A. (1984) Protoplasma 120,
  209-215
Bush, D. & Jones, R. (1987) Cell Calcium 8, 455-472
Collinge, M. & Trewavas, A.J. (1988) To be submitted
Drakeford, D. & Trewavas, A.J. (1985) in Molecular and Cellular
  Aspects of Calcium in Plant Development (Trewavas, A.J., ed.),
  p. 423, Plenum Press, New York
Gilroy, S., Blowers, D.P. & Trewavas, A.J. (1987) Development 100,
  181-184
Gilroy, S., Hughes, W. & Trewavas, A.J. (1985) in Molecular and
  Cellular Aspects of Calcium in Plant Development (Trewavas, A.J.,
  ed.), pp. 373-374, Plenum Press, New York
Gilroy, S., Hughes, W.A. & Trewavas, A.J. (1986) FEBS Lett. 199,
  217-221

Gilroy, S., Hughes, W.A. & Trewavas, A.J. (1987) FEBS Lett. 213, 133-137

Gilroy, S., Hughes, W.A. & Trewavas, A.J. (1988) Submitted to Planta

Glossmann, H., Gol, A., Ferry, D.R. & Rombasch, M. (1984) Biochem. Soc. Trans. 12, 941-943

Graziana, A., Fossett, M., Ranjeva, R., Hetherington, A. & Lazdunski, M. (1987) Biochemistry 27, 764-768

Hetherington, A. & Trewavas, A.J. (1982) FEBS Lett. 145, 67-71

Hetherington, A. & Trewavas, A.J. (1984a) Plant Sci. Lett. 35, 109-113

Hetherington, A. & Trewavas, A.J. (1984b) Proc. Phytochem. Soc. Europe 24, 181-195

Hetherington, A. & Trewavas, A.J. (1984c) Planta 161, 409-418

Hodgkin, A.L. & Keynes, R.D. (1957) J. Physiol. 138, 253-281

Larsson, C. (1983) in Isolation of Membranes and Organelles from Plant Cells (Hall, J.L. & Moore, A.L., eds.), pp. 277-309, Academic Press, London

Lisman, J.E. (1985) Proc. Nat. Acad. Sci. U.S. 82, 3055-3057

Rasmussen, H., Kojima, I. & Barrett, P. (1986) in New Insights into Cell and Membrane Transport Processes (Poste, G. & Crooke, S.T., eds.), pp. 145-174, Plenum Press, New York

Reuter, H. (1983) Nature 301, 569-574

Trewavas, A.J. (1982) in The Molecular Biology of Plant Development (Smith, H. & Grierson, D., eds.), pp. 7-27, Blackwells, Edinburgh

Trewavas, A.J. (1985) in Molecular and Cellular Aspects of Calcium in Plant Development, pp. 1-451, Plenum Press

Tsien, R.Y. & Poenie, M. (1986) Trends Biochem. 11, 450-455

Transport across membranes

John A. RAVEN

Department of Biological Sciences, University of Dundee, Dundee DD1 4HN, U.K.

SYNOPSIS

Transport across membranes is discussed in the context of the transport-catalysing functions of proteins and the role of lipids in permitting or preventing the diffusive transport of solutes and water. Particular emphasis is given to cases in which a low permeability to solute (or water) is achieved by a membrane which retains its function in catalysing transport.

INTRODUCTION

The polar lipid (bilayer forming) components of membranes constitute the continuous phase of the fluid mosaic lipoprotein membrane. They thus restrict the movement of many solutes, especially the larger and/or more hydrophobic ones, between the aqueous phases separated by the membranes (Stein, 1986). This increased the activation energy for movement across the bilayer and puts the transbilayer flux of many important solutes into the same category as the chemical interconversion of many important organic compounds: rates are very small unless there is catalysis by proteins. Catalysis also permits <u>regulation</u> of the specifity and rates of a flux.

The extent to which the bilayer restricts the flux of solutes down free energy gradients is not completely predicted by considerations of Born energies (Deamer & Bramhall, 1986) in the case of small electrolytes; the 10^5-fold greater permeability for Na^+ (and Cl^-) is suggestive of ion and sugar fluxes through aqueous 'defects' in the bilayer. Such 'defects' are apparently, less necessary to account for water (solvent) permeability, and are generally not needed to account for the permeability coefficient of bilayers to organic solutes with relatively high oil/water partition coefficients (Bangham & Hill, 1986; Deamer & Bramhall, 1986; Stein, 1986).

It is widely accepted that the transmembrane fluxes of many low molecular weight, non-ionized solutes under 'physiological' conditions can be accounted for by the non-mediated flux across the membrane. Examples are O_2 entry in respiration and efflux in photosynthesis, CO_2 exit in respiration and entry in C_3 photosynthesis, H_2O fluxes for net cell growth, and N_2 entry for (symbiotic) diazotrophy (see Raven, 1982, 1984a). Less clearcut is the influx of indoleacetic acid in "chemiosmotic" polar auxin transport (Goldsmith, 1977; Raven & Rubery, 1982). Similarly, the impermeability of lipid bilayers to ions (including H^+) and polyhydroxy compounds is generally (albeit implicitly) assumed to be adequate to prevent significant energy or nutrient losses by leakage (see Raven, 1984a). This paper sets out to

examine cases where the effective prosecution of plant metabolism
involves membrane properties that run counter to these two
generalisations since *in vivo* permeabilities lower than those found *in
vitro* in lipid bilayers are involved.

TRANSPORT OF CO_2, O_2, N_2 AND H_2O IN RELATION TO PHOTOTROPHY AND DIAZOTROPHY

The problems

A number of plants transport inorganic C from the external medium
to the site of RUBISCO (ribulose bisphosphate carbosylase-oxygenase)
activity by a mechanism other than the diffusive flux of CO_2 across
membranes and of CO_2/HCO_3^- in aqueous phases. These plants maintain a
CO_2 concentration and CO_2/O_2 ratio at the site of RUBISCO activity
which is higher than that found in the medium (or water in equilibrium
with it in the case of terrestrial plants). A "CO_2 pump" which
achieves this CO_2 concentration is based on an auxilliary
carboxylation-decarboxylation sequence together with specialized
anatomy and symplastic transport of organic and/or amino acids in the
C_4 plants. Most aquatic plants have a 'CO_2 pump' involving active
influx of CO_2 and/or HCO_3^- across some membrane(s) between the medium
and RUBISCO. The operation of the aquatic plant type of 'CO_2 pump'
would clearly be compromised by a CO_2 permeability of the pumping
membrane(s) which was as high as that of membranes in photosynthetic
cells of C_3 plants (see Raven, 1984a; Raven & Lucas, 1985).
    N_2 fixation presents a rather different problem. N_2 entry is
invariably diffusive; concommittant diffusive O_2 entry would, if
intracellular steady-state O_2 concentrations exceeded a few μmol m^{-3},
irreversibly inactivate nitrogenase (Appleby, 1984). The problem here
is, *inter alia*, one of balancing resistances to O_2 and N_2 transport to
permit sufficient N_2 entry without excess O_2 entry, and *also* allowing
the other metabolite exchanges essential to the N_2-fixing cells
(Sprent, 1987).

CO_2, O_2 and H_2O fluxes during phototrophy

We deal first with the backflux of CO_2 during operation of the
'CO_2 pump' in organisms (mainly aquatic) with active influx in one (or
more) inorganic C species at a membrane(s) between the medium and
RUBISCO. In these organisms it is generally assumed that the barrier
is in the same membrane as the inorganic C pump, although there is
some dispute as to where the pump is located in eukaryotes (see Marcus
et al, 1984; Moroney *et al*, 1984; Raven, 1985; Suzuki *et al*, 1987).
    Direct measurements, and modelling, suggest a P_{CO_2} of about 10^{-7}
$m s^{-1}$ for the pumping membrane (Ogawa & Kaplan, 1987, Yokota *et al*,
1987, and references therein). This is a much lower P_{CO_2} value than
the $3.5.10^{-5}$ $m s^{-1}$ found for a phospholipid-cholesterol bilayer *in
vitro* (Gutknecht *et al*, 1977): Table 1.
    In addition to the problem of how the membrane can have such a
low P_{CO_2} there is the difficulty with O_2 efflux. A low P_{CO_2} is likely
to be paralleled by a low P_{O_2}. This would lead to a higher steady-
state concentration of O_2 around RUBISCO during photosynthesis
(Samish, 1975; Raven, 1977a; Berry & Farquhar, 1978). Such a build-
up of O_2 would diminish the possible selective advantages of the CO_2
pump in terms of increasing the achieved *in vivo* carboxylation
activity of RUBISCO and thus increasing the rate of photosynthesis per
unit biomass, per unit incident (rate-limiting) photons and per unit

Table 1

Comparison of permeability coefficients (corrected as far as possible
for unstirred layer effects) of lipid bilayers in vitro, plant and
cyanobacterial plasmalemma, and various extracellular structures.
Temperature in the range 20-25°C.

Permeability coefficient $(m\ s^{-1})$ in

Molecule	Lipid bilayer in vitro	Plasma-lemma (a)	Extra-cellular barriers	References
CO_2	3.10^{-3}	10^{-7}		(1,3,6,14)
O_2, N_2			$\sim 4.10^{-7}$ (b,c)	(12)
H_2O	$>10^{-5}$	10^{-5}	4.10^{-10} (d)	(4,11)
NH_3	$1.3.10^{-3}$	6.10^{-6}		(7,8,13)
glycerol	$8-500.10^{10}$	10^{-12}		(4)
H^+	$\sim 10^{-6}$	2.10^{-7}		(1,4,5,11)
Na^+	$\sim 2.10^{-12}$	2.10^{-11}		(4,5)
Ca^{2+}	$\sim 2.5.10^{-13}$	3.10^{-10}		(4,5,9)

(a) Plasmalemma permeability values are the minimum recorded. Even
     lower values may occur for the lipid part of the membrane, since
     mediated fluxes of the molecules may account for part of the
     flux. P_{CO} at the plasmalemma of many C_3 plants is probably as
     high as in lipid bilayers in vitro (Raven & Glidewell, 1981).

(b) = Walls (with lipid lamellae) of heterocysts of
     cyanobacteria.

(c) Note that the permeability coefficient of O_2 and N_2 in protein
     monolayers in prokaryotic gas vacuoles is greater than $3.3.10^{-4}$ m
     s^{-1} (Walsby, 1984).

(d) = Cuticles with epicuticular wax.

References: (1) Gutknecht (1984); (2) Gutknecht et al, (1977); (3)
Ogawa & Kaplan (1987); (4) Raven (1984a); Raven (1987a); (6) Raven &
Glidewell (1981); (7) Ritchie (1987); (8) Ritchie & Gibson (1987); (9)
Rossignol et al, (1985); (10) Schonfeld & Schickler (1984); (11)
Schonherr & Merida (1981); (12) Walsby (1985); (13) Walter &
Gutknecht (1986); (14) Yokota et al, (1987).

of resource (C, N, energy) committed to the photosynthetic machinery
(Raven, 1977a; Farquhar, 1983). Furthermore, O_2 build-up can have
deleterious effects unrelated to the carboxylation reactions (Shick &
Dykens, 1985).

The recent model of Reinhold et al (1987) for cyanobacteria has
the main barrier to CO_2 diffusion at the membrane of the carboxysome
(the site of most of the RUBISCO activity: Codd, 1988) rather than at
the site of inorganic C pumping (the plasmalemma). This ingenious
model makes a number of testable predictions (e.g. a high $P_{HCO_3^-}/P_{CO_2}$
ratio at the carboxysome membrane and localization of a low carbonic
anhydrase activity in the carboxysomes), and, although the concept of
a P_{CO_2} of only 10^{-7} m s^{-1} for a proteinaceous membrane like that of
the carboxysome strains credulity (see Walsby, 1984) even more than a
similar value for a lipoprotein bilayer, the model does have the
advantage of not producing a high O_2 concentration at the site of
RUBISCO.

The other kind of "CO_2 pump" is found in C_4 plants and depends on
co-operation of two morphologically and biochemically distinct cell
types linked by plasmodesmata. Here PEPc (phosphoenolpyruvate
carboxylase) is mesophyll cells fixes HCO_3^- derived from atmospheric
CO_2. The resulting C_4 dicarboxylic acid diffuses symplastically to
the bundle sheath cells where decarboxylation yields CO_2 (Jenkins et
al, 1987; Raven, 1972) and a C_3 moiety. The CO_2 is fixed by RUBISCO
while the C_3 compound diffuses symplastically back to mesophyll cells
for recoversion to the substrate of PEPc. Prevention of leakage of
CO_2 from the site of RUBISCO activity to the cell wall-airspace
interface is a function largely of the lengthy (many um) aqueous phase
pathway involved, perhaps coupled to low carbonic anhydrase activity
in bundle sheath cells which prevents substantial HCO_3^- involvement in
the inorganic C fluxes (Raven, 1977a; Berry & Farquhar, 1978; Raven
& Glidewell, 1981; Farquhar, 1983; Furbank & Hatch, 1983).
Apoplastic (cell wall) lipid barriers may also be significant
(Hattersley & Browning, 1981; Canny & McCully, 1986; Eastman et al,
1988a,b); a major role for cell membranes of low P_{CO_2} is not,
apparently, necessary.

The contrasting requirements of the "CO_2 barrier" in plants with
the CO_2 concentrating mechanism to minimize CO_2 loss yet maximise O_2
loss are not constant during fluctuations in the environment. Using a
decrease in the incident photon flux as our example, a sub-saturating
photon flux density implies that less CO_2 is pumped to the site of
RUBISCO, less CO_2 can be fixed by RUBISCO, and less O_2 is produced
near RUBISCO, than at saturating light. Accordingly, maintenance of
the optimal CO_2/O_2 ratio at the RUBISCO site at the lower photon flux
density would be attained with an increased resistance to CO_2 leakage
and to O_2 leakage. There does not seem to be any obvious way in which
a lipid bilayer's resistance to CO_2 and O_2 diffusion could be
significantly yet reversibly changed in the time scale of seconds to
minutes over which natural light fields can change. Aqueous barriers
could, by contrast, change more rapidly. If, for example, a reduced
photon flux density permitted reversible water infiltration of the
intercellular gas spaces between mesophyll cells at the bundle sheath
end, then CO_2 and O_2 leakage could be reduced (cf. discussion of
diazotrophy below). However, there is no evidence that such water
infiltration occurs in C_4 plants.

Another conflict in membrane functioning in photosynthesis occurs
in terrestrial plants where CO_2 uptake from the atmosphere must occur
yet, at least under some circumstances, it may be selectively
advantageous to reduce H_2O loss in transpiration. As with "CO_2
pumping" plants, cell membranes are not apparently capable of
accomplishing the large changes in P_{CO_2} and P_{H_2O} which are required to
permit homoiohydry at the level of single cells in a C_3 plant nor even
in "CO_2-pumping" plants where the required maximal P_{CO_2} is lower. The
variable resistance to gas exchange is only achieved at a
supracellular level, most notably by the operation, in series with

cell membranes, of the epidermal cuticle plus stomata system (Raven, 1977b). The cuticle constitutes a lipid (wax) barrier to CO_2, O_2 and H_2O transport in parallel with a variable resistance to gas movement between the atmosphere and the intercellular gas space system. Schonherr & Merida (1981) found a P_{H_2O} of 4.10^{-10} m s^{-1} for the astomatous cuticle of _Allium_ bulb scales using labelled water in an aqueous medium with correction for unstirred layers. Schonherr & Merida (1981) point out the possible reasons why micrometeorological measurements on leaves in air yield higher gas permeability values. One such reason is the frequent presence of stomata which do not form a complete seal when closed, and we note that the lowest P_{CO_2} values for leaf cuticle pertains to _Stylites_ (_Isoetes_) _andicola_, an astomatous plant which obtains all of its CO_2 via the roots (Keeley, et al, 1984). The cuticle plus epicuticular wax may also be important in conserving a variety of nutrients in the face of efflux to water droplets on the leaf surface (Chapin, 1980). We note that the impossiblity of operating the homoiohydric system at the single cell level imposes a minimal size constraint on homoiohydrically effective plants (cf. Raven, 1984b).

N_2 and O_2 fluxes during diazotrophy

We have seen that a major problem in diazotrophy is that of balancing resistances to O_2 entry and to N_2 entry, with the former being potentially harmful at air-equilibrium O_2 concentrations, and the latter beneficial. As with the CO_2/O_2 and CO_2/H_2O pairs considered under phototrophy, no cell membrane, or other diffusion barrier, discriminates sufficiently between O_2 and N_2 to obviate the problem of O_2-sensitivity of nitrogenase. In contrast to the low CO_2 permeability of the "pumping membrane" of cells with an active transmembrane flux of inorganic C species, there are no well authenticated cases of the cell membrane per se making a substantial contribution to O_2 exclusion from cells containing nitrogenase. If the 'pumping membrane' in organisms with a 'CO_2 pump' can have a P_{CO_2} as low as 10^{-7} m s^{-1}, it is not immediately obvious why the plasmalemma of an N_2-fixing cell should not have a P_{O_2} as low as the 4.10^{-7} m s^{-1} estimated by Walsby (1985) for the diffusion barriers in the cell walls of cyanobacterial heterocysts (Table 1).

The barrier to O_2 (and to N_2) in N_2-fixers operate in multicellular systems (both free-living and symbiotic), as do the C_4 plant barriers to CO_2 and O_2, and the terrestrial plant barriers to CO_2 and H_2O. Some of these barriers are lipid in nature; material analogous to epicuticular wax, or to wall lamellae of some C_4 plant bundle sheath cells, is found in cell walls overlying the site of nitrogenase in filamentous structures such as heterocystous cyanobacteria (nitrogenase in heterocysts) and symbiotic _Frankia_ (nitrogenase in vesicles): Raven (1980), Walsby (1985), Parsons et al (1987), Newcomb & Wood (1987). The number of lipid monolayers in the cell wall varies with the O_2 concentration, being greater (and thus imposing more resistance) at higher O_2 levels. However, this is a relatively long-term expedient, taking hours for the changed permeability to become manifest.

A different mechanism is involved in legume (and _Parasponia_?) symbioses with _Rhizobium_. Here the barrier to gas diffusion is, apparently, a layer (or layers) of cells within the symbiotic nodule. Permeability changes can be effected over periods of minutes, probably by exchanging gas and liquid in intercellular spaces, thus maintaining the steady-state O_2 and N_2 concentrations around the N_2-fixing bacteroids at a near-constant values over a range of environmental conditions (Witty et al, 1986). The "faggots" of the marine, non-

heterocystous cyanobacterium _Trichodesmium_, with central N_2-fixing
cells in the bundle of filaments, may well fulfill a similar, albeit
less well regulated, function to the nodule in restricting the access
of O_2 to nitrogenase (see Raven, 1980).

A final point which relates to uptake and assimilation of NH_4^+ and
NO_3^- as well as to that of N_2, and to NH_4^+ cycling in the
photorespiratory nitrogen cycle, is that of NH_3 leakage from the
NH_4^+/NH_3 pool (Raven, 1976, 1980, 1984a, 1988; Kleiner, 1985a,b).
Minimizing NH_3 leakage would appear to be advantageous, at least to
free-living organisms, and it is of interest that P_{NH_3} of the
plasmalemma of a eukaryotic alga and a cyanobacterium is lower
($\sim 6.10^{-6}$ m s^{-1}) than is found for lipid bilayers _in vitro_, i.e.
$1.3.10^{-3}$ m s^{-1} (Ritchie, 1987; Ritchie & Gibson, 1987; Walter &
Gutknecht, 1986): Table 1.

Conclusions

     In the "CO_2-pumping" phototrophs, in terrestrial phototrophs in
general, and in diazotrophs, an "anomalously" low permeability of the
plasmalemma, or some other lipoprotein bilayer, to O_2, CO_2, N_2 and H_2O
only seems to suffice in the case of "CO_2-pumpers" as the CO_2 barrier
in the membrane across which active transport of inorganic C occurs.
In all other cases a high transport resistance is achieved by an
extracellular and supracellular barrier composed of lipid or of a
substantial thickness of water. Such extracellular contrivances
restrict access of other solutes to the cells they are "protecting",
with varying implications for the organism. Whether a low P_{CO_2} in the
plasmalemma of (e.g.) _Dunaliella_ restricts other transport capacities
in unclear; certainly P_{H_2O} (and L_p) are not anomalously low, and
mediated transport apparently works effectively in _Dunaliella_ (see
Chapter 7 of Raven, 1984a).

TRANSPORT OF GLYCEROL AND OF H_2O IN RELATION TO VOLUME REGULATION IN
EFFECTIVELY WALL-LESS CELLS

The problems

     The range of ionic concentrations in "N" phases (Mitchell, 1979),
i.e. the cytosol, plastid stroma and mitochondrial matrix of
eukaryotes, is relatively limited. This limitation on the ionic
concentration which enzymes can tolerate means that extremely
halotolerant organisms such as many _Dunaliella_ spp. must use a
"compatible solute" to make the intracellular osmolarity equal to the
external osmolarity. The use of "cheaper" (low M_r) compatible
solutes, such as glycerol, reduces the energy costs of synthesis of
the solute _provided_ this is not offset by substantial leakage to the
medium. How does _Dunaliella_ keep its low M_r compatible solute,
glycerol, from excessive leakage (Raven, 1984a)?
     The opposite problem is faced by some freshwater flagellate
microalgae such as _Chlamydomonas_ spp. Here even the minimum
concentration of essential organic metabolites and of the ions (such
as K^+ and Mg^{2+}) used as enzyme activators and modulaters gives an
intracellular osmolarity of not less than ~ 150 kPa greater than that
in the medium. This leads to net osmotic H_2O influx which cannot be
countered by the cell wall since the flagella represent 'unprotected'
plasmalemma. Active water efflux _via_ contractile vaucoles is
energetically significant, particularly (as with glycerol leakage) in
a slow-growing, energy-limited cell. What is the minimum water
permeability (better as hydraulic conductivity, Lp) compatible with

*net water influx in growth which would thereby minimize the water
flux, and thus the energy cost, of active water efflux, and how does
this relate to the observed L_p of microalgal plasmalemma and of polar
lipid bilayers (Raven, 1976, 1982, 1984a)?*

Volume regulation and glycerol leakage in <u>Dunaliella</u>

    *Raven (1984a) points out that glycerol synthesis costs, per
osmol, only about half as much energy as such compatible solutes as
mannitol or sorbitol, one-third that of galactosylglycerol, and one-
quarter that of sucrose. However, glycerol also has a much higher
permeability coefficient than do the larger-M_r compatible solutes
(Raven, 1984a; Bangham & Hill, 1986; Stein, 1986) in bilayer
membranes and many plasma membranes, i.e. 7.10^{-10} - 5.10^{-8} m s^{-1} for
glycerol as compared to some 10^{-12} m s^{-1} for the larger M_r compatible
solutes (Table 1). Raven (1984a) comments that the measured $P_{glycerol}$
of the* <u>Dunaliella</u> *plasmalemma is only some 2.10^{-12} m s^{-1} i.e. more
similar to the larger-M_r compatible solutes than to the $P_{glycerol}$ of
other biological, and* <u>in vitro</u> *lipid bilayer, membranes. How this
abnormally low $P_{glycerol}$ is achieved in* <u>Dunaliella</u>*, which also has a
low P_{CO_2} (see above), but has a 'normal' water permeability,
regardless of whether this is determined as the diffusive permeability
P_{H_2O} or as the hydraulic conductivity L_p (see Chapter 7 of Raven,
1984a), is unclear.*

Volume regulation and active H_2O extrusion in <u>Chlamydomonas</u>

    We have seen that it is unlikely that <u>Chlamydomonas</u> *has an
intracellular osmolarity which is less than 150 kPa in excess of that
in the medium. It is of interest that one of the potential
constraints on minimizing intracellular osmolarity, i.e. the high
glutamate concentration needed to achieve a high NH_4^+ affinity for
glutamate synthetase (Raven, 1980, 1982) is less severe in*
<u>Chlamydomonas</u> *(Cullimore and Sims, 1981). Even with the lowest L_p
value reported for* <u>Chlamydomonas</u>*, i.e, $1.7.10^{-15}$ m s^{-1} Pa^{-1}, the net
water influx would be $1.7.10^{-10}$ m^3 (m^2 membrane$)^{-1}$ s^{-1}. For a 5 μm
radius cell with 0.75 m^3 water per m^3 cell, and requiring 0.20 m^3
water per m^3 intracellular volume for phototrophy, the quoted net
water influx corresponds to a specific rate of increase of cell water
(corrected for water consumed in metabolism) of $8.1.10^{-5}$ m^3 water (m^3
cell water$)^{-1}$ s^{-1}. The maximum reported specific growth rate of
unicellular algae of this size at $20^{\circ}C$ is some 3.10^{-5} s^{-1}, so that the
minimum water entry overprovides that needed for growth by at least
two-fold even at the maximum specific growth rate. A reduction in L_p
of only 2.3 fold would make water entry just limiting for the maximum
growth rate of algal cells. In view of the errors in determining
osmolarity difference and L_p, it would seem that L_p in some freshwater
algae is close to growth-limiting. However, growth at any lower rate
would yield an excess of water in flux over that needed for growth,
and active efflux via contractile vacuoles would be needed to maintain
cell volume. The minimum active efflux per unit time must be higher
at the lower specific growth rate as less of the water entering is
used in growth. Even if it were possible to further decrease L_p of
the plasmalemma in effectively wall-less cells growing at lower
specific growth rates, environmental fluctuations which reduce growth
rate over a shorter time scale (seconds or minutes) than that needed
to alter membrane composition and further reduce L_p.*
    The mechanism of contractile vacuole operations (cf. the

mechanism for active water transport in insect rectum of Kuppers et
al, 1986) presumably involves water movement into the vacuole with
solutes moving by a mediated mechanism in the "isosmotic filling"
phase to a greater extent than it moves out of the vacuole in the
"solute withdrawal" phase, with solute movement via a different
mediated mechanism. At least one direction of solute movement is
active. Clearly a low L_p for non-mediated water movement would make
water efflux more energetically efficient, and constraints on the
minimum non-mediated L_p value of the plasmalemma of microalgal cells
would not apply to the contractile vacuole membrane provided the
transient fusion of the contractile vacuole membrane with the plasma
membrane does not lead to extensive exchange of membrane components.
        Space does not permit a detailed discussion of the relative
energetic merits of cell walls (which precludes flagellar motility) as
the organ of volume regulation relative to that of a contractile
vacuoles in cells in freshwater (see Raven, 1982, 1984a). We note,
however, that some freshwater algae with non-flagellate vegetative
cells and cell walls retain contractile vacuoles; an example is
Hydrurus foetidus, a member of the Chrysophyceae (Fritsch, 1935).

NON-MEDIATED MOVEMENT OF H^+, Na^+ AND Ca^{2+} ACROSS BIOLOGICAL MEMBRANES
AND ITS IMPLICATIONS

The problems

        The three cations under consideration are very generally moved by
active transport from "N" phases (e.g. cytosol, plastid stroma,
mitochondrial matrix) to "P" phases (e.g. external medium,
vacuoles/lysosomes, intrathylakoid space): see Mitchell (1979) and
Raven (1984a). The maintenance of substantial (tens of kJ per mol
ion) electrochemical potential differences for the three cations
between "N" and "P" phases, and the use of mediated transmembrane
fluxes of these cations in the energetic and informational business of
plants, is well documented (Raven, 1984a, 1987b). We propose to
briefly investigate the extent to which the non-mediated flux of the
three cations across lipid-bilayers influences the energetic and
informational roles of the cations.

Leakage of H^+: magnitude and consequences.

        Much diversity of opinion existed as to the value of P_{H^+} for lipid
bilayers; recent work (Gutknecht, 1984; Deamer & Bramhall, 1986)
suggests a greater concensus with a value of not less than 10^{-6} m s^{-1}
at pH values within a unit or two of neutrality (Table 1). Taking
this as a plausible value for P_{H^+}, we predict a 50% reduction in the
photon yield of photophosphorylation at ~10^{-4} of full sunlight, i.e.
at an incident photon flux density of some 0.2 µmol m^{-2} s^{-1} (Raven and
Beardall, 1981, 1982; Richardson et al, 1983). The occurrence of H^+
leakage through the membrane at which photophosphorylation occurs is
only an energy cost in the dark if the H^+ free energy difference is
maintained in the dark (e.g. via 'chlororespiration' in thylakoids:
Bennoun, 1982; Caron et al, 1987), or the occurrence of the membrane-
associated energy transduction reactions of photosynthesis in the
plasmalemma as in the cyanobacterium Gloeobacter (Stanier & Cohen-
Bazire, 1987). H^+ fluxes through other lipoprotein membranes across
which an H^+ electrochemical potential difference is maintained (i.e.
essentially all membranes other than those, like the outer membranes
of mitochondria and plastids with proteinaeous pores allowing
relatively free movement of solutes of M_r < 500-800) represent an

essentially continuous energy cost under both growth and maintenance conditions.

A phototroph is thus subject to a number of sequential, multiplicative energy loses due to H^+ leakage through membranes. The primary energy conversion in plastids involves H^+ leakage during ATP synthesis. Use of photosynthate (synthesised using, inter alia, photophosphorylation) in mitochondrial oxidative phosphorylation also involves H^+ leakage, the use of the ATP generated in oxidative phosphorylation by H^+-ATPases in plasmalemma, lysosomes/vacuoles and (some) microbodies (Mellman et al, 1986; Douma et al, 1987; Labarca et al, 1986), in generating $\overline{H^+}$ electrochemical potential differences also implies losses through H^+ leakage (see Raven, 1984a, 1986). These losses are most relatively significant under energy (light)-limited growth conditions.

The energy losses via H^+ leakage through bilayers are, of course, in addition to other energy losses which are proportionately more important at low photon flux densities. Examples of these additional energy losses include the "slippage" losses from the 'S-states' of photosynthetic O_2 evolution (Kok & Radmer, 1977; Raven & Beardall, 1982; Richardson et al, 1983) and from other catalytic systems (Raven, 1984a), and the leakage of solutes other than H^+ through membranes (see elsewhere in this paper), and the resynthesis of degraded macromolecules (Raven, 1984a).

The sum of the leakage, slippage and repair costs yield a predicted photon flux density of not less than 0.1 μmol photon m^{-2} s^{-1} to provide the energy for the maintenance processes, with net growth requiring a higher photon flux density (see Raven, 1984a, 1986). However, the available data on a number of microalgae and cyanobacteria do not, within the precision of the measurements, yield a finite photon flux density to satisfy maintenance requirements (Geider et al, 1985, 1986; Geider & Osborne, 1986; Osborne & Geider, 1987). This is difficult to reconcile with the in vivo occurrence of all of the energy-dissipating reactions listed above at the rates expected from data on, for example, leakage through lipid bilayers in vivo. This integrated approach which sums the achieved energy losses in vivo confirms various individual findings discussed earlier which showed that leakage of solutes such as CO_2 and glycerol may be less significant in vivo than was suggested by data on bilayers in vitro.

Laboratory data consistent with reduced leakage in vivo is paralleled by field data. Littler et al (1985) found a photosynthetically competent coralline red alga growing attached to the substratum at 268 m in the tropical ocean where the maximum incident photon flux density was little more than 10 nmol photon m^{-2} s^{-1}, or about 10^{-5} of full surface light. Although the data presented by Littler et al (1985) are incomplete, (e.g. the need for the reasonable extrapolation of the photon flux density-depth relationship to arrive at the value for 268 m, and the reasonable supposition of photolithotrophic growth), their observations are consistent with the view that ion (specifically H^+) leakage in vivo is less than that found in lipid bilayers in vitro.

Leakage of Na^+: magnitude and consequences

The origin of Na^+ exclusion from, and Na^+ extrusion by, cells was probably related to volume regulation in wall-less cells in saline media (Raven & Smith, 1982; Raven, 1984a, 1986). The crucial point is the provision of sufficient "impermeant" solute (excluded, or entering but then extruded) in the medium to osmotically balance the "impermeant" metabolites (e.g. proteins, nucleic acids) inside the cells. Na^+ exclusion is a qualitatively plausible (because simpler) precursor of Na^+ exclusion, but is the P_{Na^+} of the lipid bilayer low enough to provide the "Na^+ exclusion" role? Raven (1986) computes the specific rates of increase of intracellular Na^+ for a marine macroalga, granted a P_{Na^+} of 2.10^{-12} m s^{-1} (the value found for planar lipid bilayers; cf. the 2.10^{-11} m s^{-1} found for marine algal plasmalemma; Raven, 1984a, 1987a; Table 1) and concludes that it exceeds the maximum specific growth rate of macroalgal cells at 20^{0}C (3.10^{-5} s^{-1}). However, some estimates of P_{Na^+} for lipid bilayers are as low as 10^{-14} m s^{-1} (Deamer & Bramhall, 1986), and with such a P_{Na^+} the Na^+ influx across the lipid bilayer would be low enough to prevent excessive Na^+ entry (i.e. in excess of that needed to account for intracellular Na^+) until the cell doubling time had increased to some 266 h (specific growth rate of $7.2.10^{-7}$ s^{-1}). Indeed, growth at a rate in excess of $7.2.10^{-7}$ s^{-1} would require a mediated, regulated entry of Na^+ (e.g. by passive uniport, or a cotransport process) whose achieved capacity increases with growth rate to maintain an intracellular Na^+ concentration of 10 mol m^{-3}. Resolution of the adequacy of Na^+ exclusion as an early volume regulation mechanism awaits resolution of the magnitude of P_{Na^+} in lipid bilayers.

Leakage of Ca^{2+}: magnitude and consequences.

Prevention of a build-up of free Ca^{2+} in "N" phases of extant organisms is imperative if inhibition of enzymes and interference with the informational role of Ca^{2+} are to be avoided (Lukacs & Kapus, 1987; Raven, 1984a, 1987a). A further important role, and perhaps a decisive one in evolution of Ca^{2+} exclusion/extrusion, is the avoidance of precipitation of inorganic and organic phosphates which occur at relatively high concentrations in "N" phases (Raven & Smith, 1982; Koch, 1986).

Would Ca^{2+} exclusion be adequate to account for the low free (and total) Ca^{2+} in early cells, granted likely driving forces and the measured $P_{Ca^{2+}}$ of lipid bilayers? Raven (1986) guessed a $P_{Ca^{2+}}$ of 10^{-12} m s^{-1} (i.e. half of P_{Na^+}) and showed that Ca^{2+} entry across a lipid bilayer was less than that needed to give a total (free + bound) Ca^{2+} of 1 mol m^{-3} in a 5 µm radius cell growing at 3.10^{-5} s^{-1}. The measured $P_{Ca^{2+}}$ for lipid bilayers of $0.25.10^{-12}$ m s^{-1} (Rossignol et al, 1982; Table 1) would, if substituted for the 10^{-12} m s^{-1} used by Raven (1986), increase the need for mediated Ca^{2+} entry in rapidly growing cells. Thus, even without assuming a pro rata reduction in $P_{Ca^{2+}}$ to accord with the P_{Na^+} quoted by Deamer & Bramhall (1986), Ca^{2+} exclusion would be adequate to account for low intracellular Ca^{2+} in early cells even if their mean rate of growth was not more than 10^{-7} s^{-1} (i.e. a generation time exceeding 80 days). We note that reported values of $P_{Ca^{2+}}$ at the plant cell plasmalemma generally exceed 3.10^{-10} m s^{-1}, presumably as a result of Ca^{2+} uniporters in parallel with the lipid bilayer part of the membrane (see Raven, 1987a).

Conclusions

Membranes in vivo may be less permeable to ions (especially H^+) that many studies on lipid bilayers in vitro would suggest. In multicellular plants additional lipoidal barriers in the apoplast can act in series or in parallel with membranes (see Chapin, 1970; Peterson, 1988).

CONCLUDING COMMENTS

Data reviewed in this paper suggest that the permeability

coefficient for solutes of lipoprotein membranes *in vivo* is frequently less than the values found for lipid bilayers *in vitro* (Table 1). Direct evidence is clearest for CO_2, NH_3, glycerol and H^+; more general considerations (e.g. minimum photon flux density for growth) suggest that the phenomenon is more general with respect to the solutes involved. The possibility that the differences in lipid composition between the model membrane *in vitro* and the *in vivo* lipoprotein membrane account for the observed differences seems implausible when the permeability coefficients for a given solute differ by several orders of magnitude (Table 1). Such an explanation is also rendered less plausible by different extents of anamalously low permeabilities for different solutes in a given organism (e.g. CO_2, glycerol and H_2O at the *Dunaliella* plasmalemma).

Regardless of how the low permeabilities are achieved, they do have important repercussions for resource (energy, nutrient) handling by plants. In a number of cases the 'anomalous' permeability appears to be a *sine qua non* of the prosecution of a particular metabolic feat, e.g. the use of glycerol as a compatible solute, the functioning of a 'CO_2 pumps' based on active transmembrane transport of inorganic C species, and the capacity to grow at very low photon flux densities.

However, in a number of cases related to the manipulation of low M_r molecules such as CO_2, O_2, N_2 and H_2O the diffusion barrier is related to a lipid layer, or a substantial thickness of water, in series with the cell membrane (e.g. C_4 metabolism, symbiotic N_2 fixation, terrestrial vascular plants photosynthesis and transpiration, and heterocystous cyanobacterial N_2 fixation).

REFERENCES

Appleby, C.A. (1984). Annu. Rev. Pl. Physiol. *35*, 443-478.
Bangham, A.D. & Hill, M.W. (1986). Chem. Phys. Lipids *40*, 189-205.
Bennoun, P. (1982). Proc. Natn. Acad. Sci. Wash. *79*, 4252-4256.
Berry, J. & Farquhar, G. (1978). in Photosynthesis 77 (Hall, D.O., Coombs, J. & Goodwin, T.W., eds), pp. 119-131. Biochemical Society, London.
Canny, M.J. & McCully, M.E. (1986). J. Microscop. *142*, 63-70.
Caron, L., Berkaloff, C., Duval, J-C., & Jupin, H. (1986). Photosynthesis Res. *10*, 131-139.
Chapin, B. (1980). Annu. Rev. Ecol. Syst. *11*, 233-260.
Codd, G.A. (1988). Adv. Microbial Physiol. *29*, 115-164.
Cullimore, J.V. & Sims, A.P. (1981). Phytochem. *20*, 597- 600.
Deamer, D.W. & Bramhall, J. (1986). Chem. Phys. Lipids *40*, 167-188.
Douma, A.C., Veenhuis, M., Sulter, G.J. & Harder, W. (1987). Arch. Microbiol. *147*, 42-47.
Eastman, P.A.K., Dengler, N.G. & Peterson, C.A. (1988a). Protoplasma *142*, 99-111.
Eastman, P.E.K., Peterson, C.A. & Dengler, N.G. (1988). Protoplasma *142*, 112-126.
Farquhar, G.D. (1983). Austr. J. Plant Physiol. *10*, 205-236.
Fritsch, F.E. (1935). The Structure and Reproduction of the Algae, Vol. 1., pp.791. University Press, Cambridge.
Furbank, R.T. & Hatch, M.D. (1987). Plant Physiol. *85*, 958-964.
Geider, R. & Osborne, B.A. (1986). Marine Biology *93*, 351-360.
Geider, R., Osborne, B.A. & Raven, J.A. (1985). J. Phycol. *22*, 39-48.
Geider, R., Osborne, B.A. & Raven, J.A. (1985). J. Phycol. *21*, 609-619.
Goldsmith, M.H.M. (1977). Annu. Rev. Plant Physiol. *28*, 439-378.
Gutknecht, J. (1984). J. Membr. Biol. *82*, 105-112.

Gutknecht, J., Bisson, M.A. & Tosteson, D.C. (1977). J. Gen.
Physiol. 69, 779-794.
Hattersley, P.W. & Browning, A.J. (1981). Protoplasma 109, 371-401.
Jenkins, C.L.D., Burnell, J.N. & Hatch, M.D. (1987). Plant Physiol.
85, 952-957.
Keeley, J.E., Osmond, C.B. & Raven, J.A. (1984). Nature 310, 694-
695.
Kleiner, D. (1985a). F.E.B.S. Letters 187, 237-239.
Kleiner, D. (1985b). F.E.M.S. Microbiol. Revs. 32, 87-100.
Koch, A.L. (1987). in Environmental Physiological Ecology (Calow,
P., ed), pp. 85-103. University Press, Cambridge.
Kok, B. & Radmer, R. (1976). in Chemical Evolution of the Giant
Planets (Ponnamperuma, C., ed), pp. 183-197. Academic Press, New
York.
Kuppers, J., Plagemann, A. & Thum, V. (1986). J. Membr. Biol. 91,
107-119.
Labarca, P., Wolff, D., Soto, V., Necochea, C. & Leighton, F. (1986).
J. Membr. Biol. 94, 285-291.
Littler, M.M., Littler, D.S., Blair, S.M. & Norris, J.N. (1985).
Science 227, 57-59.
Lukacs, G.L. & Kapus, A. (1987). Biochem. J. 248, 609-613.
Marcus, Y., Volokita, M. & Kaplan, A. (1984). J. exp. Bot. 35, 1136-
1144.
Mellman, I., Fuchs, R. & Helenius, A. (1986). Annu. Rev. Biochem.
55, 663-700.
Mitchell, P. (1979). Direct chemiosmotic ligand conduction
mechanisms in proton motive complexes. in Membrane Bioenergetics
(Lee, C.P., Schatz, G. & Ernster, L. eds), pp. 361-372. Addison-
Wesley Publishing Company, Reading, Mass.
Moroney, J.V., Kitayama, M., Togasaki, R.K. & Tolbert, N.E. (1987).
Plant Physiol. 83, 460-463.
Newcomb, W. & Wood, S.M. (1987). Int. Rev. Cytol. 109, 1-88.
Ogawa, T. & Kaplan, A. (1987). in Progress in Photosynthesis
Research (Biggins, J., ed). Vol. 4, pp. 297- 300. Nijhoff/Junk,
Dordrecht.
Osborne, B.A. & Geider, R. (1987). Plant Cell Envir. 10, 141-150.
Parsons, R., Silverster, W.B., Harris, S., Gruijters, W.T.M. &
Bullivant, S. (1987). Plant Physiol. 83, 728-731.
Peterson, C.A. (1988). Physiol. Plant. 72, 204-208.
Raven, J.A. (1972). New Phytol. 71, 995-1014.
Raven, J.A. (1976). In Encyclopedia of Plant Physiology, New Series
(Luttge, U. & Pitman, M.G., eds), Vol. 2A, pp. 129-188. Springer
Verlag, Berlin.
Raven, J.A. (1977a). Cur. Adv. Pl. Sci. 9, 579-590.
Raven, J.A. (1977b). Adv. Bot. Res. 5, 153-219.
Raven, J.A. (1980). Adv. Microbial Physiol. 21, 47-226.
Raven, J.A. (1982). New Phytol. 92, 1-20.
Raven, J.A. (1984a). Energetics and Transport in Aquatic Plants.
A.R. Liss, New York.
Raven, J.A. (1984b). Bot. J. Linn. Soc. 88, 105-126.
Raven, J.A. (1985). in Inorganic Carbon uptake by Aquatic
Photosynthetic Organisms (Lucas, W.J. & Berry, J.A., eds), pp. 67-82.
American Society of Plant Physiologists, Rockville, Maryland.
Raven, J.A. (1986). in Photosynthetic Picoplankton (Platt, T. & Li,
W.K.W., eds), pp. 1-70. Can. Bull. Fish. Aq. Sci. 241.
Raven, J.A. (1987a). in Solute Transport in Plant Cells and Tissues
(Baker, D.A. & Hall, J.L., eds), pp. 166-219.
Raven, J.A. (1987b). in The Cell Surface and Signal Transduction
(Greppin, H., Millet, B. and Wagner, E., eds), pp. 205-235.
Springer, Berlin.

Raven, J.A. (1988). New Phytol. 109, in press.
Raven, J.A. & Beardall, J. (1981). F.E.M.S. Microbiol. Letters 10, 1-5.
Raven, J.A. & Beardall, J. (1982). Plant Cell Environm. 5, 117-124.
Raven, J.A. & Glidewell, S.M. (1981). in Physiological Processes Limiting Plant Productivity (Johnson, C.B., ed), pp. 109-136. Butterworths, London.
Raven, J.A. & Lucas, W.J. (1985). in Inorganic Carbon uptake by Aquatic Photosynthetic Organisms (Lucas, W.J. & Berry, J.A., eds), pp. 305-324. American Society of Plant Physiologists, Rockville, Maryland.
Raven, J.A. & Rubery, P.H. (1982). in The Molecular Biology of Plant Development (Smith, H. & Grierson, D., eds), pp. 28-48. Blackwell Scientific Publications, Oxford.
Raven, J.A. & Smith, F.A. (1982). Solute transport at the plasmalemma and the early evolution of cells. Biosystems 15, 13-26.
Reinhold, L., Zviman, M. & Kaplan, A. (1987). in Progress in Photosynthesis Research (Biggins, J., ed), Vol. 4, pp. 289-296. Nijhoff/Junk, Dordrecht.
Richardson, K., Beardall, J. & Raven, J.A. (1983). New Phytol. 93, 157-191.
Ritchie, R.J. (1987). J. exp. Bot. 38, 67-76.
Ritchie, R.J. & Gibson, J. (1987). J. Membr. Biol. 95, 131-142.
Rossignol, M., Uso, T. & Thomas, P. (1985). J. Membr. Biol. 67, 269-275.
Samish, Y.B. (1975). Photosynthetica 9, 372-375.
Schonfeld, M. & Schickler, H. (1984). F.E.B.S. Letters 167, 231-234.
Schonherr, J. & Merida, T. (1981). Plant Cell Environm. 4, 349-354.
Shick, J.M. & Dykens, J.A. (1985). Oecologia 66, 33-41.
Sprent, J.I. (1987). The Ecology of the Nitrogen Cycle. University Press, Cambridge.
Stanier, R.Y. & Cohen-Bazire, G. (1977). Annu. Rev. Microbiol. 31, 225-274.
Stein, W.D. (1986). Transport and Diffusion Across Cell Membranes. Academic Press, Orlando.
Suzuki, E., Tsuzuki, M. & Miyachi, S. (1987). Plant Cell Physiol. 28, 1377- 1388.
Walsby, A.E. (1984). Proc. Roy. Soc. Lond. B. 223, 177-196.
Walsby, A.E. (1985). Proc. Roy. Soc. Lond. B. 226, 345-366.
Walter, A. & Gutknecht, J. (1986). J. Membr. Biol. 90, 207-217.
Witty, J.R., Minchin, F.R., Skot, L. & Sheehy, J.E. (1986). Oxford Surveys of Plant Molecular and Cell Biology 3, 275- 314.
Yokota, A. (1987). Plant Cell Physiol. 28, 1363-1376.
Yokota, A., Iwaki, T., Miura, K., Wadano, A. & Kitaoka, S. (1987). Plant Cell Physiol. 28, 1363-1376.